OKLAHOMA CITY

A Better Living. A Better Life.

Produced in partnership with the
Greater Oklahoma City Chamber of Commerce

123 Park Avenue
Oklahoma City, Oklahoma 73102
405/297-8900
405/297-8908 fax
www.okcchamber.com

OKLAHOMA CITY

A Better Living. A Better Life.

By Susan Wallace • Corporate Profiles By Tamara J. Hermen

Featuring the Photography of Jack Hammett • Joe Ownbey • Fred W. Marvel • Erick Gfeller

Oklahoma City: A Better Living. A Better Life.

Produced in partnership with the
Greater Oklahoma City Chamber of Commerce
Charles H. Van Rysselberge, CCE, *President*
123 Park Avenue
Oklahoma City, Oklahoma 73102
(405) 297-8900 • fax (405) 297-8908
www.okcchamber.com

By Susan Wallace
Corporate Profiles by Tamara J. Hermen with Kathy Draper
Featuring the Photography of Jack Hammett, Joe Ownbey, Fred W. Marvel and Erick Gfeller

Community Communications, Inc.
Publishers: Ronald P. Beers and James E. Turner

Staff for *Oklahoma City: A Better Living. A Better Life.*
Publisher's Sales Associates: Robbie Wills, Dahlia Davis, Paula Haider, and John Tew
Executive Editor: James E. Turner
Managing Editor: Mary Shaw Hughes
Design Director: Camille Leonard
Designer: Rebecca Hockman Carlisle
Photo Editors: Rebecca H. Carlisle and Mary Shaw Hughes
Production Managers: Corinne Cau and Cindy Lovett
Contract Manager: Katrina Williams
Editorial Assistant: Kari Collin
Sales Assistant: Annette Lozier
Proofreader: Opal Banish
Accounting Services: Sara Ann Turner
Printing Production: Frank Rosenberg/GSAmerica
Community Communications, Inc.
Montgomery, Alabama

James E. Turner, Chairman of the Board
Ronald P. Beers, President
Daniel S. Chambliss, Vice President

CONTENTS

Part One

OKLAHOMA CITY
A Better Living. A Better Life.

A GREAT PLACE TO CALL HOME

Oklahoma City is the good life. A city with heart, it offers quality living, a comfortable pace, diversity of opportunities, and plenty of sunshine. Combined with people who place great importance on family, values, and spiritual faith, and who are the friendliest on earth, Oklahoma City is a great place to call home.

PIONEERS AND HEROES

Oklahoma City grew out of the determination and hope of many strong individuals. Since its unique beginnings of the Oklahoma Land Run in 1889, the character and hard work of its people have made this city great. Even today, Oklahoma City's most important asset is the spirit of its people.

WE MEAN BUSINESS

The American work ethic is alive and well in Oklahoma City. Many world-class businesses and organizations have seen tremendous success running their operations from right here in the heart of America, a place with a pro-business atmosphere and a superior labor force.

HIGH FLYING HISTORY… HIGH FLYING FUTURE

Aviation has historically held a presence in the Oklahoma City economy and continues to do so today. One of the nation's finest flying schools opened here in 1924. Oklahoma City is the base for many national airlines as well as Tinker Air Force Base.

THE FINER THINGS IN LIFE

Oklahoma City offers a diverse selection of cultural opportunities that rivals any U.S. city. From the Oklahoma City Philharmonic Orchestra, Ballet Oklahoma, and fine collections of art, to jazz festivals and the Red Earth Festival, Oklahoma City has much to choose from throughout the year.

OKLAHOMA CITY
A Better Living. A Better Life.

CONTENTS

Part One

Cover photo by Jack Hammett.

CONTENTS

Part Two

OKLAHOMA CITY'S ENTERPRISES

OKLAHOMA CITY'S ENTERPRISES

CONTENTS

Part Two

Photos without credits are courtesy of the Greater Oklahoma City Chamber of Commerce.

Foreword

O klahoma City. A better living. A better life.

We know it! And *now* we are telling the world!

Our big sky and amazing sunsets are the most beautiful you've ever seen. It is greener and grander than you think. The people are the nicest and most genuine you'll find anywhere in the world. It is surprising to many who visit here. But for those of us who live here, we are accustomed to its beauty and friendliness.

Our advantages are many—both in making a living and having a life. We are more arts-abundant, technologically advanced, economically advantaged, and culturally diverse than most people imagine. And our cost of living is extremely affordable. In fact, Oklahoma City ranks among the most affordable housing in the country.

In 1993, the citizens of Oklahoma City voted for a self-imposed sales tax supporting the most comprehensive downtown revitalization effort in the country. The Metropolitan Area Projects (MAPS) is the future of Oklahoma City. Our new ballpark, riverwalk, riverfront development, library/learning center, and indoor arena, along with the renovation of the Myriad Convention Center, Civic Center Music Hall, and State Fairgrounds, put Oklahoma City at the forefront of exciting, up-and-coming cities.

We also have the pro-business advantage. Our aggressive efforts to attract new business and expand those already here shows, as we have record numbers of new jobs and a below national average unemployment rate. We have entered the global marketplace, without question.

Our city has seen incredible changes over the past few years. Our growth has been exciting and significant. As you stroll through this book, you will come to know the city we live in and love. When we say "Oklahoma City. A better living. A better life." we mean it!

Charles H. Van Rysselberge, CCE
President
Greater Oklahoma City Chamber of Commerce

(on the previous page) **Photo by Joe Ownbey.**

Preface

....

This book is intended to provide a snapshot of contemporary Oklahoma City, both through photographs and words. But if I were to choose a word that best describes the city right now, that word would be change. Oklahoma City is undergoing dramatic and exciting changes as we approach the 21st century. Changes that will not only make a difference in the look and feel of the area, but will also bring more people to the city, some who are visiting and many who will call it home.

Because Oklahoma City is undergoing these dramatic changes, it was difficult to provide such a "snapshot." In fact, we were making changes and additions to the text of this book right up until the last minute. For me, it was an exciting time to be a part of such a project.

I would like to thank Dixie Reding of the Greater Oklahoma City Chamber of Commerce and the many other Chamber staff members who spent time making the book the best it could be, in spite of their already heavy workloads. Thanks also to Paul Strausbaugh and Cynthia Reid for their assistance and insight.

I am proud to call Oklahoma City home and feel fortunate to be raising my children here. As I complete this project, I am watching another kind of change that happens here every year— the dramatic and colorful change of seasons from winter to spring. What an exciting time to live in this fine city!

Susan Wallace

Photo by Jack Hammett.

Part One

A Better Living.
A Better Life.

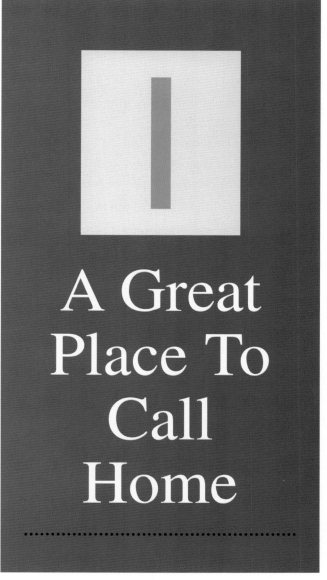

A Great Place To Call Home

Oklahoma City offers the perfect harmony of land, spirit, and progress.

Oklahoma City has a unique personality. A young city full of youthful enthusiasm, it has shown remarkable maturity in its ability to overcome adversity throughout its history. A little more than a century old, Oklahoma City has proven its metal over and over, building a quality of life that is the envy of metropolitan cities across America.

In Oklahoma City, gorgeous, never-ending skies mingle with skyscrapers and sophisticated aircraft. The music of nature, a breeze brushing the leaves and the songs of birds, mingle with the fine talent of a philharmonic orchestra and beautiful voices. Oklahoma City offers the perfect harmony of land, spirit, and progress.

Yet, when you listen closely, you hear the rhythm of a strong, distinct beating heart . . . the beating heart of people as they grow, learn, teach, work, pray, build, heal, play, and love. For with all its natural splendor, sunshine, and green grass, it is the spirit of Oklahoma City's citizens that makes this a great place to call home.

This is Oklahoma City—the men, women, and children, pioneers in their own right, who seek the American dream and hold a strong hope for their children and their children's children. Hardworking, unquestionably friendly, and deeply spiritual, the people of Oklahoma City are sometimes too modest, keeping the secret of their beautiful state all too well, protecting it as their own. Oklahoma Citians are proud of what they've built here and proud of who they are. They are proud and protective of all 625 square miles of this sprawling city and its natural beauty.

Oklahoma City has more days of sunshine each year than most any city in the United States with an average 3,000 hours annually. Each of the four distinct seasons brings an opportunity for residents to enjoy the beauty of the great outdoors. Oklahoma City has a mild climate all year, but frequent weather changes keep local meteorologists on their toes. It's this variety that so many residents love about the changing seasons in Oklahoma.

It's the spirit of Oklahoma City's citizens that make this a great place to call home. (above) Photo by Joe Ownbey. (right) Photo by Erick Gfeller.

Each of the four distinct seasons brings an opportunity for residents to enjoy the beauty of the great outdoors. (left) Photo by Fred Marvel, Oklahoma Tourism and Recreation Department. (above) Photo by Erick Gfeller.

Winters are kind to Oklahoma Citians, with just enough snow to make residents sigh at its beauty and bring children out of their homes for snowball fights and building tall snowmen. But snow doesn't stay around long here, and a sunny day is always around the corner. The average snowfall in Oklahoma City is nine inches. Springs are awesome in Oklahoma City, heralded by the blooming of thousands of Redbud trees (our official state tree), dogwood trees, azalea bushes, and spring flowers. Warm, breezy summers give way not long after Labor Day to the crisp coolness of autumn, when the breathtaking color changes of trees signal the time for football games, fall mums, and pumpkins. Oklahoma City truly has the best of all the seasons.

The Oklahoma City metropolitan area has an abundance of space to play, with 138 parks, 62 public swimming pools, 7 jogging trails, and 5 lakes. With plenty of room to spread out, residents can take advantage of long, quiet walks or toss a ball among green grass, away from the noise of traffic. And fresh air is a given. In fact, Oklahoma City is the largest U.S. city that meets EPA clean-air standards.

Oklahoma City's large land size makes living easy here. Residents enjoy the uncrowded feeling and lack of traffic jams that this community's 625 square miles offer. Just a short drive in any direction of downtown Oklahoma City is one of many smaller communities this metropolitan area has to offer, each with their own rich history and unique characteristics. They include Arcadia, Bethany, Chickasha, Choctaw, Crescent, Deer Creek, Del City, Edmond, El Reno, Guthrie, Harrah, Langston, Lindsay, Midwest City, Moore, Mustang, Nichols Hills, Norman, Piedmont, Purcell, Shawnee, The Village, and Yukon. These communities offer the advantages of small-town life along with the benefits of a large metropolitan area.

The hard-earned money of those living and working in Oklahoma City goes a lot farther here than it would in most other cities. Oklahoma City has one of the lowest costs of living in America, consistently ranking 10 percent below the national

From the redbud trees in the spring to the crisp coolness of autumn, Oklahoma City has the best of all the seasons. (above) Photo by Jack Hammett. (left) Photo by Erick Gfeller. (right) Photo by Fred Marvel, Oklahoma Tourism and Recreation Department.

(above) Feed the Children, an international, nonprofit organization, has distributed more than 225 million pounds of relief commodities to the world's needy. Photo by Jack Hammett.

(left) World Neighbors, founded by John L. Peters, has lifted millions from poverty in 43 countries. Photo by Erick Gfeller.

(right) The site of the Alfred P. Murrah Building stands as a reminder of the April 19, 1995, bombing. Photo by Erick Gfeller.

average, according to American Chamber of Commerce researchers.

In all areas of Oklahoma City, outsiders are sure to comment on the friendliness of the people. When asked what they like best about their new home, newcomers to Oklahoma City most often say its simply how nice the people are. People here are known for their neighborliness in good times and bad. They aren't hesitant to speak to one another and reach out to a neighbor in need. When the Alfred P. Murrah Federal Building was bombed in 1995, the world was amazed at the outpouring of compassion and giving from local residents during the days and months following the tragedy. Oklahoma City pulled together in a way that was only natural to them, but awe inspiring to outsiders.

Oklahoma City has long been known for its spirit of giving and compassion. Charitable organizations are in abundance here and receive great support of time and money from residents, working to care not only for their own, but those in need around the world. Two highly effective world-wide organizations were founded here and continue to base their operations in Oklahoma City. World Neighbors, a people-to-people movement founded by John L. Peters in 1951, has lifted millions from poverty in 43 countries of Asia, Africa, Latin America, and the Pacific Islands. The organization began in Oklahoma City, and its world headquarters remains here. Feed The Children, founded by Larry Jones and headquartered in Oklahoma City, is an international, non-profit Christian organization providing food,

clothing, educational supplies, medical equipment, and other necessities to people who lack these essentials because of famine, drought, flood, war, or other calamities. Since the organization was created in 1979, Feed The Children has distributed more than 225 million pounds of relief commodities to the world's needy. Oklahoma Citians give strong support to other national charitable organizations at work here, such as Habitat for Humanity, the Salvation Army, the American Red Cross, the YMCA, and YWCA. The spirit of giving here has also given birth to a number of unique local mission organizations, such as The Education and Employment Ministry (TEEM),

(below) Canterbury Choral Society of Oklahoma City presents a three-concert series from October to March. Photo by John Jernigan at David G. Fitzgerald & Associates, Inc., courtesy of the Canterbury Choral Society.

(left) The St. Elijah Antiochian Orthodox Christian Church was founded in Oklahoma City 75 years ago and now occupies a new Byzantine-style structure completed this year. The central feature of the church is the dome. Photo by Jack Hammett.

(right) The people of Oklahoma City are hard-working, friendly, and as evidenced by the many beautiful places of worship, a deeply spiritual people. Photo by Fred Marvel, Oklahoma Tourism and Recreation Department.

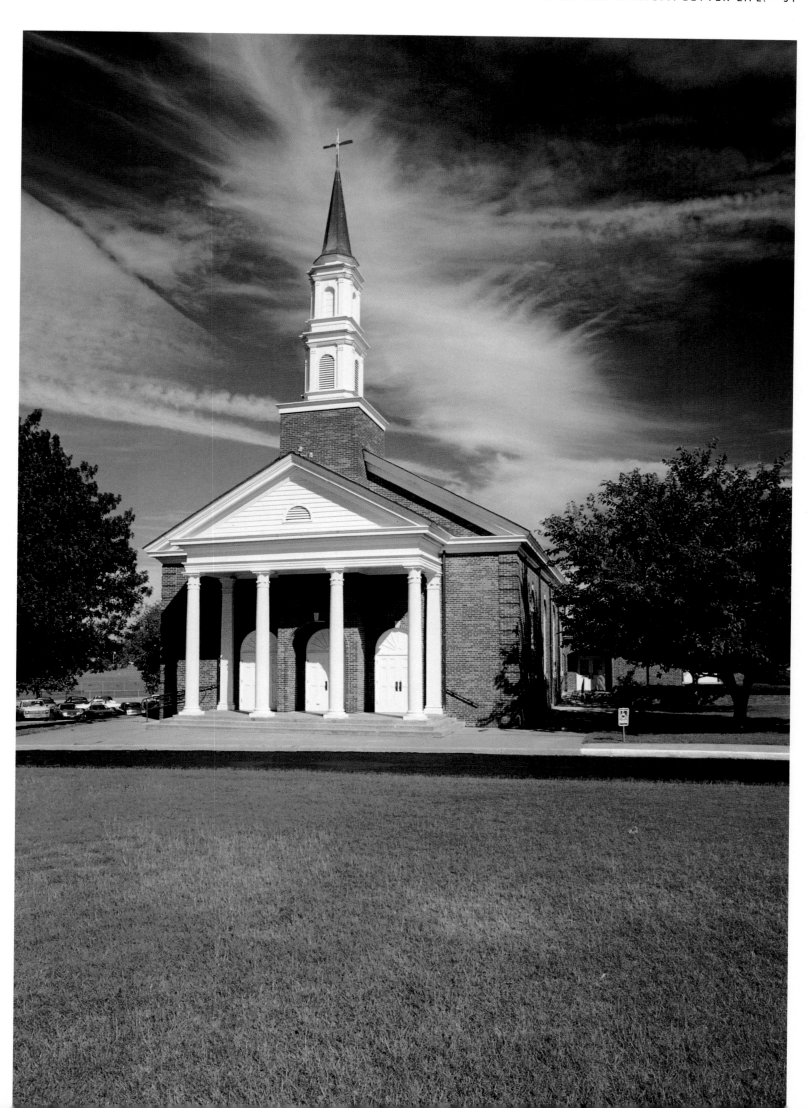

Neighborhood Services Organization, Skyline Urban Ministries, Jesus House, and the City Rescue Mission.

Much of this charitable support is drawn from the strength of local churches and synagogues. Oklahoma Citians are typically very spiritual and many attend church regularly. In Oklahoma City, there are 644 Protestant churches, 24 Catholic churches, and 2 Jewish synagogues, as well as places of worship for those practicing the Bahai and Muslim faiths.

Oklahoma City has a number of beautiful, historic churches near the downtown area that reflect the attitude of spirituality from this town's very beginnings. A number of these were tragically damaged in the 1995 bombing of the Federal Building, but are in the process of rebuilding and are coming back stronger than before. Shortly after the bombing, a banner was hung from atop the First United Methodist Church, located just across the street from the Federal Building, which reads "Our God Remains, We Will Remain." This church, with support from across the nation, is in the process of building a new sanctuary.

Other historic churches in the area include St. Paul's Cathedral, built in 1903 (four years before statehood); St. Joseph's Old Cathedral, dedicated in 1904; Old First Christian Church; First Lutheran Church; First Baptist Church; and First Unitarian Church. These churches stand as monuments amidst the skyscrapers of downtown to the people's longtime faith and values.

(left) St. Joseph's Old Cathedral was dedicated in 1904 and is one of many historic churches remaining in downtown Oklahoma City. It recently underwent extensive repairs after the 1995 Murrah building bombing. Photo by Erick Gfeller.

(below) The strong Jewish faith present in Oklahoma City is represented at two local Jewish Synagogues. Photo by Jack Hammett.

Oklahoma City reflects and benefits from the richness of diversity brought by persons from many cultural backgrounds.

Oklahoma City has built a quality of life that is the envy of metropolitan cities across America. Photo by Joe Ownbey.

Certainly Oklahoma City has a strong history of agriculture, ranching, and oil. These were the industries that shaped the early days, that Oklahoma City is most known for, and they remain strong components of the economy. But the Oklahoma City of today, although very proud of its farming and ranching base, offers a great diversity of career opportunities in a variety of industries, from aviation and manufacturing to medical research and communications.

Oklahoma City is a true melting pot for racial diversity. Asian Americans and Hispanics make up the largest group of new residents to Oklahoma City, and about eight percent of area residents speak a language other than English in the home. Oklahoma is already well-known for its rich Native American culture, of which it is very proud. The state name is derived from two Choctaw words meaning "home of the red man." But the Oklahoma City of today reflects and benefits from the richness of diversity brought by persons from many cultural backgrounds.

With a great quality of life, low cost of living, mild climate, excellent economy, diversity of opportunities, and people that make fine neighbors,

Oklahoma City is a great place to call home. It isn't surprising that *Places Rated Almanac* recently called Oklahoma City one of the "Best Places to Live in North America." This publication rated 343 metropolitan areas in 10 categories from living costs to climate and put Oklahoma City in the elite top 35 as 26th best overall.

The best may be yet to come in Oklahoma City. Residents here have big plans for the future to make Oklahoma City an even better place to live and a great place to visit. Oklahoma City truly offers a better living . . . and a better life. ✪

Downtown Oklahoma City is an interesting montage of architectural styles. Built in 1970, Stage Center (below) received the 1972 Honors Award from the American Institute of Architects. This unique structure was renovated in 1992 and houses two theaters, a Cabaret room, and dance studios.

(left) Even amid the downtown skyscrapers you don't have to look too far to enjoy the beauty of art all around.

(above) The historic Farmers Market today houses a variety of shops specializing in antiques, books, and flea market items. An outdoor market offers fresh fruits, vegetables, and flowers. Photo by Erick Gfeller.

Photo by Jack Hammett.

2

Pioneers & Heroes

....

Leonard McMurry's sculpture representing an '89er staking his claim stands in Kerr Park in downtown Oklahoma City. Photo by Joe Ownbey.

To look at Oklahoma City today, a 625-square-mile metropolis that is home to more than a million people, you would never know that just a little more than a century ago, this land was almost uninhabited prairie grass, soft rolling hills and trees, gently flowing creeks, and red earth. This was the Unassigned Lands. The U.S. government had forcibly relocated Indian tribes from all over the country into the area known as Indian Territory. The Unassigned Lands was the one parcel never given to an Indian tribe.

And then the people came and Oklahoma City was born—in one day. The people came as part of the Oklahoma Land Run in 1889 in hopes of staking claim to a new home and a promising future. There was no certainty of prosperity here, no promise of riches or stable jobs, just a land to call their own and to build on. Oklahoma City truly has one of the most unique beginnings of any American city.

On March 2, 1889, U.S. President Benjamin Harrison signed legislation that opened up the Unassigned Lands for settlement. Some eager pioneers, the "Boomers," had already slipped into the area without authorization but had been led out by the authorities. The very night before the run, another group known as the "Sooners" snuck into the area. Then on April 22, 1889, signaled by the roar of a cannon at high noon, 50,000 homesteaders charged over the boundary line to stake their claim. They came on horseback, in wagons and buckboards, on foot, and even on bicycle. Some even arrived by railroad, for tracks had been laid and a depot erected two years before on the north bank of the North Canadian River, a watering point known as Oklahoma Station. It was not long before nearly 10,000 people had staked out claims near the Oklahoma Station, known today as Oklahoma City.

Soon a tent city had been erected near Oklahoma Station. It was mass confusion at first, with boundary fights breaking out and tents being thrown up in haphazard fashion. Ten thousand people had arrived in one day's time and meant to lay their claim and protect it. Martial law was

A "tent city" sprouted as homesteaders staked their claims near the Oklahoma Station. Photo courtesy of the Greater Oklahoma City Chamber of Commerce.

Local residents celebrate the Run of '89 through a reenactment. Photo courtesy of the Oklahoma City Convention and Visitors Bureau.

declared, and the next day the people began to select a provisional government in order to restore order. Elections were held on May 1 to select permanent officials.

A city was born . . . a boomtown really. Just a month after the run, an organization of business leaders was founded to promote the city and its businesses. The Board of Trade, later known as the Commercial Club, would eventually become the Chamber of Commerce. This group decided that the best way to ensure economic stability in Oklahoma City was to attract the railroads, becoming a merchandising center of Oklahoma Territory. The freight and passenger trains began arriving, and Oklahoma City's economic prosperity was assured.

When statehood came for Oklahoma on November 16, 1907, Oklahoma City's population had grown to more than 14,000. The population continued to grow and by 1910 was nearly 64,000. By this time, downtown was lined with elegant shops and restaurants and fine brick

buildings. Numerous multistory structures were erected, changing the skyline of Oklahoma City. The first true skyscraper in Oklahoma City, the Pioneer Building, was built at Third and Broadway. Other structures included the Colcord Building, the Baum Building, the Skirvin Hotel, the Huckins Hotel, and the Hales Building.

The Chamber had attracted and funded two packing plants to the area for $300,000 each. This was the first significant industrial development in Oklahoma. A number of packing plants were located in what is now known as Stockyards City. These plants employed more than 4,000 people in 1910.

At this time, the State Capitol was located in Guthrie, 30 miles north of Oklahoma City. Many in Oklahoma City thought it should be relocated to their town. Oklahoma City eventually won this right in a popular vote and Governor Charles Haskell declared the Lee-Huckins Hotel as the temporary capitol building. A permanent state capitol was erected at Lincoln Boulevard and

(above) Oklahoma City was born in a day when on April 22, 1889, 50,000 settlers swarmed into the "unassigned lands" of central Oklahoma. Photo courtesy of the Oklahoma City Convention and Visitors Bureau.

(left) The famous Mary Sudick Number One oil well in Oklahoma City blew out and went wild on March 28, 1930. It took a long time to cap and made headlines throughout the world. Photo courtesy of the Greater Oklahoma City Chamber of Commerce.

(right) Stockyards City contains the largest and oldest cattle auction in the world.

(below) Oklahoma City truly has one of the most unique beginnings of any American city. Photo courtesy of the Oklahoma City Convention and Visitors Bureau.

23rd Street and was dedicated in 1917. The neo-classical structure was to have had a dome originally, but due to a shortage of funds, it was never added. The Capitol remains today as one of the few state capitols without a dome.

For many decades, oil played an important role in the economic situation in Oklahoma City. It was a significant day when oil was discovered here on the corner of S.E. 59th Street and Bryant on December 4, 1928. It was discovered by the Indian Territory Illuminating Oil Company, which later became Cities Service, and spewed 110,496 barrels of oil before the great gusher could be capped 27 days later. Oklahoma City became an

oil boomtown. In 1928, more than 100 oil firms had offices or headquarters here. By 1930, the city's skyline was dotted with 135 completed wells and 173 drilling rigs. Oil would pump millions of dollars into the economy. Many of Oklahoma City's finest structures and civic facilities came about as a result of the philanthropy of successful oil businessmen. Oil would continue to be the city's most important financial source until the 1980s. Today, oil plays a smaller role in Oklahoma's economic makeup; however, it still plays a major role in the global energy picture.

It was the spirit of the people that caused this city to be built in just a day and the spirit

of the people that would continue to make it stronger throughout its history. Here are the stories of some of those pioneers and heroes, some who contributed greatly to the growth of Oklahoma City and some who represent it well today through their own fame around the world.

Henry Overholser made the Run of 1889 on the first train from the north. Nearly 50 at the time he came to Oklahoma City, he would become a principal financier and promoter in those early days. Just a few days after the run, he erected eight two-story prefabricated buildings. He later acquired land and built many buildings and is attributed with ushering in the "age of brick" to the city. Henry Overholser was the first president of the Oklahoma City Chamber of Commerce from 1889 to 1891 and played a leading role in bringing the Frisco Railroad to the city. He also made many cultural contributions to Oklahoma City and built the Overholser Opera House. His son, Edward G. Overholser, served as mayor of Oklahoma City from 1915 to 1918 and was a longtime president/manager of the Chamber. Edward was the leader in the building of a city reservoir in 1918, Lake Overholser, which was named for him.

Charles F. Colcord drove in his team and wagon on the day of the run and traded them for a lot on Reno near the railroad tracks where a shanty stood. He became the first police chief under the provisional government and the first elected sheriff. He invested heavily in real estate and organized the Commercial Nations Bank of Oklahoma City. His later success in oil drilling made him a rich man, and he made many contributions to the growth of the city. He was heavily involved in building in Oklahoma City. He built the 12-story Colcord Building in 1910, which is still standing at Reno and Robinson in downtown Oklahoma City. He was also instrumental in building the Commerce Exchange Building and the Biltmore Hotel.

C.G. "Gristmill" Jones came to Oklahoma City in 1890 and established the first flour mill in Oklahoma territory. He became active in the political and economic life of the city. As mayor, he was attributed with luring the Frisco Railroad in 1897. One of his most significant contributions was organizing the first State Fair of Oklahoma in 1907, helping to establish Oklahoma City as a territorial center of ranching and agriculture.

Edward K. Gaylord bought an interest in The Daily Oklahoman in 1903 and was a longtime publisher in Oklahoma City until his death in 1974. He published *The Daily Oklahoman*, *Oklahoma City Times*, and *Farmer Stockman*. He was also president of WKY radio and TV and

(above) The Overholser Mansion is part of the legacy of entrepreneur Henry Overholser. Open for tours, the mansion features original furnishings and hand-painted, canvas-covered walls. Photo courtesy of Oklahoma City Convention and Visitors Bureau.

(right) The Oklahoma State Capitol was designed by Solomon Layton and was built in 1914-17. The working oil well south of the Capitol steps is nicknamed "Petunia" because drilling began in a flowerbed. Photo by Fred Marvel, Oklahoma Tourism and Recreation Department.

Charles F. Colcord drove his team and wagon on the day of the run. With his later financial success, he built the 12-story Colcord Building in 1910.

(on previous page) A series of majestic murals and portraits adorn the State Capitol rotunda.

involved in numerous other business interests. As an active member of the Chamber of Commerce, he was instrumental in luring many packing plants to Oklahoma City. Most notable was his agreement with the packinghouse of Nelson Morris and Company, who agreed to construct a $3 million plant in Oklahoma City in 1908.

Kate Barnard moved to Oklahoma Territory soon after the run and arrived in Oklahoma City after the turn of the century. In 1907, she became one of the first women in the nation to be elected to a state office, taking the office of commissioner of Charities and Corrections. She fought many battles for prison reform and, after two terms, continued to crusade for the poor, sick, and needy.

Frank Buttram was one of many successful oilmen to hit it big after the discovery of oil in

Oklahoma City in 1928. Buttram endowed many civic projects in Oklahoma City. He and his wife, Merle, were strong patrons of the arts, including the Oklahoma City Symphony Orchestra. He was a civic leader as well and served as president of the Chamber in 1939. He was also chairman of the University of Oklahoma Board of Regents and chairman of the Oklahoma branch of the Federal Reserve Bank.

Anton Classen was one of three principal owners of stock in the Metropolitan Street Railway Company along with Charles Colcord and Henry Overholser. The company was granted a charter to build an electric streetcar system, which would bring a boom in housing development on the city's outskirts. Much of this land was owned by Classen, who would prosper from developing it into stately neighborhoods. He platted the addition

that would later become Heritage Hills.

G.A. Nichols is attributed with developing some of Oklahoma City's most distinctive neighborhoods. Nichols spent his youth on a farm near Guthrie and came to Oklahoma City as a dentist in 1904. Capitalizing on the development of a streetcar system that would bring people to the fringes of the city, he bought land and built homes first in the University Addition (now Mesta Park) and then the Winan's addition, which were more than two miles from the downtown business district, a significant distance in that day. He built mainly medium-sized bungalows in these areas. In the early 1920s, he developed the housing additions of Military Park, Central Park, Nichols University Place, Gatewood, and Harndale. Later he built more expensive homes in Lincoln Terrace, south of

Downtown Oklahoma City is home to the anchor from the **USS Oklahoma.** *A plaque at the base of the anchor reads: "Eternal Vigilance is the Price of Liberty"* **USS Oklahoma** *Pearl Harbor, 7 December 1941. Photo courtesy of the Oklahoma City Convention and Visitors Bureau.*

(left) Tinker Air Force Base was named in honor of Maj. Gen. Clarence L. Tinker of Pawhuska, Oklahoma, who lost his life leading a strike against Japanese forces during the early months of World War II. Photo courtesy of Tinker Air Force Base.

(above) Stanley Draper was largely responsible for luring Tinker Air Force Base to Oklahoma City.

the State Capitol. His most famous project, Nichols Hills, began in 1928 as a 2,780-acre development of large mansions that remains today as one of the city's most exclusive neighborhoods.

Stanley Draper was born in North Carolina and came to Oklahoma in 1919 after serving in World War I. He began his career with the Chamber as manager of the dining room and later applied for a membership job with the Chamber and got it. He later served an impressive career as managing director and executive director of the Chamber of Commerce from 1927 until 1968, and was largely responsible for luring Tinker Air Force Base to Oklahoma City. He is considered a legend in the history of Oklahoma City because of his diligent work to secure defense contracts for the city. Under his direction, a 14-member Industries Foundation of Oklahoma was organized to acquire land for defense installations. In February, 1941, the city bid for a major supply depot. Draper and his team convinced the people of Oklahoma City to approve a major bond issue. Tinker soon became a vital link in the national

defense system and would create thousands of jobs for the area. Draper was also influential in the creation of the Urban Action Foundation, which led to a long-range plan for redeveloping the downtown district.

Fred Jones built and operated the Fred Jones Motor Company in Oklahoma City in 1922. By 1926, it had become the largest Ford agency in the Southwest and, by 1929, was the fourth-largest dealership in America. He later opened the Fred Jones Manufacturing Company for rebuilding automobile components. By 1966, it had become the largest Ford-authorized remanufacturer in the country. During World War II, Jones served as a member of the National Defense Advisory Commission.

Sylvan Goldman got involved in chain grocery operations at a young age. He came to Oklahoma City in 1930 and operated a chain of six stores as Standard Food Markets. This chain grew and he later bought the Humpty-Dumpty chain. He is perhaps best known for his invention of the shopping cart in 1936. Goldman endowed many civic

L.A. Macklanburg (right) knew the importance of dependable service. He is shown here helping a hardware dealer set up a display featuring Numetal, the weatherstrip invented and manufactured by Macklanburg in 1920. Photo courtesy of Macklanburg-Duncan.

projects in Oklahoma City, including health care institutions such as the Oklahoma Blood Institute and the Dean A. McGee Eye Institute.

Robert A. Hefner was twice drafted as mayor of Oklahoma City and served from 1939 until 1947. His administration is most remembered for the completion of the $7 million Lake Hefner project and a $15 million public improvement program. Hefner was also a Supreme Court justice and an oilman. His former home at N.W. 15th and Robinson was donated to house the Oklahoma Heritage Association and the Oklahoma Hall of Fame.

Dean A. McGee teamed up in 1937 with Robert S. Kerr of Kerlyn Oil Company to help build what is known today as the Kerr-McGee Corporation. In 1953, he headed a Chamber committee to promote additional scientific thinking and education. This committee would develop into an organization called Frontiers of Science, which attracted international attention. McGee was very active in civic and cultural affairs. He led fund-raising for the Myriad Botanical Gardens in downtown Oklahoma City and was the principal donor for the Dean A. McGee Eye Institute located in the Oklahoma Health Center.

A.S. "Mike" Monroney served Oklahoma in the U.S. House and Senate for more than 37 years. An avid promoter of aviation, he authored the Federal Aviation Act of 1958, which established the agency now known as the Federal Aviation Administration. The Mike Monroney Aeronautical Center in Oklahoma City is named for him, a support facility for the FAA that employs more than 5,000 Oklahomans.

John and Eleanor Kirkpatrick have contributed greatly to cultural, educational, and humanitarian causes in Oklahoma City through the Kirkpatrick Foundation. John E. Kirkpatrick was born in Oklahoma City in 1908, the son of a dentist. He married Eleanor Blake in 1932, the daughter of an early Oklahoma City merchant. A graduate of the Naval Academy, John served in World War II and was awarded two bronze stars. Following the war, he entered the oil business with Hubert E. Bale, forming the company of Kirkpatrick and Bale, which would become Kirkpatrick Oil Company in 1950. Since that time, the Kirkpatricks have contributed millions of dollars to local endeavors. Their first major bequest was to the Oklahoma Science and Arts Foundation, which would later give birth to the Omniplex

Science Museum. Their contributions funded the museum complex of galleries that includes the Omniplex, the Air and Space Museum, the planetarium, and other galleries. They were also major contributors to Oklahoma City University and the National Cowboy Hall of Fame & Western Heritage Center. Other programs that have benefited from the Kirkpatrick Foundation include Lyric Theater, Oklahoma City Zoo, Allied Arts Foundation, Camp Fires, Girl Scouts and Boy Scouts, drug and alcohol rehabilitation programs, numerous colleges and universities, local hospitals, health foundations, churches, shelters for the homeless, black education, and many others.

Edward L. Gaylord has served as chairman and publisher of the Oklahoma Publishing Company since taking over the reins from his father, Edward K. Gaylord, in 1974. He is also chairman of Gaylord Entertainment. Mr. E.L. Gaylord has been strongly involved in the Chamber and in numerous civic activities, serving as chairman of the Oklahoma Industries Authority, president of the Oklahoma City Chamber of Commerce, president of the State

Fair of Oklahoma, vice chairman and trustee of the Myriad Gardens Authority, and numerous other positions with various charitable, educational, hospital, and cultural organizations.

Through history, there have been a number of Oklahoma Citians who were leaders in the retail business world, establishing businesses that were a mainstay for city shoppers for years. These include William J. Pettee, founder of Pettee's Hardware Store; Stephen F. Veazey, founder of the Veazey Drug Store Chain; John A. Brown; C.R. Anthony; and B.C. Clark. Raymond A. Young founded TG&Y Stores that at one time had more than 1,000 stores.

Many Oklahoma Citians have gained national fame through their literary, musical, artistic, scientific, and athletic talents. Among them was author Ralph Ellison, who was born in Oklahoma City in 1914. His first book, *Invisible Man*, was highly acclaimed and won the Russwurm Award, the National Book Award, and the National Newspaper Publishers Award.

John Peters founded World Neighbors in Oklahoma City in 1951. His vision changed the destiny of more than 25 million people and

Early civic leader Robert Hefner is remembered for the completion of the $7 million Lake Hefner project.

(above) The Oklahoma Air and Space Museum was built by contributions from John and Eleanor Kirkpatrick. Photo by Fred Marvel, Oklahoma Tourism and Recreation Department.

(right) By 1926, the Fred Jones Motor Company in Oklahoma City had become the largest Ford agency in the Southwest.

Entire Fire Dept.
OKLA. City. 1923.

STONE PHOTO.

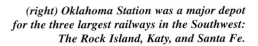

(left) Anton Classen owned much of the land that was developed into stately neighborhoods and would later become Heritage Hills. Photo by Fred Marvel, Oklahoma Tourism and Recreation Department.

(above) The pioneers of yesterday and today include hardworking people such as teachers, bankers, ministers, firefighters, and others who work hard to improve the quality of life. Photo courtesy of Greater Oklahoma City Chamber of Commerce.

(right) Oklahoma Station was a major depot for the three largest railways in the Southwest: The Rock Island, Katy, and Santa Fe.

earned him the nomination of the Nobel Peace Prize. He served as a U.S. Army chaplain during World War II and was inspired to move the world to peace. Later as a United Methodist minister, he inspired the formation of this movement that has lifted millions from poverty in 43 countries.

John Sabolich is founder of NovaCare Sabolich Prosthetics in Oklahoma City and is a pioneer in the prosthetics industry. Today, NovaCare Sabolich is a world-renowned provider of orthotic and prosthetic services. Sabolich developed the patented Sabolich Socket System and has improved the lives of patients from around the world.

The determination of these individuals and countless others whose achievements may not have earned them mention in the history books is what built Oklahoma City and continues to make it strong. The pioneers and heroes of today include businessmen, lawyers, bankers, teachers, ministers, police officers, firefighters, builders, and individuals from every profession who work hard every day to improve their own lives and the lives of others. ✪

3

We Mean Business

Oklahoma City has a diversified economy with more than 1,000 manufacturers, some 6,000 retailers, and numerous distributors, corporate headquarters, and high-tech companies.

Oklahoma City was built out of the hard work of driven individuals. That strong work ethic still thrives today. As a result, many first-rate companies have seen tremendous success in Oklahoma City, and many more are taking a look at the numerous advantages of doing business here.

A perfect location, superior work force, and a great quality of life—combined with a highly competitive incentives package, abundant resources, and an efficient transportation network—make Oklahoma City one of the best choices for locating a business. Pro-business leadership working for a pro-business city located in a pro-business state. . . this has been the underlying spirit in Oklahoma City that has spawned the success of many area businesses, both large and small.

So many companies have reaped the benefits of doing business in Oklahoma City, such as American Floral Service, the second-largest floral wire service in the world, which was born here in Oklahoma City in the founder Herman Meinders' garage. And Macklanburg-Duncan, one of the largest home improvement products manufacturers in the world, has been in business in Oklahoma City for more than 75 years. NovaCare Sabolich, founded by the ingenuity of Oklahoman John Sabolich, develops the most advanced prosthetics in the world and has served patients from all parts of the globe.

(above) The Oklahoma City Business Network works on behalf of the Chamber to coordinate all existing business assistance efforts in the Oklahoma City metropolitan area.

(right) The art-deco style city hall in downtown Oklahoma City was built at the turn of the century. Photo by Joe Ownbey.

Downtown Oklahoma City's unique architecture makes an aesthetically pleasing environment for businesses to locate.

The list goes on and on, making Oklahoma City's economy extremely diverse. No longer does the oil and gas industry primarily drive the state and local economy. Industry here includes agribusiness, aviation and aerospace, construction and real estate, communications, convention and tourism, distribution, energy, government, health care, manufacturing, military, mining, and wholesale and retail trade. The varied industry base in Oklahoma City draws its strength from several key advantages.

LOCATION—THE MID-AMERICA ADVANTAGE

Oklahoma City's location makes it an ideal place for companies of many types to conduct business. Located in the geographical center of the United States, companies can easily conduct business on both coasts. It is an ideal location for companies involved in distribution operations. These companies can reach numerous U.S. locations from Oklahoma City quickly and at a great savings.

Eight federal highways intersect in Oklahoma City, including I-35, I-40, and I-44. With these major interstates converging in Oklahoma City, a large percentage of the United States can be reached by truck overnight, and much of the region, from south Texas to South Dakota and from eastern Utah to western Tennessee, can be reached the same day. Any continental U.S. destination can be reached within a two-day delivery by truck.

Getting in and out of Oklahoma City is easy for distribution traffic as well. The city has a well-engineered system of highways designed for quick and efficient movement. Oklahoma City is known for minimal traffic jams, even during the busiest times of the day.

Oklahoma City's central location is home to a number of successful distribution firms such as Fleming Companies, Wal-Mart, Hobby Lobby Creative Centers, Mac Tools, United Parcel Service, and Federal Express. Numerous motor freight carriers also have terminals in Oklahoma City, including CRST, Donco Carriers, Roadway Express, Waggoners Trucking, Yellow Freight System, and others.

(left) The Hertz Corporation is one of Oklahoma City's top 25 employers with a local work force of 3,000.

(right and below) Oil, gas, and agricultural industries were the primary driving forces behind the local and state economy for many years. Photos by Jack Hammett.

Oklahoma City's location makes for efficient air travel. The city's international Will Rogers World Airport offers an average of 144 scheduled passenger flights daily. It is among the five largest airports in the United States in land area and is easily accessible, just 10 minutes from downtown Oklahoma City. Major airlines operating at Will Rogers include American, Continental, Delta, Northwest, Reno Air, Southwest, Trans World Airlines, United, and Western Pacific. The airport also services two regional carriers, American Eagle and Atlantic Southeast Airlines, as well as a growing number of charter services. Will Rogers is served by all major U.S. air cargo carriers, providing complete air cargo and freight services to U.S. and international destinations.

Will Rogers World Airport is a Port of Entry and has U.S. Customs designation as a Port of Exportation. A Foreign Trade Zone is located on the airport property.

A number of public-use airports in the Oklahoma City area serve private and corporate aviation needs. Wiley Post Airport, located in northwest Oklahoma City, provides a base for more than 300 aircraft ranging from single and twin engine planes to turboprop and jet aircraft. This airport serves numerous business and corporate air travelers and many licensed private pilots. Other air parks serving the Oklahoma City business community include Clarence E. Page

Airport, Downtown Airport, Expressway Airpark, and Westheimer Airport in Norman.

Rail access connects Oklahoma City to all major U.S. cities, markets, and ports. Burlington Northern, Santa Fe, and MK&T offer major rail service to and from the city. A statewide network of more than 4,000 miles of track is a valuable resource to Oklahoma City's manufacturing and distribution companies.

Oklahoma City offers an efficient location for companies doing business in Mexico, with I-35 serving as the corridor into the southern region. Companies doing business from Oklahoma City can easily service the growing Mexico market.

Because of Oklahoma City's location in the central time zone, companies doing business on both coasts benefit from convenient hours in which to conduct phone business. This central location offers measurable savings to companies that use long-distance service extensively.

PRODUCTIVE LABOR FORCE PROVIDES BACKBONE

Oklahoma City's abundant work force has been ranked among the nation's most productive. Workers here are well-known for their strong work ethic and are highly motivated and well-trained. Local residents enjoy a lower-than-average cost of living and therefore, industrial wage rates are consistent with or below the national average. Absenteeism,

(below) Oklahoma City's central location is an important factor to many successful businesses such as Federal Express. Photo by Jack Hammett.

(right—top and bottom) Sonic Corporation is one of many quality corporations that have chosen Oklahoma City as their headquarters. (bottom right) Photo by Jack Hammett.

(left) Kerr Park in downtown Oklahoma City offers a lovely resting place for downtown workers and visitors.

work stoppages, and turnover levels are also below average.

Oklahoma City's work force was recognized by *World Trade* magazine as one of the top 10 nationwide for global competitiveness and as one of the top 10 in the nation for international business.

Several Oklahoma City companies have received top-quality awards thanks to the area's excellent and committed work force. Local divisions of Lucent Technologies and Xerox played a significant role in their companies receiving the coveted and highly acclaimed Malcolm Baldrige Award. Lucent Technologies' plant in Oklahoma City has achieved certification from

the International Standards Organization and was also the first manufacturer of electronics equipment in the United States to be designated as a STAR work site by the U.S. Occupational Safety and Health Administration. The local General Motors plant, one of Oklahoma City's top 10 employers, was designated in 1993 as the number one North American plant for quality. The plant was also recognized by a major national survey firm for building the highest-quality car in North America. Autocraft Industries, formerly Fred Jones Industries, became the first company in Oklahoma to receive the Big Three automakers strictest quality certification, QS-9000. This automotive remanufacturing and electronics

(right) Macklanburg-Duncan has been manufacturing building products in Oklahoma City for more than 75 years. One of the largest home improvement products manufacturers in the world, its products are sold across the United States and in 23 foreign countries. Photo by Erick Gfeller.

(below) Hobby Lobby Creative Centers have grown into an operation of 138 stores in 16 states since its beginnings in Oklahoma City in 1972. Photo by Erick Gfeller.

supplier was also one of the first in the country to earn the prestigious designation.

INCENTIVE PACKAGES EXEMPLIFY PRO-BUSINESS ATTITUDE

Oklahoma offers one of the most unique and powerful incentive programs in the country. The Oklahoma Quality Jobs Program offers significant incentives to new or expanding businesses in Oklahoma. It provides quarterly cash payments to qualifying companies of up to five percent of new taxable payroll for a 10-year period. For most companies, this means receiving a minimum of more than a million dollars over a decade. Many companies receive several million dollars. The program was created through an act in the Oklahoma legislature to spur economic growth by attracting new companies and encouraging existing companies to expand. Quality Jobs illustrates Oklahoma's proactive commitment to the business community.

Since the inception of the Quality Jobs Program in 1993, more than 100 companies have signed on, creating more than 6,000 new jobs in the metro area alone. Companies benefiting from payments now include America Online, API, Autocraft Industries, The Boeing Company, Seimans, American Paging, Southwest Airlines, DataCom Sciences, Hertz, Freymiller Trucking, Fleming Companies, Lucent Technologies, and MATRIXX MARKETING.

In addition, the state of Oklahoma offers a five-year ad valorem tax exemption, investment/new jobs tax credit, and sales tax exemptions for manufacturing, as well as state and local industrial financing programs.

Aggressive, quality-minded programs such as these have assisted the Greater Oklahoma City Chamber of Commerce in attracting new business and assisting existing businesses to bring new jobs to the area. With the ability to offer the most competitive incentives package in the United States today, the growth possibilities for Oklahoma City are astounding.

LAND OF ABUNDANCE OFFERS PLENTY FOR LESS

Oklahoma City businesses benefit from the state's rich supply of resources. Oklahoma offers abundant reserves of natural gas, petroleum, gypsum, iodine, coal, lumber, helium, and silica sand. The land has been good to Oklahoma residents, providing more than adequate commodities such as winter wheat, grain sorghum, cotton, soybeans, peanuts, corn, alfalfa, beef cattle, hogs, and poultry. Oklahoma also has more horses per capita than any state in the United States.

The FAA's Mike Monroney Aeronautical Center is an important component of the dynamic aerospace industry in Oklahoma City. Photo by Erick Gfeller.

National defense has played a role in the making of Oklahoma City. Tinker Air Force Base has proven the importance of its strategic position in the geographical center of the nation.

Water is plentiful in Oklahoma City. City leaders throughout history have provided a well-designed system of lakes and reservoirs, providing more than enough water to meet both residential and business needs.

Businesses also benefit from utility rates that are among the nation's lowest. Oklahoma's industrial electric and gas rates are highly cost-competitive and energy is in good supply.

SMALL BUSINESSES OFFERED BIG SUPPORT

Community leaders recognize the vital role played by small business in the Oklahoma City economy. As a result, a number of unique services have been made available to help small business succeed.

The Oklahoma City Business Network works on behalf of the Chamber to coordinate all existing business assistance efforts in the Oklahoma City metropolitan area. This network closely coordinates the activities of more than 40

organizations working toward the success of the small business owner. The network has its own facility in downtown Oklahoma City and houses the Business Information Center and the Small Business Development Center.

The Business Information Center is a partnership program with the U.S. Small Business Administration, the Greater Oklahoma City Chamber of Commerce, and the University of Central Oklahoma. It is a one-stop shop serving as a self-service, no-fee multimedia information source for small businesses. Resources include hundreds of volumes of books on business ownership, business start-up manuals, computer workstations loaded with the most advanced applications, videos and reference manuals, and a trained staff. The center also offers counseling from volunteers with the Service Corps of Retired Executives (SCORE). The SCORE staff is experienced in a variety of areas including accounting, law, manufacturing, wholesaling, and government agencies.

The Small Business Development Center offers low-cost seminars and workshops designed to guide participants through all phases of business ownership, as well as one-on-one consultations.

DIVERSE BUSINESS CLIMATE OFFERS STABLE ECONOMY

Fifteen years ago, Oklahoma City's economy was highly vulnerable to the fluctuations in the energy industry. But today, Oklahoma City has a much more balanced mix of industries from the more traditional wholesale and retail trade, mining, services, and government to highly technological industries such as cellular, satellite technology, communications, online services, and medical technology. As a result, Oklahoma City's economic fluctuations tend to follow national trends more closely.

Communications is one area of recent growth for Oklahoma City. Many companies have chosen to locate reservations centers here, including Hertz and Southwest Airlines. A number of large inbound and outbound marketing firms such as Teleservice Resources, ITI, and MATRIXX MARKETING are taking advantage of the cost savings afforded by Oklahoma City's central location. America Online chose Oklahoma City as home to its new telephone customer support center. Industry giants such as Southwestern Bell Telephone and Lucent Technologies are top employers in the area and contribute to making Oklahoma City a base for state-of-the-art technologies affecting business around the world.

Oklahoma City has also seen positive growth in the area of manufacturing with quality, forward-thinking companies such as General Motors, Seagate, Xerox, Seimens, Lucent Technologies, and Autocraft Industries.

FORWARD OKLAHOMA CITY— THE NEW AGENDA

In 1995, the Greater Oklahoma City Chamber of Commerce launched a campaign to raise $10 million toward an aggressive economic development plan. The campaign exceeded its goal in

Shamu III was dedicated in 1994 to commemorate Southwest's status as the Official Airline of Sea World. Southwest Airlines has benefited from the Oklahoma Quality Jobs Program.

(above) Many companies such as the American Floral Service have reaped the benefits of doing business in Oklahoma City. Photo by Erick Gfeller.

(right) America Online chose Oklahoma City as home to its new telephone customer support center. Photo by Erick Gfeller.

A Lucent Technologies employee at the Oklahoma City plant tests circuit packs for the Subscriber Loop Carrier (SLC®) access transmission equipment produced for telephone network providers around the world. The factory provides switching and transmission equipment in more than 50 countries. Photo by Jim Argo, courtesy of Lucent Technologies, Oklahoma City.

TOP 25 LOCAL EMPLOYERS

COMPANY	EMPLOYED
State of Oklahoma	37,400
Tinker Air Force Base	22,000
Oklahoma Health Center	12,900
Oklahoma City Public Schools	5,500
INTEGRIS Health	5,442
FAA Aeronautical Center	5,000
City of Oklahoma City	4,297
General Motors Corporation	4,200
Lucent Technologies	4,100
University of Oklahoma	3,500
Hertz Corporation	3,000
U.S. Postal Service	2,095
Southwestern Bell Telephone	2,033
Seagate Technology	2,000
St. Anthony Hospital	1,858
Mercy Health Center	1,750
Norman Regional Hospital	1,657
Dayton Tire Company	1,500
Oklahoma Gas & Electric	1,500
County of Oklahoma	1,472
Homeland Stores	1,402
Fred Jones Autocraft Industries	1,311
Deaconess Hospital	1,240
Fleming Companies	1,200
Unit Parts Company	1,200

early 1996 with the support of 130 Oklahoma City area companies. *Forward Oklahoma City— The New Agenda* will generate well over 28,000 new jobs, $450 million in new capital investment, $490 million in new payroll, and $356 million in new retail sales over a five-year period. Each company made a commitment ranging from $5,000 to $1.2 million.

Forward Oklahoma City will position Oklahoma City for unprecedented growth through seven initiatives. These include marketing Oklahoma City to the world, enhancing the city's pro-business environment, promoting private sector development with the Metropolitan Area Projects program, expanding existing business through a coordinated business assistance network, expanding international trade and recruiting foreign investment through a World Trade Council, expanding Tinker Air Force Base's role and preserving Air Logistics Center missions, and developing infrastructure for technology transfer.

Forward Oklahoma City results from the long-term vision and strategic planning of local business leaders that will benefit the entire community. ✺

In 1980, Seagate Technology produced the computer industry's first 5.25-inch hard disc drive. Today, the company's data storage products include the broadest range of hard drives available, including the award-winning Hawk™ family designed by Seagate's Oklahoma City Operations. Photo courtesy of Seagate Technology.

By 1966, Fred Jones Manufacturing Company had become the largest Ford-authorized remanufacturer in the country. Photo by Jack Hammett.

(above) Millions of American consumers enjoy the benefits of shopping at a Fleming-supported supermarket. Oklahoma City-based Fleming is a leading provider of marketing and distribution services for supermarket enterprises. Photo courtesy of Fleming Companies.

(left) The Oklahoma City General Motors plant is one of their top sites in the United States and the home of the Chevrolet Malibu. With more than 4,200 employees, it is one of Oklahoma City's top employers. Photo courtesy of General Motors.

Turn-of-the-century architecture mingles with the more modern glass structures in downtown Oklahoma City.

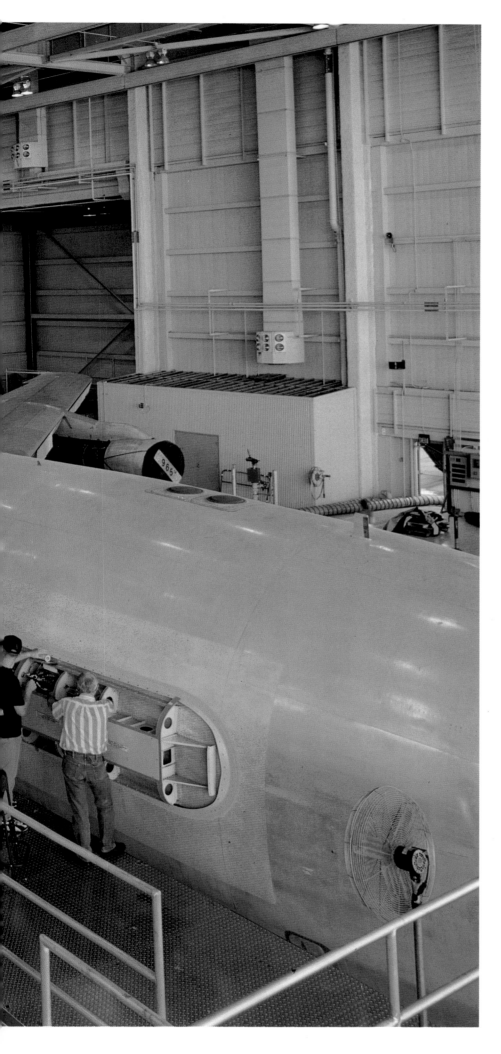

4

High Flying History... High Flying Future

Tinker Air Force Base is made up of 5,020 acres with 771 buildings, including 136 acres of indoor maintenance.

Aviation has historically held a strong presence in the Oklahoma City economy and continues to do so today. The city's first airport, Municipal Aviation Park, was dedicated in 1924. One of the first flying schools in the nation would open here that same year, the Berl Tibbs Flying School. In 1928, Oklahoman Paul Braniff, co-founder of Braniff International Airlines along with his brother Thomas, established the first air passenger service from Oklahoma City to Tulsa.

To meet the growing air travel needs in Oklahoma City, the city purchased the tract of land in 1930 that would later expand and become Will Rogers World Airport, a large business complex making a significant economic impact on Oklahoma City and the entire state. Today, the 67 airport tenants include the Mike Monroney Aeronautical Center, the Federal Transfer Center, Southwest Airlines Reservation Center, and Metro Tech Aviation Career Center.

The Mike Monroney Aeronautical Center (MMAC) is a vital service and support facility for the Federal Aviation Administration (FAA) and the U.S. Department of Transportation. It has been located in Oklahoma City for more than 50 years. With more than 5,000 employees, it has the largest concentration of employees in the U.S. Department of Transportation (DOT). It is located on the west side of Will Rogers World Airport in an area that housed the U.S. Army Air Corps during World War II. Today, it occupies 20 major buildings and 35 smaller structures on nearly 1,000 acres. The center is named for Oklahoma Senator A.S. "Mike" Monroney, an avid promoter of aviation who authored the Federal Aviation Act of 1958, establishing the agency now known as the FAA. All area FAA facilities were financed by the Oklahoma City Airport Trust.

Almost all U.S. aviation activities are tracked by the MMAC's operations. In addition, all of the nation's air traffic controllers are trained here. Other functions include Aircraft and Airman Registry and the payroll functions of the DOT.

Other aviation related firms located in the city include Gulfstream, a manufacturer of aircraft parts and subassemblies; Commander, an aircraft manufacturer; and Chromalloy, a company that repairs and overhauls aircraft engines and parts.

(right—top and bottom) Will Rogers World Airport is one of the 10 largest in the country with abundant room for expansion.

*(below) Presidential flights in aircraft with ties to Oklahoma City began years ago and continue today. Here, John F. Kennedy checks the flight deck of an early twin engine **Commander** manufactured in Oklahoma City. President Clinton enjoys the use of an Oklahoma City Gulfstream jet. Photo courtesy of Gulfstream Aerospace Technologies.*

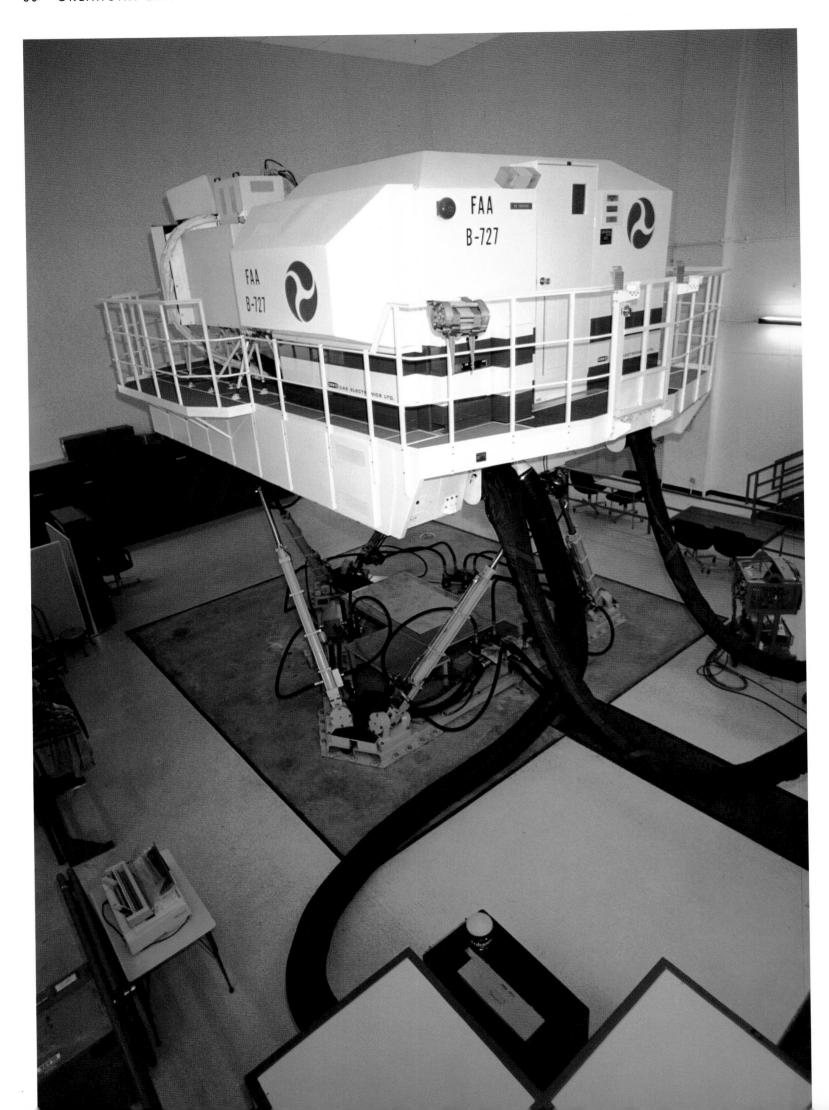

(left) The Federal Aviation Administration's Mike Monroney Aeronautical Center trains pilots in flight simulators like this one. Photo by Erick Gfeller.

(above) The Russian Anatov AN-225 "Mechta" meaning dream in Russian, is the world's largest plane (290 feet from wingtip to wingtip). The plane departed from Aerospace America '90 with over 1,000,000 pounds of food and medicine from Oklahoma City's Feed The Children for the children of Chernobyl. This was the first U.S. appearance of this huge aircraft.

(right) With more than 5,000 employees, the Mike Monroney Aeronautical Center has the largest concentration of employees in the U.S. Department of Transportation. Photo by Erick Gfeller.

TINKER AIR FORCE BASE SERVES AMERICA AND OKLAHOMA CITY WELL

Tinker Air Force Base has not only played a vital role in the defense of our nation during its more than 50-year history, but in the Oklahoma City and state economy as well. Today, Tinker employs nearly 22,000 and with an annual payroll of $806.5 million has an annual economic impact of $1.9 billion (October 1996 figures). Many feel that the establishment of the base was the most significant turning point in Oklahoma City's economic history.

It was the efforts of the Greater Oklahoma City Chamber of Commerce that brought the facility to Oklahoma City in the early 1940s. The Chamber, directed by Stanley Draper, formed the Industries Foundation in 1940 for the purpose of raising funds and acquiring land to attract military settlement. The Foundation soon learned that the Army Air Corps was considering locating an air depot in the Midwest, and Oklahoma City was one area being considered. The Industries Foundation, under the chairmanship of E.K. Gaylord, quickly went to work to pursue the depot. Over the next year, the Foundation and the Chamber would work diligently to meet all the demands of the Army Air Corps, promising adequate land, workers, and community support. Oklahoma City seemed the perfect place for an

Maj Gen C L Tinker
United States Army

(top left) Tinker Air Force Base was named for Maj. Gen Clarence L. Tinker, a native Oklahoman who lost his life while leading bombers on a long-range strike against Wake Island in World War II. Photo courtesy of Tinker Air Force Base.

(bottom left) Tinker's largest organization is the Oklahoma City Air Logistics Center, one of five depot repair centers in the Air Force Materiel Command. Photo courtesy of Tinker Air Force Base.

(above) It was the efforts of the Greater Oklahoma City Chamber of Commerce, directed by Stanley Draper, that brought Tinker to Oklahoma City in the early 1940s. Photo courtesy of the Greater Oklahoma City Chamber of Commerce.

air base, with an average 328 flying days each year, flat terrain, and relatively mild winters. Official announcement that the depot would be located in Oklahoma City was made by the War Department on April 8, 1941. Oklahoma City voters soon approved a $928,000 bond issue for the purchase of the land. City and county voters would later approve an additional $685,000 in bonds for water lines and $1,225,000 for access roads. The facility was originally named Midwest Air Depot, Oklahoma City.

Before completion of the depot, its size would change dramatically from the original concept. In November of 1941, it was announced that the Army Air Forces would make additional expenditures of $6 million to make it one of the largest and most modern facilities in the United States. The U.S. House of Representatives had just approved the expenditure days before the Japanese struck Pearl Harbor on December 7, 1941. When the U.S. Senate approved the bill on

December 11, 1941, America was at war.

The total construction cost of the depot exceeded $21 million. This alone was a boon for the Oklahoma City economy following a decade of economic depression. Meanwhile, as the base began hiring civilian employees, its payroll impact would have an even greater positive effect on the growth of our city.

In February of 1942, the facility's new commander, Lt. Col. William Turnbull, changed the name to Oklahoma City Air Depot. The base was officially activated on March 1, 1942. Later, upon the suggestion of the Oklahoma City Chamber, the base was renamed Tinker Field in honor of native Oklahoman Maj. Gen. Clarence L. Tinker who lost his life in service during early World War II while leading his bomber command at Wake Island in Japan. Tinker Field officially become Tinker Air Force Base in 1948.

Throughout its history, Tinker has played important roles during wartime and military

Tinker's physical plant has become by many standards a "megafacility."

B-2 engine, avionics, and software work is completed at Tinker Air Force Base.

crises. For the remainder of World War II, personnel repaired B-24 and B-17 bombers and fitted B-29s for combat. During the Korean War, Tinker contributed support by keeping planes flying and funneling supplies to the Far East. The base played a key role during the Cuban crisis and throughout the Vietnam War, providing logistics and communications support to Air Force units in Southeast Asia. Tinker also provided frontline support to the forces engaged in Operation Desert Shield and Desert Storm during the early 1990s.

Tinker has survived two rounds of base closings in the past decade thanks to the excellent service it provides, superior teamwork, and a strong commitment from the surrounding communities. Today, it is the largest single-site employer in Oklahoma and has the potential to become even stronger. The base is home to the Oklahoma City Air Logistics Center, the 72nd Air Base Wing, and seven other associate organizations. The Air Logistics Center (ALC), Tinker's

largest organization, is one of five depot repair centers in the Air Force Materiel Command. The ALC provides logistics support to aerospace defense weapon systems, managing a wide range of aircraft, engines, missiles, and commodity items worldwide. Under its management are 2,267 aircraft, including the B-1, B-2, B-52, C/KC-135, E-3, VC-25, VC-137, and 25 other Contractor Logistics Support aircraft. The center also manages more than 13,000 jet engines, including B-2 engines such as the F118. The ALC manages missile systems as well, including the Air Launched Cruise Missile, Short Range Attack Missile, Harpoon and Advanced Cruise Missiles. The 72nd Air Base Wing hosts many critical base functions for Tinker. Its assigned organizations include 72nd Medical Group, 72nd Support Group, 72nd Civil Engineer Group, 72nd Operations Support Squadron, 72nd Logistics Directorate, the Base Chapel, and the offices for Plans, Social Actions, International Military Students and Arms Control.

Other tenant organizations include the 552nd Air Control Wing, the Navy's Strategic Communications Wing ONE, 38th Engineering Installation Wing, Oklahoma's Air Force Reserve Flying Unit—the 507th Wing, the 3rd Combat Communications Group, the Defense Distribution Depot Oklahoma, and the Tinker Information Processing Mega Center. The 552nd Air Control Wing flies the E-3 AWACS aircraft.

Tinker Air Force Base has strong bonds with its surrounding communities in the Oklahoma City metropolitan area. The base reaches out to community organizations by providing volunteer hours and financial support. In April of 1995, within minutes of the Murrah Federal Building bombing, Tinker workers were on the scene aiding in rescue and relief efforts. Since that time, Tinker employees devoted more than 46,000 hours of professional and volunteer support to activities related to the bombing.

Tinker Air Force Base will continue to bring more jobs to the city. In 1996, the Senate Armed Services Committee made a decision to allow Tinker and other bases to compete for military maintenance work contracted out to the private sector. In addition, The Boeing Company announced that it would create more than 1,100 new jobs at Tinker over the next six years. Northrop Grumman Corporation will also bring new jobs to Tinker— more than 400 during the next three years. They will open 15 software laboratories at the base to support the B-2 radar-evading bomber. All of these additional jobs will make a $20 billion impact on the community over the next 20 years. ❂

More than 2,000 people attended the naming ceremony for the Spirit of Oklahoma *B-2 Bomber.*

Dignitaries help celebrate the **Spirit of Oklahoma** *B-2 Bomber on display outside a hangar near a Tinker runway.*

5

The Finer Things In Life

The seven-story Crystal Bridge Tropical Conservatory features a fascinating collection of palm trees, flowers, and exotic plants from across the globe. Photo by Jack Hammett.

While Oklahoma City may be best known for its Native American culture and cowboys and rodeos, the art community of Oklahoma City reflects the melting pot that exists here. The tastes and priorities of a town are reflected in where its people choose to commit their time and resources. Residents here have long been committed to a variety of art forms and give strong support to their Philharmonic, ballet, modern dance, theaters, museums, and choirs.

The variety of opportunities to enjoy the arts contributes to the high quality of living in Oklahoma City. Throughout the year, residents flock to the theater, take their families to art museums, stroll through one of many excellent arts festivals, or treat themselves to a special afternoon or evening at the Civic Center Music Hall to enjoy some of the world's most talented performers. When it comes to fine arts, Oklahoma City has it all.

Orchestral music has held a 70-year tradition in Oklahoma City. Today, the Oklahoma City Philharmonic Orchestra continues that tradition as it begins its ninth season. The Philharmonic offers an annual schedule that includes nine Classics Series concerts, 14 Pops Series concerts, 3 Family Series concerts, and 4 Youth Concerts for grade-school children in the metropolitan area and provides orchestra services to Ballet Oklahoma and Canterbury Choral Society. The Philharmonic also participates in a number of special programs each season, including concerts around the state of Oklahoma and the award-winning classroom music education program, "We've Got Rhythm."

Each season, the Oklahoma City Philharmonic Orchestra features special performances by the world's leading artists. The Philharmonic is under the direction of Maestro Joel Levine, who has been associated with the

The National Cowboy Hall of Fame and Western Heritage Center, covering 220,000 square feet on 20 acres, preserves the rugged individualism and spirit of the frontier.

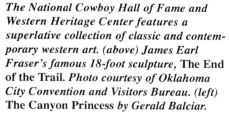

The National Cowboy Hall of Fame and Western Heritage Center features a superlative collection of classic and contemporary western art. (above) James Earl Fraser's famous 18-foot sculpture, **The End of the Trail.** *Photo courtesy of Oklahoma City Convention and Visitors Bureau. (left)* **The Canyon Princess** *by Gerald Balciar.*

The Myriad Botanical Gardens is a 17-acre oasis in the heart of downtown Oklahoma City with beautifully landscaped rolling hills surrounding a lake.

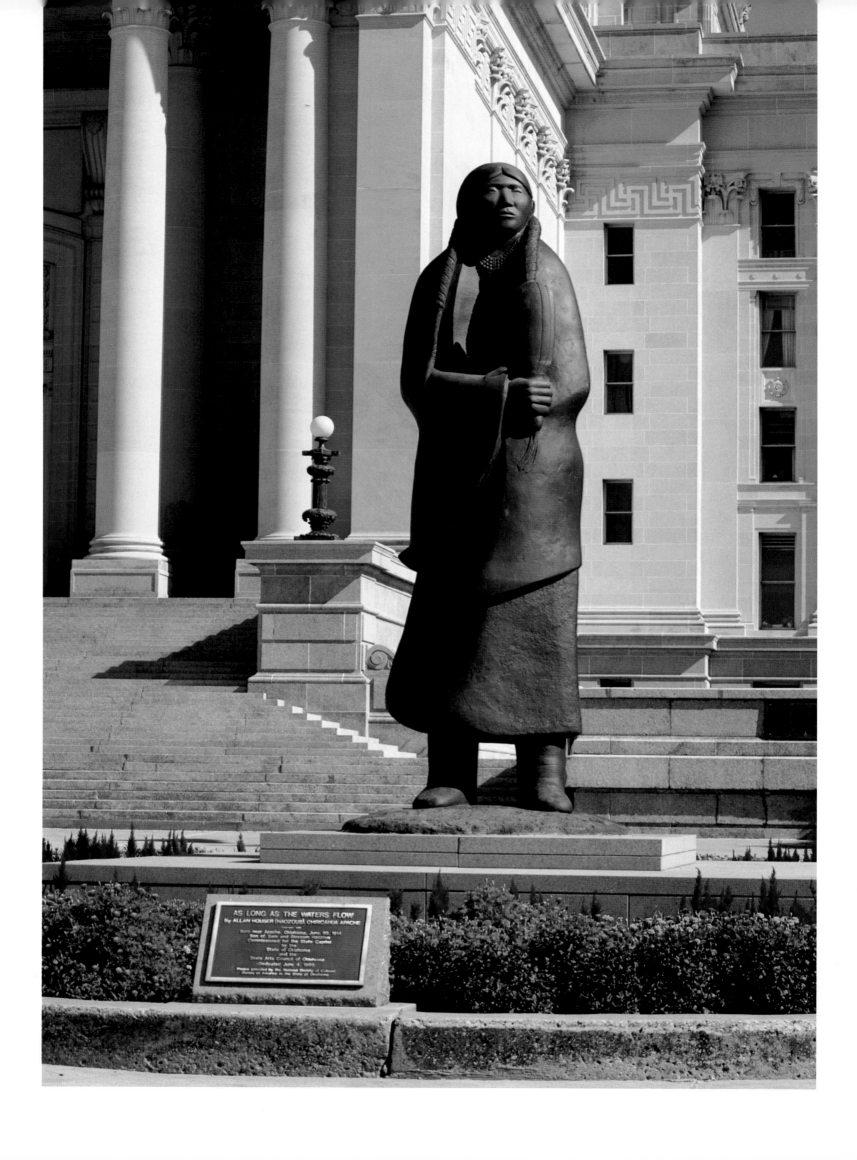

AS LONG AS THE WATERS FLOW
By ALLAN HOUSER (HAOZOUS) CHIRICAHUA APACHE

(left) Indian heritage is celebrated on the South Capitol Plaza with a monumental sculpture by noted Indian sculptor Allen Houser. Photo by Fred Marvel, Oklahoma Tourism and Recreation Department.

(above) The Governor's Mansion, restored through the efforts of First Lady Cathy Keating and hundreds of private donors, offers an elegance of a bygone era. Photo by Fred Marvel, Oklahoma Tourism and Recreation Department.

performing arts in Oklahoma City for more than 20 years. Levine has conducted extensively throughout the United States, Mexico, and Europe for a wide variety of dance, theater, and symphonic performances. He is considered to be one of the most versatile and active conductors in America today. Oklahoma Citians love Maestro Levine and attest to that through record attendance at Philharmonic concerts season after season in the Civic Center Music Hall. Levine was named "Oklahoma Musician of the Year" for 1992 and was a recipient of the 1989 Governor's Arts Award.

Ballet Oklahoma, Oklahoma City's resident professional ballet company, recently celebrated its 25th anniversary season. Directed by Bryan Pitts, its company of 15 full-time dancers perform four programs each season for a total of 17 performances at the Civic Center Music Hall, which includes 8 performances of the *Nutcracker*, an annual tradition for many Oklahoma City families. Ballet Oklahoma also offers a School of Ballet, which provides preprofessional training as well as recreational dance instruction. Its Arts Education Live Performance program includes an eight-week series of classes that culminate with a live performance at the Civic Center Music Hall.

An exceptional collection of more than 3,000 works of art are on display year-round at the Oklahoma City Art Museum. This privately funded museum, located at the State Fairgrounds, features mainly 19th and 20th-century American Art. Also a part of the collection are European and American prints and drawings dating back to the 16th century, including works by unknown Italian masters as well as Camille Corot and Pablo Picasso. The highlight of the museum's offering is the Washington Gallery of Modern Art collection, which focuses on contemporary American art of the 1950s and 1960s. This collection includes works by Ellsworth Kelly, Richard Diebenkorn, and Robert Indiana. The museum also features a number of special exhibits throughout the year that are regional, national, and international in scope. The Art Museum building itself features a very unusual circular architecture from the late 1950s.

The artistic talent of African-Americans from Oklahoma and around the world are showcased through the Black Liberated Arts Center. B.L.A.C., Inc. celebrates the Black American culture through a series of plays and musical events held each year at the Civic Center Music Hall. Nationally and internationally known guest artists

The finest Native American artists, dancers, and singers from more than 100 tribes across the United States and Canada can be seen each June at the Red Earth Festival in Oklahoma City. Photos by Fred Marvel, Oklahoma Tourism and Recreation Department.

participate in a variety of performances. The season ends with the Deep Deuce Jazz Festival on 2nd Street each May, an event that honors jazz great Charlie Christian and celebrates the rich heritage of jazz greats from Oklahoma City. Christian, a famous guitarist who pioneered Jazz guitar, was featured with the Benny Goodman orchestra in his day.

The 150-voice Canterbury Choral Society combines the talent of some of the finest vocalists in the Oklahoma City metropolitan area. Their repertoire ranges from folk to the major choral masterworks with accompaniment by the Oklahoma City Philharmonic Orchestra. Music Director and Conductor Dennis Shrock leads the chorus in presenting a three-concert series each year at the Civic Center Music Hall. The group also performs at numerous community events throughout the year.

Oklahoma City is home to a number of first-class theaters offering a variety of plays and musicals throughout the year. Carpenter Square Theatre

is located downtown in a converted retail store and presents a number of works from contemporary playwrights, including comedy, drama, and musicals by an outstanding theatrical cast and crew.

The Jewel Box Theatre has been a favorite for nearly 40 years and is Oklahoma City's oldest community theater. This theater-in-the-round has been the recipient of the Governor's Arts Award for providing continuing excellence in live theater. In addition to its theater performances presented August through May, the Jewel Box presents two outdoor performances in the summer.

Oklahoma City's only professional musical theater, Lyric Theatre, has just completed its 35th successful season. Its summer season each year includes a number of Broadway hits and is the delight of musical lovers in Oklahoma City.

The Oklahoma Opera and Music Theater at Oklahoma City University has offered many unforgettable theatrical performances over the years. Its season includes magnificent music, lavish costumes, and spectacular sets performed by some of the country's most talented performers.

Country and western music is a part of Oklahoma City's culture, and many Nashville greats have gotten their start in Oklahoma. The Oklahoma Opry, located in the historic Capitol Hill area, features regular Saturday night live country and gospel performances in its unique 800-seat auditorium. The Opry has grown to be one of the most successful and respected country music shows in the Southwest since it opened in 1977. The stars of tomorrow are showcased, backed by the popular Oklahoma Opry Band.

Prairie Dance Theatre is considered one of the nation's most unique modern dance companies. This talented group weaves traditions of the Indian with contemporary concerns and comedy, drawing lessons from the first Americans on saving our planet and our spiritual selves. Their performances are a delight to all ages. They perform three times a year in Oklahoma City and tour through nine states the remainder of the year, presenting modern dance, concerts, and special children's shows. They also provide children's workshops. Prairie Dance Theatre has been performing for 14 years and was Oklahoma's first modern dance company.

The City Arts Center, an arts education facility, offers exhibitions by local and regional artists featuring a variety of art mediums. Exhibits change every four to six weeks, and the Gallery Shop offers unique handcrafted gifts. Children can take a class in storytelling or participate in special art projects. On Friday Family Days, the whole family can enjoy a pizza dinner and then take a choice of classes in the arts.

The City Arts Center is also home to Oklahoma Children's Theatre. Children enjoy delightful performances of favorite stories such as *Puss in Boots, Cinderella,* and the *Princess and the Pea*. On weekends, children have the opportunity to meet the cast after the show.

The Paseo is Oklahoma's only artists' community, boasting dozens of artists, studios, restaurants, and shops, as well as an annual arts festival. (left) Photo by Joe Ownbey. (above) Photo by Fred Marvel, Oklahoma Tourism and Recreation Department.

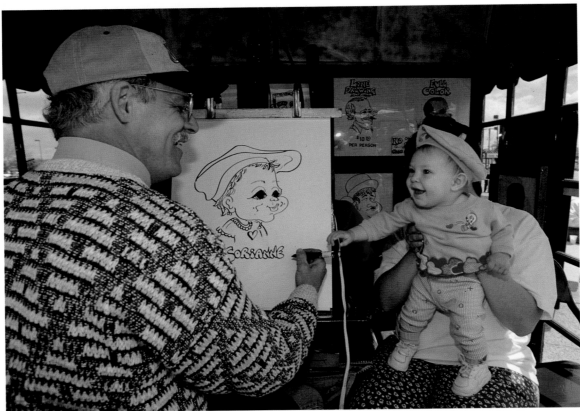

Drawing an audience of more than 750,000 annually, the Festival of the Arts has become the premier event on Oklahoma City's cultural calendar. Photo of musician by Fred Marvel, Oklahoma Tourism and Recreation Department. Photo of portrait artist by Jack Hammett.

Children have a special place at the Festival of the Arts located in the Myriad Gardens.

Festival organizers have helped the festival achieve the status of "Top 10 Outdoor Festivals" in the United States. Photo by Jack Hammett.

The festival features artwork created by more than 140 visual artists and six days of performances on three stages simultaneously.

Oklahoma City is fortunate to have a number of very talented local artists who show their work at any of a number of local galleries and festivals throughout the year.

The Oklahoma City Festival of the Arts is a rite of spring here. Held in downtown Oklahoma City each April among the beauty of the Myriad Botanical Gardens, the festival features 144 of the finest artists from around the country. It is considered to be one of the top 10 outdoor festivals in the United States and features six days of art exhibitions, continuous entertainment on three outdoor stages, arts and crafts demonstrations, and delicious international foods from 21 food booths.

The Festival del Paseo is held every Memorial Day weekend in the historic Paseo artist district. This unique neighborhood of Spanish architecture was built in the 1920s and has several blocks of white stucco and rounded porticos. It has become Oklahoma City's only artist's community and is home to a number of small artists' studios and galleries, as well as restaurants and cafes. The artist's colony includes such talent as the internationally known Mike Larsen, as well as a number of well-known regional artists. In addition to the annual festival, the artists host open gallery walks on the second Saturday of every month.

Each Labor Day weekend, south Oklahoma City is home to the Arts Festival Oklahoma held

(left) The Oklahoma Children's Theatre has year-round performances at the multifunctional City Arts Center. Photo by Jack Hammett.

(below) Children of all ages enjoy the special Prairie Dance blend of music, literature, art, rhythm, movement, and humor. Photo by Steve Sisney, courtesy Prairie Dance Theatre.

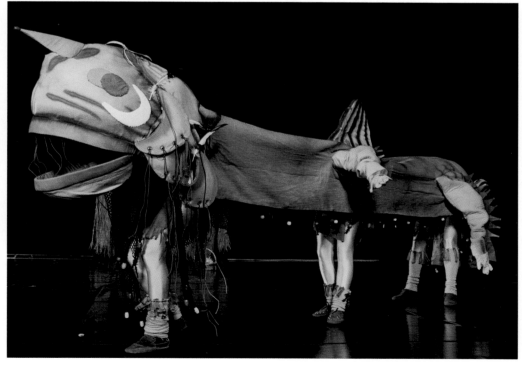

on the campus of Oklahoma City Community College. The festival includes more that 160 arts and crafts booths, a special children's area, music and dance performances, and festival food booths. On Saturday and Sunday evenings of the festival, participants enjoy fireworks and a special performance by the Oklahoma City Philharmonic Orchestra.

Oklahoma City is proud to be home to the National Cowboy Hall of Fame and Western Heritage Center, a monument to the men and women who pioneered the West. The museum opened in 1965 and has grown extensively throughout the years, recently adding a new 60,000-square-foot wing and renovated galleries. An additional 130,000-square-foot addition was completed in 1997. Its expansive collection includes classic and contemporary western art representing the history and cultural diversity of the 17 western states. The museum's permanent collection includes the James Earle and Laura G. Fraser studio collection, the Albert K. Mitchell Russell-Remington collection featuring works by Charles Russell and Frederic Remington, the Joe Grandee western history collection, the John Wayne collection, and the

Doubleday and the Helfrich photographic rodeo collection. Of special interest are Gerald Balciar's 15-ton marble sculpture, *Canyon Princess*, James Earle Fraser's well-known 18-foot-high plaster sculpture, *The End of the Trail*, and five western landscape triptychs by Wilson Hurley, which span 50 linear feet and are 16 feet tall. The museum also includes an authentic old western street complete with sod house, marshal's office, and stage depot, as well as the Rodeo Hall of Fame and beautifully maintained gardens.

The National Cowboy Hall of Fame hosts a number of annual events, including the Prix de West art show and sale, the Bolo Ball, the Chuck Wagon Gathering, the Cowboy Poetry Gathering, the National Children's Cowboy Festival, and the Western Heritage awards. The Western Heritage awards event draws stars from the fields of movie, television, music, and literature to celebrate the great American West during an Oscar-awards type program. This national awards program has been around since 1960. Past award recipients have included John Wayne, James Stewart, Clint Eastwood, John Ford, Barbara Stanwyck, Louis L'Amour, and James Michener.

Home to many country-western music stars, Oklahoma City also boasts the Oklahoma Opry, featuring Saturday night performances. Photo courtesy of the Oklahoma Opry.

*The Oklahoma Firefighters Museum features
extraordinary turn-of-the-century fire engines
that were once used in Oklahoma communities.
Photo courtesy of Oklahoma City Convention
and Visitors Bureau.*

The Omniplex has been called the
Smithsonian of the Southwest. This expansive
complex, located in northeast Oklahoma City
near the Oklahoma City Zoo and the Remington
Park racetrack, houses five museums, a planetarium, galleries, gardens, and greenhouses in a
10-acre complex. The International Photography
Hall of Fame and Museum offers a spectacular
360-degree panoramic laserscape view of the
Grand Canyon. It also features a number of photography exhibitions of some of the world's most
influential photographers such as Mathew Brady,
George Eastman, and Ansel Adams. The
Kirkpatrick Galleries feature fine and contemporary art as well as historical collections and
cultural artifacts from around the world. The
Red Earth Indian Center preserves the American
Indian culture through art and artifacts, educational displays, and hands-on exhibits. The
Hands-On Science Museum, the Kirkpatrick
Planetarium, and the air and space museum, are
also housed at the Omniplex, and featured in
chapter five.

The Red Earth Indian Center also offers the
Red Earth Festival each June in downtown
Oklahoma City. During this unique festival, the
Myriad Convention Center comes alive with
color, tradition, music, and dance as more than
1,500 Native American dancers representing
tribes from across North America gather for the

largest competition festival in North America.
Red Earth gives persons of all cultures an opportunity to experience the Native American spirit
through dance, music, a juried art show, and
food. The festival honors the ceremonial protocol
of the Powwow, yet it differs from the smaller
regional Powwows because it highlights the
dance competition, while the Powwow emphasizes the fellowship aspect. Both Indians and
non-Indians who have contributed to the revitalization of the Native American spirit are recognized during the gathering. The Red Earth
Festival was the recipient of the National Cultural
Heritage Award in 1994.

St. Gregory's College in Shawnee offers the
Mabee-Gerrer Museum. This museum and art
gallery offers a unique collection of Egyptian
artifacts, paintings, and sculpture from the
Renaissance to the early 20th century.

An outdated firehouse in Norman has been
converted into a wonderful art center, the
Firehouse Art Center. The walls and halls feature
rotating exhibits including paintings, sculpture,
photography, bonsai, and more. This unique center also offers a variety of classes to children and
adults with everything from mask making to
jewelry making and photography.

Also in Norman on the University of Oklahoma
campus is a fine collection of contemporary art
treasures housed at the Fred Jones, Jr., Museum of

(left) A taste of pioneer life can be found at the authentic Harn Homestead and Barn, located on the site of an 1889 Land Run claim. Photo by Fred Marvel, Oklahoma Tourism and Recreation Department.

(above) **Ballet Oklahoma has brought the beauty of ballet into the lives of Oklahomans for more than 25 years. They also tour extensively and have earned a reputation that continues to increase their reach. Shown here, A Midsummer Night's Dream.** *Photo by Keith Ball Photography.*

(top right) **The African Art Exhibit is just one of the unique museum offerings housed at the Omniplex. (bottom right) The International Photography Hall of Fame and Museum also offers a spectacular 360-degree panoramic landscape view of the Grand Canyon.** *Photos by Joe Ownbey.*

Art. Established in 1936, it offers contemporary, American, and European art. The permanent collection of nearly 6,000 objects of art includes Mexican masks, Native American paintings, graphic arts, ceramics, and photographs.

In the heart of downtown Oklahoma City is a beautiful oasis called the Myriad Botanical Gardens and Crystal Bridge Tropical Conservatory. The Botanical Gardens opened in 1988 as a product of a downtown revitalization project and offers visitors an amazing collection of exotic plants. Its 17 acres include beautifully landscaped rolling hills around a lake stocked with Goldfish, Golden Orfes, and Imperial Japanese Koi. The highlight of the gardens is the Crystal Bridge, a seven-story conservatory made from 3,028 translucent acrylic panels. Visitors experience the deep green foliage of a rainforest, a gently cascading waterfall, a realistic mountain, and a collection of palm trees, flowers, and exotic plants from around the world. A bird's eye view of the conservatory is offered atop a skywalk. It truly is a tropical paradise complete with chameleons and Zebra Longwing butterflies.

Of course, there is no better way to experience Oklahoma City's rich culture and art than through its historical buildings, preserved for the generations.

The historic Harn Homestead sits on land that was claimed during the Run of 1889 by William Fremont Harn. He later donated seven and a half acres to the state for capitol grounds, but his family continued to live at the home on N.E. 16th for many years. Today, this prestatehood farmhouse stands restored and open to visitors. The six-building homestead includes the farmhouse with furnishings of the period, a restored stone and cedar barn with relics of prestatehood days, the Shepherd House, the 1897 one-room Stoney Point School, a territorial farm, and a 1904 dairy barn. Visitors can get a taste of what it was like living in Oklahoma Territory during the years between the run and statehood.

The Overholser Mansion, built by Oklahoma City pioneer entrepreneur Henry Overholser, was the first mansion in Oklahoma City. Built in 1902 and 1903, it now stands as a museum in the historic Heritage Hills neighborhood. The Victorian-era-style home is open several days a week to

(left) Oklahoma City's Deep Deuce Jazz Festival is a two-day music extravaganza commemorating jazz guitarist Charles Christian. Photo by Fred Marvel, Oklahoma Tourism and Recreation Department.

(below) Revelers enjoy celebrating the German heritage during a local Germanfest celebration. Photo courtesy of the Oklahoma City Convention and Visitors Bureau.

tours and looks much as it did in the years after it was built. It includes original furnishings and hand-painted canvas-covered walls.

The Oklahoma Heritage Center is the former residence of Judge Robert A. Hefner. The 1917 mansion is open to the public and contains elegant Louis XVI furnishings and a collection of bells, canes, Meissen china, and fine art and tapestries. The Oklahoma Hall of Fame Gallery is housed on the third floor, which was originally the home's ballroom. The Oklahoma Heritage Center also includes a library, chapel, and formal gardens.

The stately governor's mansion, east of the State Capitol, has been the official residence of Oklahoma's chief executive since 1928. It reflects the same Dutch Colonial style as the State Capitol building. The current governor and his wife, Frank and Cathy Keating, recently undertook a major refurbishment of the mansion with the help of private donations. A special housewarming was held, giving all Oklahomans the opportunity to help accessorize the home with their own gifts. The mansion is open for tours one day a week and, since its refurbishment, stands as a great source of state pride.

Oklahoma City's diverse selection of cultural opportunities rivals any U.S. city. Oklahoma Citians truly know how to enjoy the finer things in life. ✪

(below) Visitors to the University of Oklahoma are able to enjoy a wide variety of fine arts events, including the newly acquired Fleischaker Collection, on permanent display at the Fred Jones Museum of Art. Photo courtesy of the University of Oklahoma.

(right) Former residence of the Judge Robert A. Hefner family, the Oklahoma Heritage Center contains Louis XVI furnishings, fine art, and houses the Oklahoma Hall of Fame Gallery. Photo by Jack Hammett.

(above) Maestro Joel Levine conducts the Oklahoma City Philharmonic Orchestra and Canterbury Choral Society in the annual Yuletide Festival. Photo by Joseph Mills, courtesy of the Oklahoma City Philharmonic Orchestra.

(left) Lyric Theatre's production of Evita. *Lyric's performances are in the Kirkpatrick Fine Arts Auditorium on the Oklahoma City University campus. Photo by Mike Baroli, courtesy of the Lyric Theatre.*

(right) From ballet to Broadway, conventions to concerts, the Civic Center Music Hall lends a special ambiance to any event.

6

A Time
for Fun

This is just one of the coaster rides to experience at Frontier City. Photo by Jack Hammett.

Having a good time is just a part of the culture in Oklahoma City. With 3,000 hours of beautiful sunshine each year, plenty of room to play, and easy commutes, Oklahoma Citians enjoy many opportunities for fun and leisure. From the high drama of professional and collegiate sporting events to family activities at the zoo or one of Oklahoma City's many parks or lakes, residents have an array of activities to choose from year-round.

Oklahoma has long been famous for its successful collegiate sports programs. But local fans have proven themselves to be terrific supporters of professional sports teams as well. Oklahoma City 89ers baseball has been around for many years, and newer on the scene is Oklahoma City Blazers hockey.

The Oklahoma City 89ers baseball team has been bringing America's favorite pastime to Oklahoma City since its formation in 1962. The Triple A baseball team is the top farm club of the Texas Rangers. Many major-league players have moved up from the 89ers over the years. In 1996, the 89ers won the American Association Championship Series. In 1998, games will be played in the new Bricktown Ballpark built as a part of the Metropolitan Area Projects initiative in downtown Oklahoma City. Oklahoma Citians love the thrill of the ballpark, the excitement of the game, and the entire American baseball experience.

When Oklahoma City's Central Hockey League team takes to the ice each October, the fans absolutely go wild. Ice hockey is a fairly

(below) Visitors are often surprised at the abundance of lakes and water activities available in central Oklahoma.

(right) Everyone has fun at White Water Bay, Oklahoma City's wet and wild getaway. Photo by Fred Marvel, Oklahoma Tourism and Recreation Department.

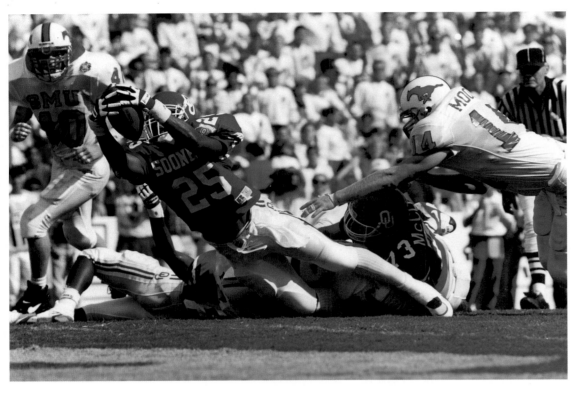

new sport to this part of the country, but Oklahoma Citians are crazy about the Blazers and continue to pack the Myriad Convention Center game after game. For attendance, the Blazers have finished in the top five out of more than 90 minor league hockey teams for the past five years. And the Blazers have rewarded the fans well, winning their second Adams Cup Central Hockey League Championship for regular season play in 1996 as well as the Levins Cup for the post-season championship.

The Oklahoma City Coyotes roller hockey is an International League team that borrows many of its team members from the off-season Blazer hockey team. Also new to Oklahoma City, this sport is quickly drawing enthusiastic support from fans of all ages.

Oklahoma City is also home to a number of annual sporting events that make for sought-after hot tickets among sports fans. These include celebrity golf tournaments such as *The Daily Oklahoman* Open and the 1988 PGA Tour. Celebrity tennis tournaments have included the IGA Tennis Classic, the Virginia Slims Tennis Tournament, the Legends Tennis Tournament, and others.

Collegiate sports hold a long tradition of excellence in central Oklahoma. For residents here, autumn Saturdays were made for Big 12 football games. Through good times and bad, University of Oklahoma Sooner fans pack Oklahoma Memorial Stadium in Norman, decked out in red and white, while Cowboy fans flock to Lewis Field in Stillwater dressed in Oklahoma State University orange and black.

Oklahoma Memorial Stadium is the 19th-largest collegiate stadium in the United States with a capacity of 75,000. The Oklahoma Sooners football team has won six National Championships in its 102-year history. The most notable championships came under the leadership of famous head coaches Bud Wilkinson and Barry Switzer.

Oklahoma State University has claimed 42 National Championships for its varsity sports program over the years. In fact, OSU has won at least four league championships in each academic year since the 1978-79 season.

A number of other nearby colleges and universities offer many more exciting collegiate sports moments, including men's and women's basketball, baseball, women's softball, wrestling, track, gymnastics, tennis, soccer, golf, and volleyball. Local sports fans can always find a winning team to cheer on in the collegiate sports arena.

(above) The University of Oklahoma continues to build upon its tradition of excellence in intercollegiate athletics. University sporting events are a great source of pride for both students and the community. Photo courtesy of University of Oklahoma.

(right) Oklahoma State University has claimed 42 national championships for its varsity sports programs over the years. Photo courtesy of Oklahoma State University.

The Oklahoma City 89ers Triple A Baseball Team offers exciting baseball and great family fun each summer. Photos by Jack Hammett.

Two days before going to press, the Oklahoma City 89ers name changed to the Oklahoma RedHawks. The RedHawks will have their first season at the new Bricktown Ballpark in 1998.

The Oklahoma City Blazers bring exciting hockey action to town. (above) Photo by Fred Marvel, Oklahoma Tourism and Recreation Department. (left) Photo by Lisa Hall.

(above) Oklahoma City has an abundance of public and private tennis courts to enjoy. Photo by Joe Ownbey.

(right) Members and guests can enjoy a great day of golf on the beautifully manicured course at Quail Creek Country Club. Photo by Jack Hammett.

(above) The Inaugural Induction Ceremony for the International Gymnastics Hall of Fame was held in June, 1997, at the Medallion Hotel in Oklahoma City. Photo by Jack Hammett. (right) Photo of Shannon Miller courtesy of Garrison Photography, Edmond.

(below) The All-College Basketball Tournament held at the Myriad Convention Center is the oldest holiday basketball tournament in the country. Photo by Fred Marvel, Oklahoma Tourism and Recreation Department.

The National Softball Hall of Fame displays the colorful history of softball and its greatest players.

Oklahoma City University Chiefs men's basketball team has won more national championships than any other National Association of Intercollegiate Athletics (NAIA) team. They have won four championships since 1991. The women's Lady Chiefs softball team won three national titles in a row in 1994, 1995, and 1996.

The Southern Nazarene University Redskins women's basketball team has won the NAIA National Championships three years in a row for a total of four championships since 1989.

Each year, Oklahoma City is host to several exciting collegiate sports special events. The All-College Basketball Tournament is held at the Myriad annually between Christmas and New Year's Day. It is the oldest basketball tournament in the country. Hosted by the University of Oklahoma, it features teams from around the country. The Oklahoma City area is also traditionally host to the Big 12 Conference Women's Softball Championship in May at the

Softball Hall of Fame Stadium, as well as the Big 12 Baseball Championship at All Sports Stadium. The Big 12 Swimming and Diving Championships will be held in February of 1998 at the Aquatic Center at Oklahoma City Community College. The Aquatic Center debuted in 1989 during the U.S. Olympic Festival. It features an 8-lane, 50-meter swimming pool with movable bulkhead and a separate 18-foot diving well with 4 springboards and 3 platforms. The facility sponsors a variety of classes, community events, and national competitions throughout the year.

Oklahoma City is home to the National Softball Hall of Fame, where visitors can glimpse at memorabilia from the history of softball and displays featuring softball stars who have been inducted into the Hall of Fame. The complex is also home to the International Softball Federation, the U.S. Amateur Softball Association, and the American Softball Association Hall of Fame Stadium, a state-of-the-art ballpark where

numerous national and world championship events are held each year.

Of course, Oklahoma City has plenty of opportunities for the amateur athlete as well. Citywide leagues for softball, baseball, soccer, basketball, hockey, and football are available for all ages.

Running and bicycling enthusiasts have numerous opportunities to participate in marathon-style competitions in Oklahoma City. One of the most popular, the Red Bud Classic, is held each spring in Nichols Hills amidst the blooming redbud trees. It includes a 10, 30, and 50-mile bicycle tour, a children's fun run, 10 kilometer and 2-mile runs, and a 2-mile walk.

World-class pari-mutuel horse racing is featured at Remington Park, a state-of-the-art racing complex. With more than half a million visitors a year, Remington Park is one of Oklahoma's top tourist attractions. The track offers thoroughbred racing in the spring and fall and quarter horse racing in the summer. Facilities include restaurants,

(left) Authentic chuck wagons and chuck wagon cooks from across the West attend the annual Chuck Wagon Festival. Photo by Jack Hammett.

(right and below) Rodeos and livestock shows are part of Oklahoma history that are still enjoyed today. Photos by Fred Marvel, Oklahoma Tourism and Recreation Department.

gift shops, and an infield park with a playground for children.

Oklahoma City is truly the horse show capital of the world, with many special events, rodeos, and other opportunities for the horse lover each year. The city is host to a number of international, national, and regional rodeo events and horse shows, more than any other city in the country. The International Finals Rodeo is held each January at the State Fair Arena. The top 15 International Pro Rodeo Association men and women compete in seven events for the world championships during this exciting rodeo. Oklahoma City is also home of the American Quarter Horse Association World Championship Show and Sale, Grand National and World Championship Morgan Horse Show, the United States Team Roping Championships, the National Appaloosa Horse Show, the International Arabian Youth Nationals, the National Reining Horse Futurity, and the World Championship Barrel Racing Futurity. Horse shows draw large numbers of visitors to the city each year and have a tremendous impact on the local economy. In 1996, these shows attracted

233,000 participants and led to the sale of 49,000 hotel room nights.

Just north of Oklahoma City in Guthrie is the Lazy E Arena, a first-class facility where many equine events and rodeos are held year-round. A few recent events have included the Pro Bull Riders Tour Challenge, U.S. Team Roping Championship, and the National Finals Steer Roping Championship.

With Oklahoma City's rich history as a center for ranching and agriculture, it's only natural that it is home to one of the top five state fairs in the nation. The State Fair of Oklahoma, held each year in late September at the State Fairgrounds, draws record crowds for 17 fun and exciting days. Fairgoers enjoy numerous activities and exhibits, including a championship rodeo, live-stock shows, entertainment of all types, art exhibits, carnival rides and games, and, of course, lots of food from Indian tacos and corn dogs to scrumptious cinnamon rolls.

The National Cowboy Hall of Fame and Western Heritage Center is a wonderful place to celebrate the history and culture of the American West. It is home to the Rodeo Hall of Fame, an

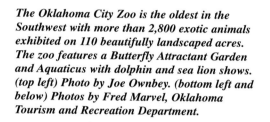

The Oklahoma City Zoo is the oldest in the Southwest with more than 2,800 exotic animals exhibited on 110 beautifully landscaped acres. The zoo features a Butterfly Attractant Garden and Aquaticus with dolphin and sea lion shows. (top left) Photo by Joe Ownbey. (bottom left and below) Photos by Fred Marvel, Oklahoma Tourism and Recreation Department.

expansive collection of art and unique exhibits for children, as well as special events such as the Chuck Wagon Gathering and the National Children's Cowboy Festival.

A major source of pride for Oklahoma Citians is the Oklahoma City Zoo, one of the finest in the nation. The 110-acre zoo is the oldest in the Southwest and is home to more than 2,800 of the world's most exotic animals. The Great EscApe exhibit houses a major collection of gorillas, chimpanzees, and orangutans amidst tropical rainforests, meadows, and streams. The gorillas can be viewed up close through large glass windows. Aquaticus is home to more than 1,500 aquatic creatures from sharks and piranhas to sea lions and dolphins. Visitors enjoy dolphin and sea lion shows and numerous aquariums full of colorful tropical fish. One of the newest exhibits at the zoo is the Cat Forest and Lion Overlook. The cats are housed in a naturalistic habitat and can be viewed from trails that wind through two-and-a-half beautifully landscaped acres.

Some of the smallest and most delicate creatures at the zoo have their own open air, 200,000-square-foot Mecca. The Butterfly Attractant Garden has created a beautiful haven, incorporating more than 15,000 plants that sustain the complete life cycle of the butterfly.

The Oklahoma Firefighters Museum offers a wonderful glimpse into the history of firefighting. Restored equipment dating back to 1736 is showcased, including turn-of-the-century fire engines once used in Oklahoma communities, Oklahoma's first fire station, and a working circa-1900 alarm box. The museum is considered one of the top firefighting museums in the country.

The International Gymnastics Hall of Fame is new to Oklahoma City. From its temporary headquarters downtown, it offers gymnastics memorabilia such as competition medals, an Olympic Torch, equipment used in competition, a fine collection of art, gymnastics videos, and photographs of gymnastics champions. Eventually, the International Gymnastics Hall of Fame will move to its permanent new facility in Bricktown.

Housed in the Omniplex, the Hands-On Science Museum is truly a learning opportunity for kids and grown-ups alike. More than 350 interactive exhibits introduce museum guests to the mysteries and wonders of science. Visitors can stand under the colossal two-story replica of a plant-eating Camarasaurus dinosaur, crawl inside a 24-foot-tall diamond molecule, take a Virtual Reality trip through the human digestive system, star in their own music video, and experience what a real earthquake feels like. Or they can visit the Kirkpatrick Planetarium and be

Oklahoma City residents celebrate the Fourth of July in a big way. Photo by Fred Marvel, Oklahoma Tourism and Recreation Department.

The annual Independence Day Festival includes parades, fireworks, food, and entertainment.

After the Land Run, Bricktown began as a military outpost from Fort Reno to keep peace in the new Oklahoma City. Later it grew as a warehouse area. Today, it is thriving as an entertainment district.

Today, Bricktown is Oklahoma City's newest dining and entertainment district.

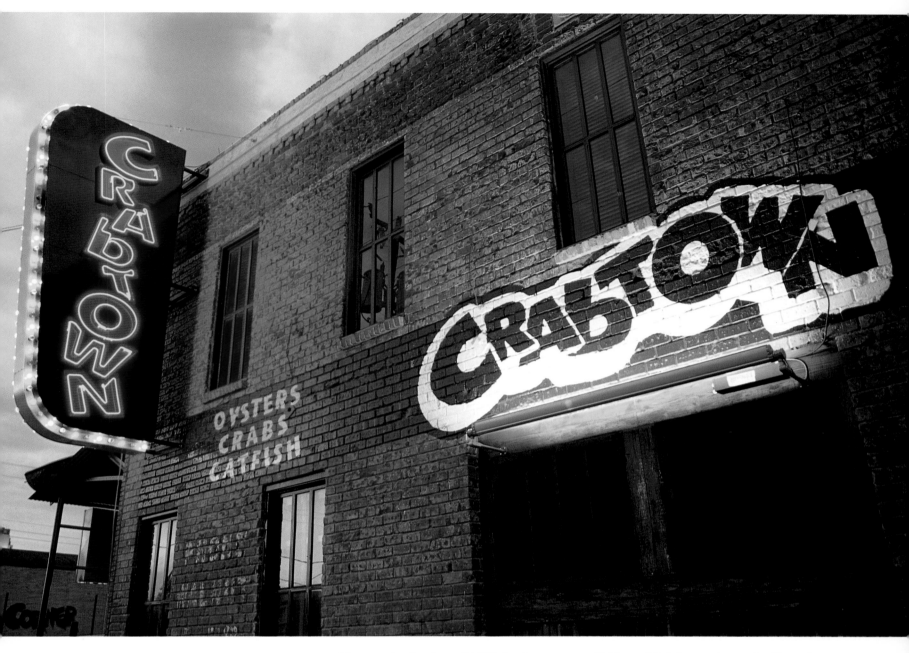

awed by a spectacular show of 4,000 dazzling stars filling a 40-foot dome.

Under construction at the Omniplex is a new $7.3 million OmniDome Theater, which will house a 70-foot domed theater screen that will completely surround the audience. The 268-seat theater will stand nearly seven stories tall and will use the world's most sophisticated motion-picture projection system to show film that is three times the size of a standard motion picture film. The OmniDome will feature a variety of dramatic and educational films exploring the wonders of science and technology.

The Air and Space Museum is also housed at the Omniplex and offers a collection that will thrill any aviation or space buff. Numerous artifacts, photographs, and exhibits celebrate the rich history of the aerospace industry in Oklahoma and the entire nation.

Oklahoma City's long and extensive history in the aviation industry has resulted in a number of high-flying entertainment and educational opportunities for the young and old who are fascinated by aviation. Aerospace America, one of the top five airshows in the country, is held each year at Will Rogers World Airport. This event has grown tremendously during its 12-year history and now has the largest contingency of military support of any civilian airshow in the country. The show features top aerobatic performers, military jet exhibitions, specialty parachute acts, and precision flying teams. On view are more than 100 modern military aircraft, more than 80 World War II-era aircraft, and a tradeshow. An evening performance includes a spectacular pyrotechnic performance and a reenactment of a World War II air battle. This performance is the only one of its kind in the

Members of the large Hispanic population in Oklahoma City celebrate their culture during a Hispanic Day celebration. Photos by Fred Marvel, Oklahoma Tourism and Recreation Department.

(left) BalloonFest has become an annual August event in northwest Oklahoma City. Photo by Joe Ownbey.

Aerospace America is one of the top five air-shows in the country and features top aerobatic performers, military jet exhibitions, parachute acts, and precision flying teams. (right) Photo by Fred Marvel, Oklahoma Tourism and Recreation Department. (below) Photo by David Fitzgerald, courtesy of Oklahoma City Convention and Visitors Bureau.

country. The three-day event draws attendance of more than 100,000 people.

Oklahoma City area residents love to celebrate the Fourth of July in a big way. A number of annual Independence Day festivals are held, including the Bricktown Fourth of July Festival with a parade, fireworks, food, arts and crafts, live entertainment, a patriotic show, and the Fourth of July Garden Party at the Myriad Botanical Gardens featuring a spectacular fireworks display. Edmond is host each year to the weeklong Liberty Fest during the first week of July. The festival offers a little bit of everything with a kite festival, Road Rally, IPRA Rodeo, bike tour, one of the best parades in the state, watermelon festival, and an Independence Day naturalization ceremony followed by fireworks.

No major city is complete without an amusement park. Frontier City, owned and operated by Oklahoma City-based Premier Parks, offers old-fashioned theme park entertainment in a western frontier town setting. The 50-acre park offers more than 60 rides, live entertainment shows, western frontier retail shops, and down-home eateries. Frontier City also hosts a number of special events, including nightly fireworks during the summer, an Easter Eggstravaganza, Old-Fashioned American Celebration on the Fourth of July, a summer concert series, Oktoberfest, and HallowScream.

On New Year's Eve, downtown Oklahoma City is the place to be for ringing in the new year. Designed as a New Year's alternative for the entire family, Opening Night brings the downtown area alive each year with numerous entertainment venues in many different downtown buildings and parks. This is such a popular event that it continues to grow every year. The evening culminates in a midnight countdown and a huge fireworks display.

Downtown is also home to the city's newest entertainment and dining district. Bricktown, a renovated warehouse district east of central downtown, currently has more than 11 restaurants and clubs, gift and antique shops, and wonderful turn-of-the-century charm. Visitors can even enjoy a horse-drawn carriage ride. Bricktown has become one of the hottest spots for nightlife in Oklahoma City and is sure to become even hotter with the completion of a new ballpark and a new

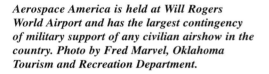

Aerospace America is held at Will Rogers World Airport and has the largest contingency of military support of any civilian airshow in the country. Photo by Fred Marvel, Oklahoma Tourism and Recreation Department.

Oklahoma City has plenty of opportunities for the amateur athlete. The Red Bud Classic is held each spring in Nichols Hills amidst the blooming redbud trees. Photos by Jack Hammett.

downtown canal riverwalk, both a part of the Metropolitan Area Projects plan (MAPS).

The Paseo district is another popular spot for dining and entertainment. Oklahoma City's artist's community, Paseo offers a number of popular restaurants, cafes, and galleries.

During the day, many Oklahoma Citians like to enjoy a more relaxing type of entertainment at one of Oklahoma City's many lakes and parks. Any warm weekend from early spring to Labor Day, heading to the lake becomes a favorite weekend pastime. The Oklahoma City metropolitan area offers 138 parks, 62 public swimming pools, 7 jogging trails and 5 lakes. Lake Hefner is a popular sailing spot and other lakes offer great locations for motor-boating, water-skiing, jet-skiing, and fishing.

For golf enthusiasts, Oklahoma City has 12 public and 11 private golf courses. Two of the finest courses in the area are Oak Tree Golf and Country Club in Edmond and Karsten Creek Golf Club in Stillwater. Oak Tree was host to the 1988 PGA Championship. Karsten Creek is ranked among the top 20 public courses in the United States.

Whether celebrating with a bang or relaxing in peace and quiet, Oklahoma City residents have much to choose from during their free time. ❂

The Myriad is Oklahoma City's multi-purpose sports, convention, and entertainment center. Photo by Jack Hammett.

(left and bottom left) The Oklahoma Centennial Horse Show is one of a few charity multi-breed shows in the nation. Photos by Jack Hammett.

(right and bottom right) Remington Park is a $97 million pari-mutuel horse racing facility with thoroughbred racing in the spring and fall, and quarter horse racing in the summer. Photos courtesy of the Oklahoma City Convention and Visitors Bureau.

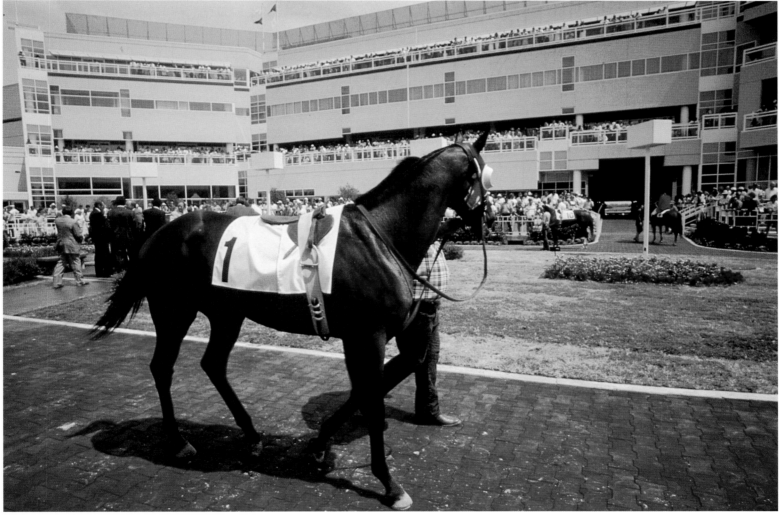

The Oklahoma State Fairgrounds is one of the largest and busiest fairgrounds facilities in the country. (top and above) Photos by David Fitzgerald, courtesy of the Oklahoma City Convention and Visitors Bureau.

The annual Spring Fair and Livestock Exposition and the State Fair of Oklahoma are just two of the many events held at the Oklahoma State Fairgrounds. (top right) Photo by Fred Marvel, Oklahoma Tourism and Recreation Department. (bottom right) Photo by Erick Gfeller.

Frontier City is a 50-acre "old-west-styled" theme park with entertainment for all ages. (top left) Photo by Fred Marvel, Oklahoma Tourism and Recreation Department. (bottom left) Photo by Jack Hammett.

(above and right) Little cowpokes can learn to make a rope, brand a wooden steer, ride a horse, and other cowboy fun at the National Children's Cowboy Festival held at the National Cowboy Hall of Fame and Western Heritage Center. Photos by Fred Marvel, Oklahoma Tourism and Recreation Department.

7

Learning & Growing

The University of Oklahoma is located just south of Oklahoma City in the community of Norman.

The key to education is opportunity and the key to opportunity is a quality education. Oklahoma City offers boundless opportunities for quality educational experiences for all ages. Approaching the 21st century, Oklahoma City's educational lineup is preparing the superior work force of tomorrow and continually better equipping the work force of today.

Lifelong education is a high priority here. With nearly half of the state's institutions of higher learning located in or near the metropolitan area, more than 10 percent of the Greater Oklahoma City adult population is enrolled in area colleges and universities. Oklahoma City is a major educational center with 17 public and private colleges in the metro area. In fact, higher education is the sixth-largest part of the Oklahoma City economy, employing more people than the oil and gas industry.

Oklahoma City is a well-educated community, with one-fourth of area residents holding a college degree. Excellent college and high-school attainment rates have resulted in a high-quality, well-educated labor force in Oklahoma City. Median school years completed in this area is 12.3 compared to the national average of 12.0. A well-educated community means strong support of local schools and an excellent start for our young people.

The University of Oklahoma is located just south of Oklahoma City in the community of Norman and enrolls more than 25,000 students systemwide each year. OU was founded in 1890, 17 years before statehood, and today is a doctoral degree-granting research university drawing students from across the nation. It currently ranks first per capita among all comprehensive public universities in the number of National Merit Scholars in the student body. OU students are enrolled in 19 colleges located on the Norman campus, the University of Oklahoma Health Sciences Center in Oklahoma City, and the OU Health Sciences Center-Tulsa Campus. At the main campus, students also take advantage of a wide variety of classes offered through the College of Continuing Education.

OU has a renowned College of Engineering specializing in aerospace, geological, petroleum, physics, and environmental engineering. *U.S. News and World Report* ranks OU's petroleum engineering program among the top three in the nation. More than 21 percent of the College of Engineering freshmen in 1996 were National Merit Scholars. The Minority Engineering Program is ranked in the top three among public universities in the United States in the enrollment of African-American National Achievement Scholars and National Hispanic Scholars. The College of Engineering has produced more than 700 corporate presidents, vice presidents, and CEOs.

OU's College of Arts and Sciences is the school's largest college and encompasses the humanities, social sciences, natural sciences, and

(left) Aviation maintenance technician students working on aircraft in the powerplant hangar of the Metro Tech Aviation Career Center. Photo courtesy of Metro Tech.

(right) Casady School is one of the leading private schools in Oklahoma City. The K-12 Episcopal day school was founded by The Right Reverend Thomas Casady in 1947. Photo by William Mercer, courtesy of Casady School.

The University of Oklahoma currently ranks first per capita among all comprehensive public universities in the number of National Merit Scholars in the student body.

some professional studies. The college offers a strong liberal arts education combined with major programs preparing students for employment in careers such as journalism, foreign service, the sciences, law enforcement, education, government, computer science, and counseling.

The College of Continuing Education's Economic Development Institute was ranked by the *Wall Street Journal* as the second most popular executive-training program in the nation.

The OU Health Sciences Center offers one of the finest education training programs for physicians, scientists, nurses, dentists, pharmacists, and public health and allied health professionals. *U.S. News and World Report* ranks the OU College of Medicine among the top 20 comprehensive medical schools in the country, and the OU College of Pharmacy is among the top 10 pharmaceutical programs in the United States.

The University of Central Oklahoma in nearby Edmond was also founded in 1890 and is the state's third-largest university with a total enrollment of more than 15,000. It offers 7 bachelor's degrees with 60 undergraduate majors in addition to 5 master's degrees with 34 majors. Enrollment at this school has boomed in recent years, and UCO has recently undergone a $54 million capital improvements project to handle the growing student population. This university prides itself in meeting the diverse needs of its students, from full-time resident students to commuter students as well as older students returning to school part-time while maintaining full-time employment in the community.

Other public colleges located in the Oklahoma City metropolitan area include Oklahoma State University-Oklahoma City, Oklahoma City Community College, and Rose State College.

OSU-OKC is a branch campus of Oklahoma State University in Stillwater and offers more than 40 associate degree and certificate programs. Its division of business technologies offers contract training to area businesses by delivering job-specific customized in-house training. This includes entry-level training for employees of new and expanding business and industry as well as retraining of employees for new technological developments.

The community of Stillwater, home of Oklahoma State University, is well within commuting distance from the Oklahoma City metro area. The state's largest university with an enrollment of more than 26,000 systemwide, OSU's student body included 1,001 high school valedictorians in 1996.

OSU is well known for its College of Engineering, Architecture, and Technology. This college was in the top eight percent in a *U.S News and World Report* survey of engineering deans who were asked where they get their best graduate students. The Industrial Engineering doctoral program was recently ranked 16th nationally in a National Research Council survey.

In the last 15 years, OSU's architecture students have won or placed in more than 125 international and national design competitions. In fact, OSU's School of Architecture ranks second in the nation for the number of Paris prizes won by an institution. OSU's fourth-year architectural students completed a comprehensive three-dimensional model showing what downtown Oklahoma City will look like in 30 years as a result of the Metropolitan Area Projects (MAPS) and the projected spin-off developments. After a combined effort of 27,000 man hours, the model is on display for public viewing.

OSU is a leader in engineering and agricultural research as well. The university's Noble Research Center houses the nationally known Center for Laser and Photonics Research and other labs that specialize in genetic engineering and other processes. The new state-of-the-art Food Processing Research Center was dedicated in November, 1996. It serves as a statewide research facility to develop Oklahoma's capabilities in adding value to food and other Oklahoma products. Also recently added was the Advanced Technology Research Center, which provides world-class laboratory facilities in the areas of energy storage and conversion, manufacturing, materials processing, hazardous and industrial waste management, and laser applications in industry and medicine.

OSU was recently chosen to house a statewide Nuclear Magnetic Resonance Center. It will include a 600-megahertz spectrometer, one of the most powerful research tools in the world. The spectrometer is used to determine the three-dimensional structures of molecules in the development and testing of new products. The tool will be available to scientists from all Oklahoma universities, research foundations, and private industry for use in such research areas as gene therapy, enzyme development, and the production and testing of new medicines.

The Oklahoma City metropolitan area also boasts a number of fine private universities. Oklahoma City University, a United Methodist

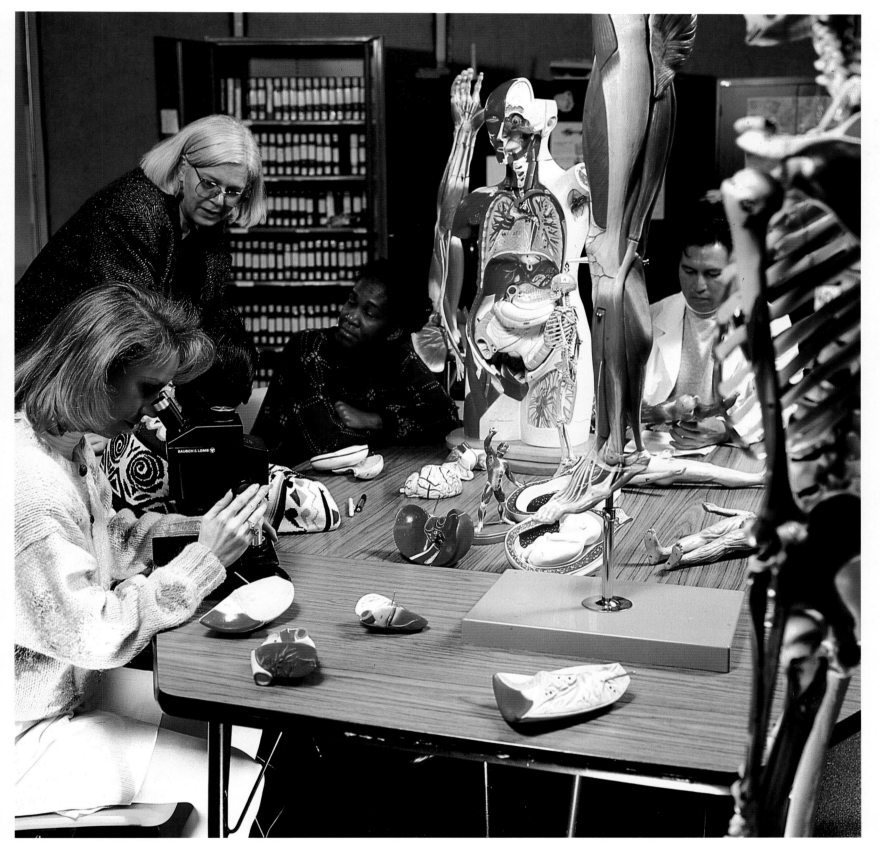

Oklahoma City Community College has the highest acceptance rate of transfer students applying to the University of Oklahoma Health Sciences Center. Photo courtesy of Oklahoma City Community College.

institution located in northwest Oklahoma City, proves to be a convenient location for commuters, but draws students from around the nation and world as well. Its students now represent 70 different countries. With an enrollment of more than 4,700, Oklahoma City University was recently named by *The National Review College Guide* as one of the best liberal arts schools in the United States and was listed in the recent *Student Guide to America's 100 Best College Buys*. The guide cited OCU's strong global awareness and student-centered education. OCU was the only Oklahoma university named in the guide. OCU's line-up also includes an accelerated MBA program that allows local business persons to complete an MBA in just 60 weeks.

OCU has one of the finest and most well-known fine arts colleges in the nation. Seven OCU alumni currently perform on Broadway. OCU students have won numerous pageants. Since 1956, the university has had three Miss Americas, one Miss Illinois, three Miss Arkansas, two Miss Colorados, and 17 Miss Oklahomas.

Southern Nazarene University, located in Bethany, offers quality academic programs in business, education, the sciences, and the fine arts. The university offers excellent adult studies programs in Management of Human Resources, Family Studies and Gerontology. SNU graduates have gone on to chair Fortune 500 companies and be honored as nominees for the Nobel Peace Prize. The university prides itself in small class sizes, maintaining a 17 to 1 student/teacher ratio.

Oklahoma Christian University of Science and Arts, located in far north Oklahoma City, was recognized in 1995 by *U.S. News and World Report* as a "Best Value" in higher education based on the quality of the education program offered at a relatively reasonable price. Affiliated with the Church of Christ, the university offers 46 degree options including pre-professional programs in Bible, Business, Communications, Fine Arts, and Engineering. Its 200-acre campus offers 30 contemporary buildings, a performing arts theater, and an olympic-size swimming pool.

Oklahoma State University, located in Stillwater, is the state's largest university with enrollment of more than 26,000 system wide. The student body included more than 1,000 high school valedictorians in 1996. "Old Central" photo by Andy Maxey, courtesy of Oklahoma State University.

Oklahoma City's vocational training program is among the finest in the world and assists in producing a high-quality work force for local business and industry. Facilities include five vocational technical schools in the metropolitan area—Metro Tech, Francis Tuttle, Moore/Norman, Mid/Del, and Eastern Oklahoma County. According to *Fortune* magazine, Oklahoma City's lineup of adult education facilities is considered the best in the United States. These schools offer specialized training in aerospace, business, health, automated manufacturing, and many other areas. Area public schools work closely with the vo-tech system through an innovative program that results in graduates well prepared for jobs ahead.

Oklahoma's vo-tech schools have a unique partnership with business through the Training for Industry Program. Since its formation in 1968, the program has provided more than 600 industries with a trained, start-up work force. Customized industrial training is provided at no cost to the employer through a statewide network of 49 sites. Management training is also available through 29 schools. This ongoing program works

Oklahoma State University-Oklahoma City trains firefighters from across the nation at its Oklahoma City training facility. Photos courtesy of Oklahoma State University, Fire Service Training.

(above) Students at the University of Central Oklahoma take the traditional walk under the Old North Tower before graduation. Photo by Daniel Smith, courtesy of University of Central Oklahoma.

(right) Not all learning takes place in the classroom. Photo by Fred Marvel, Oklahoma Tourism and Recreation Department.

Built in 1922, the Clara E. Jones Administration Building was the first structure on the Oklahoma City University campus. Photo courtesy of Oklahoma City University.

Oklahoma City University cast members sing the final number in their production of "Elixir of Love." Each year the university provides top-notch entertainment with four fully staged productions and more than 100 concerts and recitals. Photo courtesy of Oklahoma City University.

to maintain the skills of the work force it produces, thereby helping companies to remain competitive.

The Oklahoma City metropolitan area's public schools are among the best in the nation. In fact, test scores are, on average, 10 percent above the national average. Oklahoma County has 15 school districts. The largest, Oklahoma City Public Schools, has 87 schools and serves more than 40,000 students. This school district has made dramatic progress in the last five years. Its grade schools now perform above the national average and at the regional average.

The Oklahoma City Public Schools magnet program has created some shining stars in the system, helping to better prepare local children for life and work in the 21st century. The magnet school program includes three secondary programs and four elementary programs. These magnet schools offer programs not available at other sites in the district. Students receive the advantage of specialized resources, small class sizes, unique course offerings, faculties selected for their expertise in a particular field, customized curriculum, and up-to-date facilities and equipment.

One of these magnet schools, the Classen School of Advanced Studies, is a public college

preparatory school for students in grades 6 through 12 who show promise of exceptional achievement in academics or the visual or performing arts. Students who meet strict admission requirements may choose to study in a demanding International Baccalaureate Diploma program or they may choose to pursue studies in the Visual and Performing Arts program.

The International Baccalaureate Diploma program is a rigorous, interdisciplinary, academic program in which students in the middle years study advanced core subjects, including foreign language study. The Visual and Performing Arts program allows students to study in the areas of visual art, dance, vocal music, drama, or instrumental music, including strings, band, keyboard, and guitar.

Southeast High School serves as a magnet school for students interested in professional fields specializing in the use of advanced technologies. This comprehensive program for grades 9 through 12 emphasizes the development of student skills to work in a technical field and/or to prepare for post-secondary studies.

Northeast High School offers a BioMedical Sciences program and has a unique partnership with the University of Oklahoma Health Sciences Center

(left) Oklahoma Christian University of Science and Art has been recognized by U.S. News and World Report as a "best value" in higher education. (above) Oklahoma Christian University is also home to Enterprise Square USA, a hands-on adventure in free enterprise, as well as a new learning center featuring the Midwest's only multimedia computer lab for public and private sector training.

(left) Students in the Biomedical Sciences Program at Northeast High School focus on a special curriculum designed for students interested in health care careers. Photo courtesy of Oklahoma City Public Schools.

(below) A volunteer mentor for HOSTS (Help One Student To Succeed) works one-on-one with students at Capitol Hill Elementary School. Photo courtesy of Oklahoma City Public Schools.

Oklahoma City Parents as Teachers program helps parents teach their children during the important preschool years. A parent educator, on left, assists a parent with teaching aids. Photo courtesy of Oklahoma City Public Schools.

and the Metro Tech Vocational Center. Students prepare for a career in science or health-related areas through one of three career paths—physical sciences, biological sciences, or health sciences.

The Oklahoma City Public Schools also offer four elementary magnet and specialty programs. Nichols Hills Elementary focuses on international studies, Cleveland Elementary focuses on arts and sciences, and Thelma R. Parks Elementary focuses on foreign language and computer technology, while Columbus Elementary is a specialty school for language.

The local Adopt-a-School program gives area businesses the opportunity to take a hands-on as well as a financial interest in a selected school. Supported by more than 600 participating businesses, this program has an impact on the district of more than $2 million.

The Putnam City school district, the area's second-largest district, is located in far northwest Oklahoma City and is a national leader academically. This district recently earned a top 100 national ranking from *Money* magazine. The magazine based its ranking on academics

as well as housing costs within the school district.

Putnam City schools are in the country's top 20 percent academically based on an average SAT score of 1,044 (the national average is 910) and an average ACT score of 21.6 (the national average is 20.8). Since 1991, Putnam City has produced 69 National Merit Scholar finalists.

The Putnam City school district participates in the School-to-Work program, which offers career awareness for elementary students, vocational exploration for middle school students, and school/work connections at the high school level.

The metro-area school districts of Putnam City, Edmond, Western Heights, and Deer Creek offer the Distance Learning Network to their students and teachers. This program provides video interaction with teachers statewide, allowing students and teachers to access career classes as well as undergraduate and graduate college courses.

OneNet is a unique telecommunications and information network for Oklahoma education and government sponsored by the Oklahoma State Regents for Higher Education and the Oklahoma Department of Finance. Through a partnership between the state of Oklahoma and private telecommunications companies, OneNet electronically links Oklahoma's public schools; vocational-technical schools; colleges and universities; public libraries; government at all levels; court systems; hospitals and clinics; and research entities. The system allows for cost-effective access and exchange of information as well as Internet access. For education, it can provide sharing of library resources as well as video conferences and distance educational programs and courses.

Oklahoma City is fortunate to be home to the Oklahoma School of Science and Mathematics (OSSM), a public, tuition-free, residential high school for juniors and seniors with exceptional mathematical and scientific abilities. This school was created through legislative action in 1983 and graduated its first class of 44 seniors from across the state in 1992. In 1996, its graduates included 60 National Merit Scholars, with 46 National Merit Commended Scholars, one National Presidential Scholar, and 21 National Presidential Scholar semifinalists.

This outstanding school has a two-fold mission: to foster the educational development of Oklahoma high school students who are academically talented in science and mathematics and who show promise of exceptional development through participation in a residential educational setting; and to assist in the improvement of science and mathematics education for the state by developing, evaluating, and disseminating instructional programs and resources to all schools and students of the state.

A nationally known expert on gifted education, Dr. Julian Stanley, has called OSSM "the most rigorous academic program of its kind in the nation." Students are offered a wide range of advanced courses, most of which are college level.

OSSM's 32-acre campus is adjacent to the Oklahoma Health Center, allowing a close relationship with professionals from the scientific and technological community.

The Oklahoma City area also offers a number of excellent private and parochial schools for elementary through high school students. These include Bishop McGuiness High School, Mount Saint Mary High School, Casady School, and Heritage Hall.

Oklahoma City's well-educated population is committed to offering excellent opportunities for those wanting to learn. This commitment to life-long education is key to the outstanding work force available to business and industry in this area as we enter the 21st century. ✪

(below) Putnam City Schools' Hilldale Elementary School teacher Amy Bixler and Putnam City High School senior Andrew Anthony assist elementary students in planting flower seeds in a school garden. The Grants for Kids project was provided by the Putnam City Public Schools Foundation, a nonprofit organization created to enhance and enrich educational opportunities in the Putnam City school district. Photo by Jessica L. Cook, courtesy of the Putnam City Schools Office of Communications.

(right) Rose State College Logo Tower. Courtesy of Rose State College.

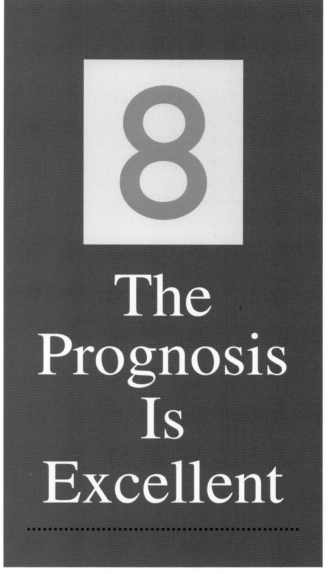

8

The Prognosis Is Excellent

....

The University Hospitals operate Oklahoma's statewide air ambulance service, Medi Flight Oklahoma. Photo courtesy of The University Hospitals, Oklahoma City.

In the everchanging world of health care, Oklahoma health providers have stayed at the forefront, offering the finest in care to Oklahoma City area residents. The Oklahoma City medical community has gone through many transitions in the past decade. Yet one thing has remained constant. The quality of care is outstanding, and those providing the care still do so with a personal, sensitive touch true to the Oklahoma spirit.

When the Alfred P. Murrah Building was bombed in 1995, the medical community of the entire metropolitan area came together to provide fast, top-notch care. The world looked on in awe as emergency medical technicians from many areas worked quickly in a coordinated effort and hospitals went into disaster mode with amazing efficiency, handling victim after victim with proficiency and compassion. The world discovered something we here in Oklahoma City knew all along—that our health care community is second to none.

Oklahoma City's medical community is made up of more than 2,000 doctors and nearly 400 dentists who work with 14 general medical and surgical hospitals, 8 specialized hospitals, and 2 federal medical installations. From any point in the metropolitan area, residents have easy access to first-rate care.

These hospitals are backed up by a number of excellent home-care agencies, outpatient care facilities, mental health providers, freestanding surgery centers, birthing centers, fitness centers, rehabilitation centers, and hospices as well as retirement homes that provide quality, dignified care to our older residents.

Like any major metropolitan area, the medical community in Oklahoma City is indeed a diversified and complex structure. Health care institutions have undergone restructuring processes and now put a stronger focus on providing outpatient care and wellness programs designed to keep persons out of the hospital. All of these programs are an effort to lower health care costs for everyone. Because of the expertise, ingenuity, and forward thinking of Oklahoma City's health care leaders, residents now have the luxury of choice among providers of all types.

The Oklahoma City medical community has remained on the forefront of technology and research. Residents have access to specialized, state-of-the-art care in the areas of cancer, burns, organ transplants, laser surgery, eye care, senior health, women's health, and many others.

(below and right) Local members of the medical profession proved to be among the highest quality in the nation during the hours, days, and months following the bombing of the Alfred P. Murrah Federal Building in downtown Oklahoma City in 1995. The Survivor Tree, located across from the building, is considered "The Tree of Life" representing "the people who bent, but did not break." President Bill Clinton, August 1997, Washington, D.C. Photos by Jack Hammett.

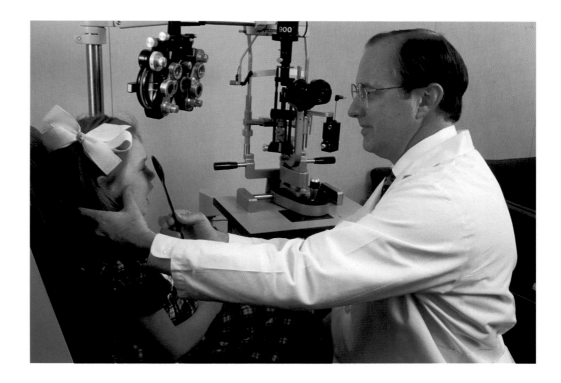

Dean A. McGee Eye Institute is a comprehensive clinical and research facility offering patient care in all subspecialty areas of ophthalmology. Photo courtesy of Dean McGee Eye Institute.

(left) Mercy Hospital is located in northwest Oklahoma City and offers specialty programs in many different areas.

(above) INTEGRIS Baptist Medical Center is one of 13 INTEGRIS Health Hospitals across Oklahoma. Photo courtesy of INTEGRIS.

Oklahoma City has more than 24,000 medical personnel. Forty percent of the state's medical community works in the metropolitan area.

The largest concentration of medical providers in the Oklahoma City area is the Oklahoma Health Center, a 275-acre complex located east of the downtown area and just south of the Oklahoma State Capitol Complex. This area is comprised of 18 different organizations, including four hospitals, seven colleges of medical education and patient care, research facilities, allied health divisions, state and county health care offices, and specialized clinics. The Oklahoma Health Center is the largest medical and academic health center in the state of Oklahoma. It is also one of the largest employment concentrations in the region with approximately 13,000 employees.

The primary teaching hospitals of the University of Oklahoma Health Sciences Center are located here in the Oklahoma Health Center. These include the University Hospital and Children's Hospital of Oklahoma. They offer a number of comprehensive specialty services. Children's Hospital is the state's only full-service

pediatric health care provider and is the only hospital in the state with an infant heart surgery and transplantation unit, an adolescent medicine clinic, a burn center for infants and children, and a pediatric intensive care unit.

The OU Health Sciences Center, which makes up the educational core of the Oklahoma Health Center, offers more than 50 degree programs and serves more than 3,500 students. It attracts some of the finest faculty members from around the world and outstanding students from across the country. Funding for its research programs helped to lead research breakthroughs in Alzheimer's disease, AIDS, blindness, immune disorders, emotional and behavioral disorders, aging, and disease prevention.

The Robert M. Bird Health Sciences Library, located on the OU Health Sciences Center campus, houses the largest collection of medical books in Oklahoma. It serves students and health professionals across the state.

Near the Health Sciences Center, the Oklahoma Medical Research Foundation (OMRF) ranks as one of the top independent biomedical research institutes in the United States.

This independent, nonprofit institution has an excellent Cardiovascular Biology Research Program that is recognized worldwide. The foundation has also received national and international recognition for its scientific research in the areas of cancer, diabetes, arthritis, AIDS, and Alzheimer's disease.

The Oklahoma Telemedicine Network, operated by the Center for Telemedicine at the Health Sciences Center, expands access to health care in rural areas through computer and telecommunication technologies. It is the largest telemedicine network in the nation and has become a world-class model. The network links rural physicians and hospitals to metropolitan hospitals, improving the quality of care for residents of Oklahoma's most remote communities.

Under construction at the Health Sciences Center is a multimillion-dollar Biomedical Research Center that will house research laboratories and scientific equipment. The facility will aid in attracting leading scientists, faculty, and students. A part of the long-term vision of the Oklahoma Health Center is the development of the

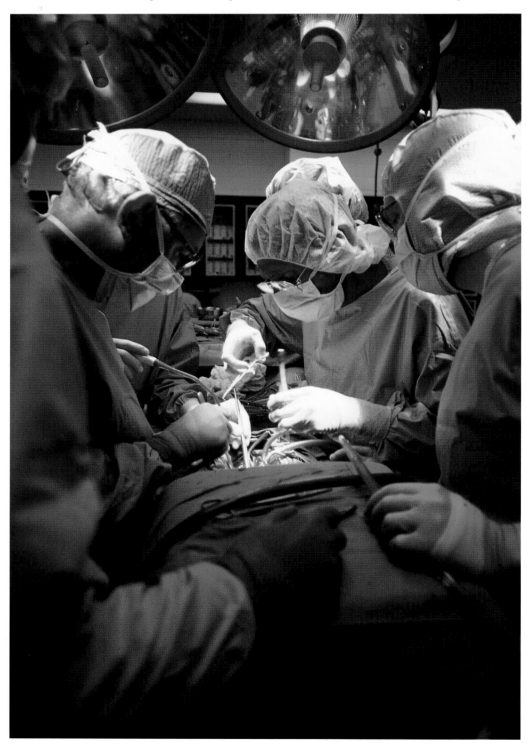

(left) The Oklahoma City medical community has remained in the forefront of technology and research.

(right) The Robert M. Bird Health Sciences Library is the largest health resource in the state and currently has 242,835 volumes. It is one of 12 health sciences libraries in the United States that serve a diverse population of health care professionals, including dentistry, allied health, nursing, public health, medicine, and pharmacy. Photo courtesy of the University of Oklahoma Health Sciences Center.

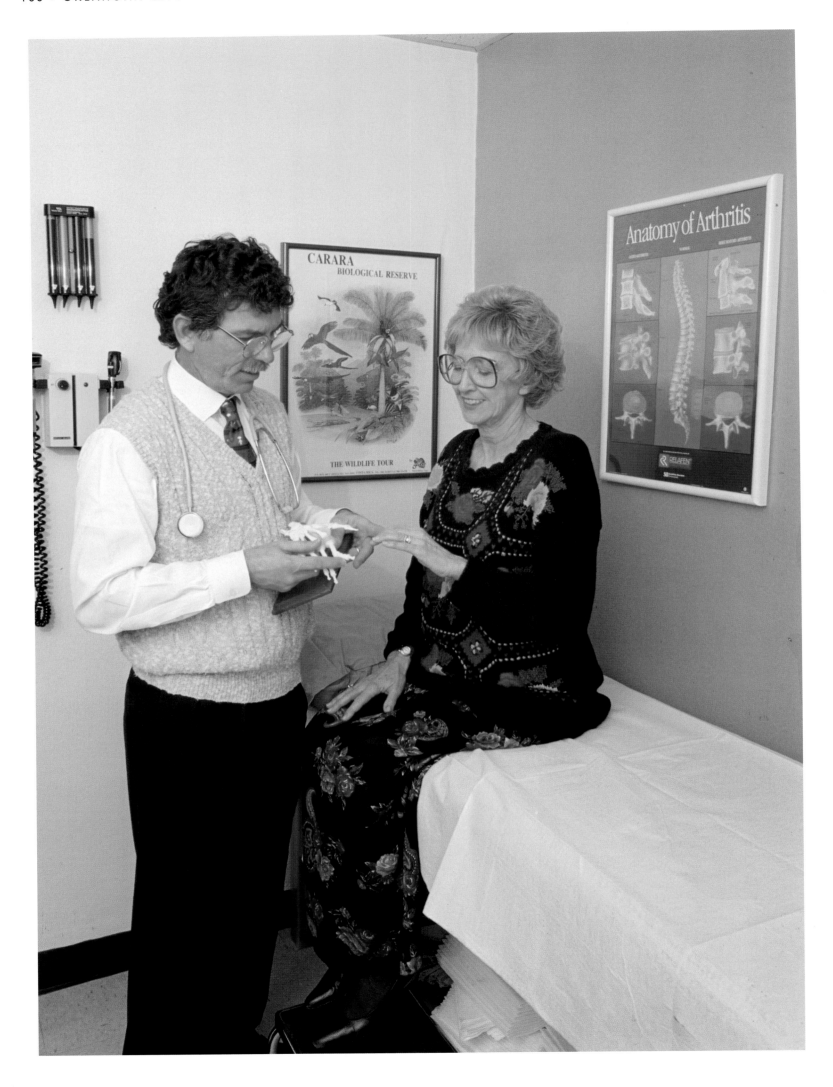

Biomedical Research Park located between the Oklahoma Health Center and the downtown business district. The park will bring together state-of-the-art facilities in an accessible campus setting to establish Oklahoma City as a center for research development activities in biotechnology, biomedical engineering, medical equipment, robotics, and communications. UroCor, the park's anchor tenant, has located its corporate headquarters and operations here. UroCor provides diagnostic analysis of urological cases and expert pathology consultation.

Also located on the Oklahoma Health Center campus is Columbia Presbyterian Hospital, a private teaching hospital. Presbyterian offers excellent programs in the areas of oncology, ophthalmology, gastroenterology, cardiology, genetics diagnostics, sleep disorders, and chemical dependency treatment. Presbyterian is also home to the Center for Athletes and a brain rehabilitation program.

The Veterans Affairs Medical Center, a modern tertiary care facility located on the Oklahoma Health Center campus, provides a broad range of medical, surgical, and psychological services to more than 300,000 veterans.

The Dean A. McGee Eye Institute, also located at the Oklahoma Health Center, is a comprehensive

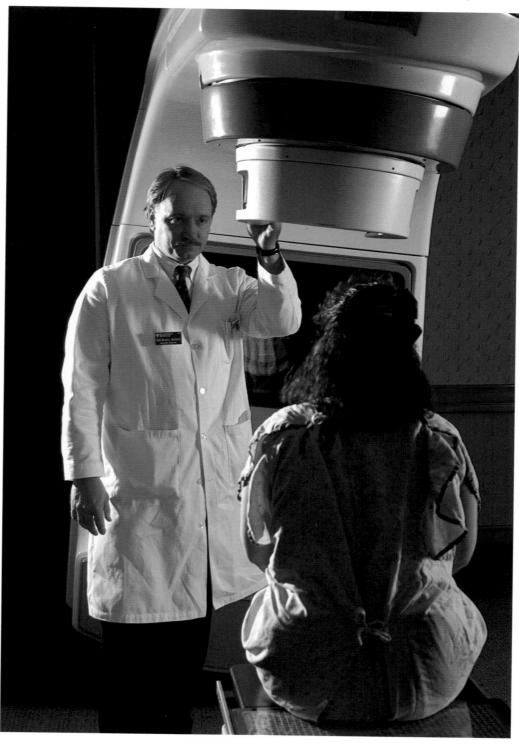

(left) Larry Willis, M.D., a rheumatologist with McBride Clinic, educates a patient about arthritis to help her better manage the disease. Photo by Brent Purdy, courtesy of McBride Clinic.

(right) Deaconess Hospital combines state-of-the-art technology with a human touch. Photo courtesy of Deaconess Hospital.

(above) University Hospital is a 200-bed adult tertiary care facility, which serves as a primary teaching hospital for the University of Oklahoma Health Sciences Center. Photo courtesy of The University Hospitals, Oklahoma City.

(left) Interior of the newly remodeled Labor and Delivery Unit of University Hospital offering a private, home-like setting. Each year, more than 2,500 babies are born at University Hospital. Photo courtesy of The University Hospitals, Oklahoma City.

Clinton A. Medbery III, M.D., radiation oncologist at the Columbia Gamma Knife Center at Columbia Presbyterian Hospital, uses images from an MRI and the treatment planning computer to calculate the dosage a patient will receive for a brain tumor. Photo courtesy of the Columbia Gamma Knife Center.

clinical and research facility offering patient care in all subspecialty areas of ophthalmology. A staff of board-certified ophthalmologists provides state-of-the-art diagnostic testing and surgery for all types of eye disease and vision problems.

Children's Medical Research raises funds to assist investigators seeking causes and treatment of children's diseases. Money raised through its Children's Miracle Network Telethon stays in Oklahoma and is used for pediatric research at the OU Health Sciences Center and to supplement Children's Hospital of Oklahoma with funding not covered by state programs.

Other residents of the Oklahoma Health Center include the American Red Cross, Office of the Chief Medical Examiner, Oklahoma Allergy & Asthma Clinic, Oklahoma City Clinic, Oklahoma Department of Mental Health and Substance Abuse Services, Oklahoma School of Science and Mathematics, Oklahoma State Department of Health, Presbyterian Health Foundation, and Sylvan Goldman Center Oklahoma Blood Institute.

Medi Flight, central Oklahoma's emergency medical air ambulance service, is publicly funded and based at Children's Hospital of Oklahoma. This service provides emergency helicopter services to about 75 percent of Oklahoma, transporting patients to Oklahoma City hospitals.

Outside of the Oklahoma Health Center, the metropolitan area boasts a number of exceptional hospitals, each with its own unique areas of excellence to offer the community.

St. Anthony Hospital, Oklahoma City's oldest hospital founded in 1898, offers specialized services through programs such as the Laser Center, Joyful Beginnings, Hand Center, Comprehensive Medical Rehabilitation, Sportsciences/ Rehabilitation, Spine Center, Oklahoma Cardiovascular Institute, and the Specialized Center for Rejuvenation and Exercise (SCORE).

INTEGRIS Health is the state's largest Oklahoma-owned health system and is the parent corporation of INTEGRIS Baptist Medical Center, INTEGRIS Southwest Medical Center, INTEGRIS Mental Health, and Baptist Healthcare of Oklahoma. INTEGRIS Baptist Medical Center is home to the nation's fourth-largest burn center. This hospital pioneered heart

The Children's Hospital of Oklahoma is the state's only full-service, freestanding pediatric hospital and is also a primary teaching hospital for the University of Oklahoma Health Sciences Center. Photo by Jack Hammett.

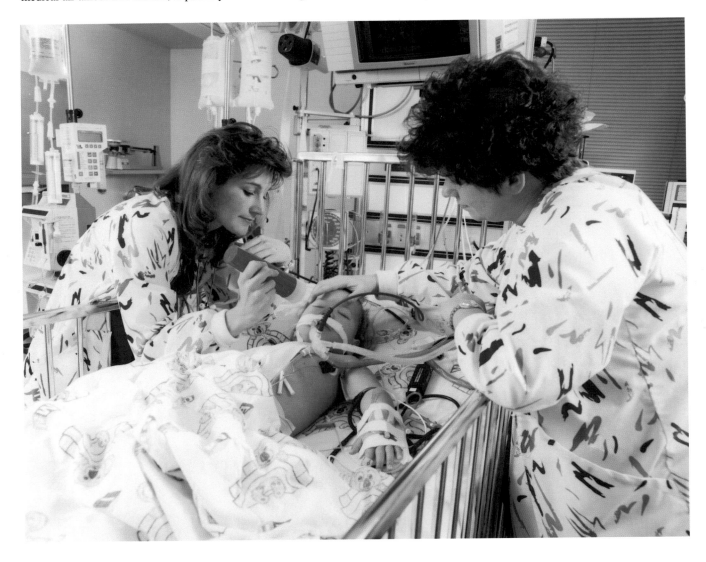

transplantation in Oklahoma City and remains one of the top institutions for transplants of all kinds. Other services include the Oklahoma Heart Center, the Troy and Dollie Smith Cancer Center, Hough Ear Institute, and Baptist Women's Center.

INTEGRIS Southwest Medical Center houses the Sleep Disorders Center, the only center of its kind to be accredited by the American Sleep Disorders Association. Other centers include the Central Oklahoma Cancer Center and the Southwest Breast Health Center. The INTEGRIS Jim Thorpe Rehabilitation Hospital is also located on this campus and is the largest rehabilitation facility in Oklahoma.

Deaconess Hospital is one of only two hospitals in the metropolitan area that remains an independent hospital. A Christian institution and a ministry of the Free Methodist Church, its specialty areas include the Deaconess Cancer Center, Cardiac Health Center, Center for Women's Health, Birth Center of New Generations, Senior Diagnostic Center, and Adult Mental Health Center. Its Oklahoma Lithotripsy Center offers

nonsurgical relief from kidney stone pain and is the only program of its kind in western Oklahoma.

The Oklahoma Urology Center opened recently on the campus of Deaconess Hospital. A joint venture of Deaconess Hospital, Oklahoma Lithotripter Associates, and Urological Surgery Center Associates, the center will bring together more than 40 Oklahoma urologists to provide the most comprehensive care available for the treatment of urological disorders.

Mercy Hospital, located in northwest Oklahoma City, offers specialty programs in cancer diagnosis and treatment, comprehensive cardiac treatment and rehabilitation, the Family Birthplace, senior health programs, neurology, and others. Mercy's NeuroScience Institute offers specialized diagnosis, treatment, and research for diseases of the brain, spine, and peripheral nerves.

Other hospitals serving the Oklahoma City metropolitan area include Hillcrest Health Center, McBride Bone and Joint Hospital, Columbia Bethany Hospital, Columbia Edmond Regional

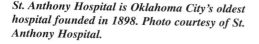

St. Anthony Hospital is Oklahoma City's oldest hospital founded in 1898. Photo courtesy of St. Anthony Hospital.

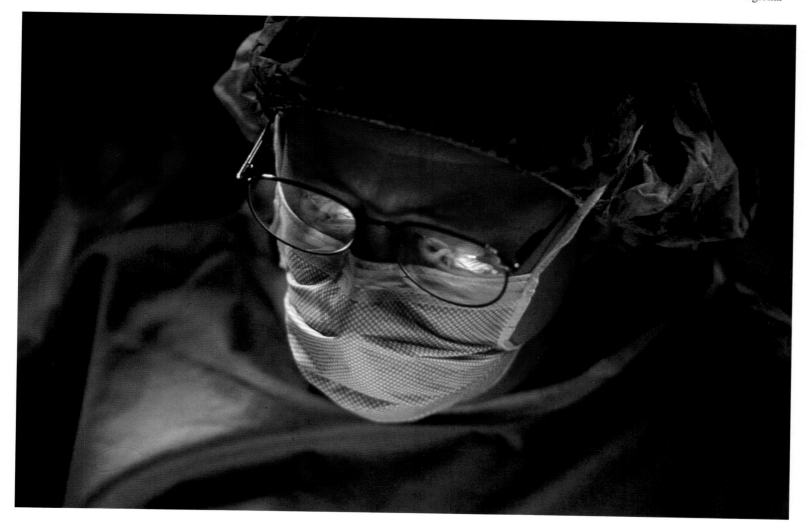

UroCor provides diagnostic analysis of urological cases and expert pathology consultation.

NovaCare Sabolich develops the most advanced prosthetics in the world and has served patients from all parts of the globe.

Medical Center, Midwest City Regional Hospital, Shawnee's Mission Hill Memorial Hospital, and Norman Regional Hospital.

Oklahoma City is the headquarters for a number of state and county professional medical organizations that work to strengthen the medical profession in Oklahoma. The Oklahoma State Medical Association has 4,300 members and works along with the American Medical Association to represent medical doctors on local, state, and national issues. The Federation of Medicine has recognized it as one of the nation's most successful state societies.

The Oklahoma Hospital Association, also located in Oklahoma City, provides numerous services to its members to meet the health care needs of their communities, educates the public and providers on health policy and issues, and promotes quality health care for all Oklahomans.

Oklahoma City's health care delivery system will no doubt continue to undergo much change into the next decade. But the community continues to be dedicated to providing a quality system that emphasizes technology, wellness programs, research, and education, all with a caring touch. ✪

The Oklahoma Health Center is the largest academic and medical health complex in the state of Oklahoma. Photo courtesy of University of Oklahoma Health Sciences Center.

9

A Sense Of Community

Photo by Jack Hammett.

Residents of Oklahoma City aren't seeking the American Dream . . . they've already found it. In Oklahoma City, the living is easy and affordable. As the boomtown grew into a thriving metropolis, this unique community retained its hometown feeling. Strong neighborhoods are still important to those making their home here and remain as the core of the community's existence.

Perhaps most fortunate for homebuyers here is the unequaled value in owning a home. For two years, Oklahoma City was named as the most affordable housing market in the United States in an annual ranking survey by Coldwell Banker. That means that more hardworking individuals and families can own the home of their dreams in Oklahoma City.

According to the Coldwell Banker Home Price Comparison Index, a buyer of a 2,200-square-foot home in Oklahoma City would pay nearly eight times the dollar amount for a similar-sized home in Beverly Hills, California, and nearly seven times the amount for the same house in Greenwich, Connecticut. The study evaluated the cost of a single-family dwelling with four bedrooms, two and one-half baths, family room, and a two-car garage. Oklahoma City topped the 1997 index with an average price of $99,333.

The city has ranked high in other nationwide studies as well. In a 1996 study by Ernst and Young Kenneth Leventhal Real Estate Group and the Koll National Real Estate Index, Oklahoma City ranked number one in the category of "The Nation's Most Affordable Housing Markets."

Oklahoma City also has consistently ranked among the nation's 30 most affordable housing markets according to quarterly survey results released by the National Association of Home Builders. This survey measures the proportion of homes sold in a specific market that a family earning the median income could afford to buy. It also takes into consideration the difference in property tax rates and property insurance rates in each community ranked. Oklahoma City has not

Oklahoma Citians retain a strong sense of community and neighborly bonds. (above) Photo by Joe Ownbey. (right) Photo by Erick Gfeller.

Gaillardia features a championship golf course designed by Arthur Hills, a Normandy French Clubhouse and a private residential community. The clubhouse is shown above. The rendering is courtesy of William Zmistowski Associates.

The Oklahoma City metropolitan area offers a wide variety of living styles in its many neighborhoods. (left) Photo by Jack Hammett.

only ranked well nationally, but in the top seven regionally.

Affordable housing, low utility costs, and a low cost of consumer goods ensure Oklahoma City's cost of living index remains below the national average year after year, according to studies produced by the American Chamber of Commerce Researchers Association.

The Oklahoma City metropolitan area offers a wide variety of living styles in its many neighborhoods, from well-maintained historic areas to new home additions and villages as well as apartment, townhome, and condominium complexes. Homes in a variety of styles and sizes meet the needs of many types of homebuyers. Sale prices of single-family homes range from an average $47,000 to $260,000, and prices for condominiums and townhomes range from an average $33,000 to $85,000. Rental prices are also lower than in many cities, with the average cost of renting a home ranging from $350 to

$850 and apartment rental rates ranging from $350 to $650.

Oklahoma City boasts many historic neighborhoods where strong preservation groups work to maintain the integrity of their neighborhoods. Heritage Hills, located north of downtown Oklahoma City, is made up of some of the finest homes built in the early 20th century ranging in styles from Queen Ann, Victorian, and Neoclassical to Colonial Revivals and Prairie-style bungalows. Near Heritage Hills, Mesta Park neighborhood homes were also built in the early 1900s.

The historic Gatewood neighborhood, bordered by North Pennsylvania, Classen Boulevard, N.W. 16th, and N.W. 23rd, is made up of Tudor Revival, Bungalow and Mission/Spanish Colonial Revival homes built between 1915 and 1935.

Near the State Capitol Complex, the historic Lincoln Terrace area offers beautiful larger homes built in the early 1900s, many of which are being restored to their original beauty.

Paseo is a unique neighborhood of Spanish architecture built in the 1920s and is Oklahoma City's artist colony as well as home to many restaurants and cafes.

Oklahoma City's waterfront areas have a nautical atmosphere and spirit.

The American Dream is attainable in Oklahoma City—and the living is easy. (bottom left) Photo by Erick Gfeller.

Magnificent homes can be found in neighbor-hoods throughout the metropolitan area at a fraction of the cost found on the East or West coasts. Photo by Jack Hammett.

Many other neighborhoods offer well-preserved historic homes, such as Edgemere, Crown Heights, and Putnam Heights.

Nichols Hills, developed by G.A. Nichols in the 1920s and 1930s, quickly became a location for some of the most luxurious homes in the city. Located north of N.W. 63rd Street surrounding Pennsylvania Avenue, the City of Nichols Hills has homes built during every decade since its development and new homes are still being built there today.

Hundreds of suburban additions in Oklahoma City and the many communities in the metro area have been developed over the years, which offer comfortable living and easy access to schools, work, and shopping.

Neighborhood associations working together throughout the metro area aim to bring neighbors together, beautify their surroundings, and increase security. The Oklahoma City Neighborhood Alliance now works with more than 300 neighborhood organizations in the Oklahoma City area.

New homebuyers have numerous locations and a long list of qualified builders to choose from in the Oklahoma City area. With a range of sizes and prices, new homes are within the reach of more residents, even young buyers purchasing their first home. New homes now under construction range from smaller two and three bedroom homes to larger executive homes, many in planned communities offering landscaped parks and recreational facilities. Many new subdivisions are platted each year in some of the most popular areas, such as Edmond, Putnam City, Moore, Norman, and Deer Creek.

Because homebuyers in Oklahoma City today frequently include first-time homebuyers, retirees, and empty-nesters, area developers are planning new concepts that will meet the needs of these groups. Oklahoma City also offers an increasing number of retirement villages designed to provide all levels of amenities to senior citizens. These communities, such as Epworth Villa, Copper Lake Manor, and The Fountains at Canterbury, have various types of housing on one campus, from apartments for independent living to full-care nursing facilities.

Wherever residents choose to live in Oklahoma City, they enjoy hassle-free commutes.

Strong neighborhoods are important to those making their home in Oklahoma City and remain the core of the community's existence.

In fact, Oklahoma City's average commute time is under 20 minutes, according to the U.S. Census Bureau. Oklahoma City is one of a few communities in the United States where commute times actually decreased between 1980 and 1990. This can be attributed to the city's continued planning to improve the highway system, with new construction and expansion of interstates making it easier to get around. Oklahoma recently implemented the Pike Pass system, allowing motorists to prepurchase automated passes for Oklahoma turnpikes. The system lowers the cost for motorists and eliminates time waiting in toll-gate lines.

The American Dream is attainable in Oklahoma City . . . the living is easy. Oklahoma Citians retain a strong sense of community and of neighborly bonds, even as the city continues to grow and offer more opportunities. ✪

(above) Even in the heart of downtown Oklahoma City, there is a sense of belonging to a friendly community.

(right) Oklahoma City also offers an increasing number of retirement villages, such as Epworth Villa, designed to provide all levels of amenities to senior citizens. Photo by Joseph Mills, courtesy of Epworth Villa.

Oklahoma City boasts many historic neighbor-hoods where strong preservation groups work to maintain the integrity of their neighborhoods. (left) Photo by Jack Hammett. (above) Photo by Fred Marvel, Oklahoma Tourism and Recreation Department.

10

The City Of The Future

....

Residents of Oklahoma City aren't seeking the American Dream . . . they've already found it.

The citizens of Oklahoma City believe in their community's potential. They believe so strongly that in 1993 they approved a five-year, one-cent sales tax to fund a sweeping plan to build for the future. When they approved the Metropolitan Area Projects (MAPS) plan, they gave a resounding yes to the grandest package of city upgrades ever undertaken by a U.S. city at one time.

MAPS is an aggressive $300 million development plan, constructing convention, tourism, educational, cultural, recreational, and sports facilities downtown and at the State Fairgrounds. It is comprised of nine major projects aimed at revitalizing the city, encouraging private development, and turning Oklahoma City into a favored convention and tourist destination.

Early in the 1990s, city leaders saw the need for major improvements to city facilities and the need to plan such projects in a coordinated effort. MAPS has been called the legacy of Mayor Ronald J. Norick, who in 1991 appointed five citywide task forces to develop recommendations in the areas of cultural, downtown hotel, riverfront, baseball park, and Myriad renovation. A MAPS committee was formed in the fall of 1992 to develop an economic development and public improvement plan based on the recommendations of these task forces. The MAPS committee's goal was to improve the economic well-being of area residents by developing a successful plan for

Mayor Ronald J. Norick appointed five citizen task forces to develop recommendations for Oklahoma City's renovation projects.

funding and constructing new and improved convention, tourism, sports, cultural, and education facilities. The committee was made up of city, county, and business leaders and took input from numerous metropolitan area organizations.

After many meetings and significant analysis, the committee put together a recommended project list that included these nine projects: a 20,000 seat Indoor Sports Arena, renovation and expansion to the Myriad Convention Center, construction of a baseball park in Bricktown, Civic Center Music Hall renovation, a new Metropolitan Library/Learning Center, upgrading State Fairgrounds facilities, a North Canadian riverfront development, a canal with a riverwalk through downtown, and a trolley system.

The MAPS committee's recommendations were taken under consideration by the City Council, who formed a committee to develop the proposed projects, funding, and other issues. On October 13, 1993, the City Council set an election date for an ordinance to fund MAPS, and on December 14, 1993, the voters of Oklahoma City approved it.

The Greater Oklahoma City Chamber of Commerce was responsible for raising $355,000 to fund and manage the "Vote Yes Campaign" called "Believe in Our Future," which passed with a 54 percent approval.

The city went to work to make MAPS happen. A team of city staff dedicated to MAPS was formed, and Frankfurt-Short-Bruza Associates, P.C., was selected to assist as the MAPS program coordinator. The City Council reviews and authorizes all phases of development, design, construction, and expenditures upon the recommendation of a Citizens Oversight Board.

The goals of the MAPS program, as stated in the project's Master Plan, are to encourage private investment by creating a climate that will attract both in and out-of-state businesses and industries; to promote tourism and convention activity that will bring about increased revenues to Oklahoma City and the state; to enhance the quality of life for both visitors and residents by offering a much broader selection of cultural, entertainment, and sporting events, as well as learning opportunities; and to revitalize downtown Oklahoma City and its surrounding environs. MAPS projects will be debt-free when construction is completed early in the 21st century.

Already, this aggressive plan has brought about a substantial amount of private sector investment and interest in the downtown area. This includes the announcement of a new 350-room hotel adjacent to the Myriad Convention Center, construction of new restaurants, clubs and retail shops, renovation of existing hotels,

Visitors flock to historic Bricktown for dining and entertainment. Photo by Jack Hammett.

(above) The 25-year-old Myriad will undergo a remodeling to upgrade the exhibit hall and expand meeting rooms, lobby areas, and the ballroom. The new exterior will include a sky-walk to a new hotel. Photo courtesy of the City of Oklahoma City.

(right) The Canal River Walk will extend from the Myriad Convention Center to the ballpark, then southeast to the North Canadian River. Courtesy of Frankfurt-Short-Bruza.

and the formation of a local sports commission and investment group.

Residents and visitors are now seeing dramatic and exciting changes, with several projects well under way.

BASEBALL PARK

The new baseball park has its inaugural season, in historic Bricktown, in the spring of 1998. The stadium is adjacent to the future canal, just across Walnut Avenue, and is designed to recall a nostalgic atmosphere and blend naturally with the turn-of-the-century charm of Bricktown.

Designed in the old-time tradition of such stadiums as Camden Yards in Baltimore, Maryland, and The Ballpark at Arlington in Arlington, Texas, the new stadium will offer a special charm inspired from the early days of baseball when the sport became America's favorite pastime.

The field will be natural turf, and the stadium will consist of four levels with facilities to accommodate 12,500. This includes room for

11,500 in stadium seating, with room for about 1,000 more persons in the picnic area and grassy berm. The facility will be complete with novelty shops, luxury suites, and a club restaurant.

The new stadium will be home to the Oklahoma City 89ers, the Triple A affiliate of the Texas Rangers. It will also host NCAA baseball tournaments, outdoor concerts, and other entertainment events.

The stadium will act as an anchor to other entertainment venues in the downtown area. Fun-seekers will be able to see a game, stroll the canal, and enjoy Bricktown dining, shopping and clubs in one outing. The stadium will also offer a dramatic view of the downtown skyline.

CANAL AND RIVERWALK

The canal will bring an even more festive flair to downtown Oklahoma City and has perhaps been one of the most anticipated projects by local residents. Originating east of the historic Santa Fe depot at California and Santa Fe streets, the

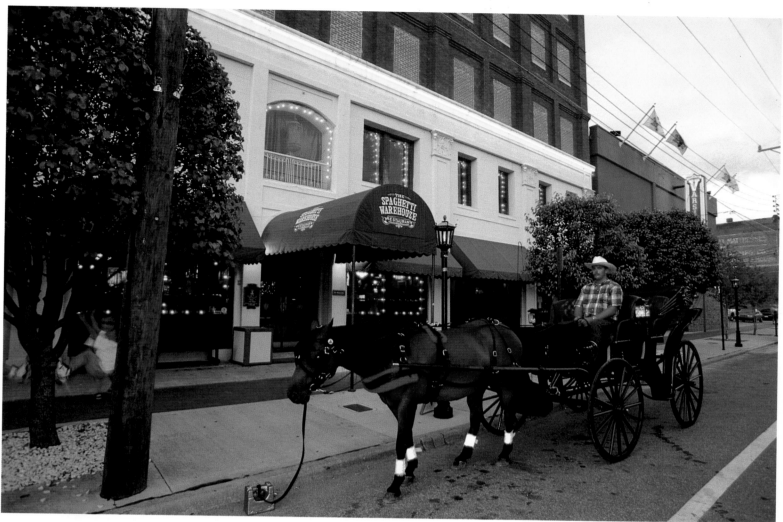

Bricktown has become one of the hottest spots for nightlife in Oklahoma City. Visitors can even enjoy a horse-drawn carriage ride.

first section along Santa Fe will be anchored with fountains at either end of a well-landscaped area, stepping down to a center plaza. At this center will be a cascading fountain and tranquil pool area, as well a boat turning basin.

As the canal flows eastward along California through historic Bricktown, pedestrians will enjoy a refreshing oasis 14 feet below street level, with access to sidewalk cafes and retail shops. Existing early 20th-century warehouse buildings along California Street will be renovated through private development along the canal.

At the east end of the California section will be a major entrance to the new baseball park. At this point, the canal turns southward and reaches toward the North Canadian River. Along this section will be space for private development of restaurants, theaters, shops, and office buildings, as well as picnic areas, walking, biking and hiking trails, and beautiful landscaping. A cascading waterfall and pedestrian overlook will enhance another boat turning basin at this part of the canal. Beyond the overlook, the canal will wind south to another turning basin. A final section will serve as a transition between the river and the canal. Several low-water dams along the river will cause water to back up into this south portion.

INDOOR SPORTS ARENA

Located just south of the Myriad Convention Center, a new 20,000-seat Indoor Sports Arena is being designed to accommodate a major league hockey or basketball franchise. It will host other entertainment and sports events as well, including professional rodeos, ice shows, family shows, and music concerts. It will also be suitable for major exhibits and conventions.

The multilevel facility will include a food court, extensive concession areas, a novelty shop, and a suite/club level with an upgraded finish, 48 luxury suites, party suites, a lounge, and a restaurant. The street frontage along the north side, Reno Avenue, will be developed as a landscaped plaza serving the entries to the Myriad Arena and a proposed 350-room hotel to be built on the northeast corner of the site.

MYRIAD CONVENTION CENTER

The 25-year-old Myriad will receive extensive remodeling, expansion, and upgrading, allowing it to better serve the convention needs of the city. A 25,000-square-foot ballroom along with a 30,000-square-foot meeting room complex will be created on the north side of the facility in an area that can be separated from the rest of the building. This complex will include upgraded finishes, a grand staircase, and an extensive lobby

Local favorite Garth Brooks and other big name entertainers will enjoy the new spacious Myriad Arena. Photo by Joe Ownbey.

area, as well as a new business center, and a banquet kitchen.

The existing exhibit area will receive extensive renovations that include new flooring, lighting, and walls. Service corridors will be provided to make food service more accessible in both the ballroom and meeting room areas. The new ballroom will be nicely finished with vinyl wall coverings, carpeting, special ceiling treatments, and light fixtures. This room can be divided into five smaller rooms or handle 1,800 people for a banquet.

A new skywalk will extend from the concourse level to a new hotel on the north.

The entire building, except for the arena bowl, will receive a new sprinkler system. The arena area will receive minor renovations, including refurbishing or replacing seating on some levels. The building will also receive a new heating and cooling plant. Two much-needed truck loading

docks will be added on the north and south ends near the exhibit hall. Underground parking garage access will be moved from the north and south to the east and west ends to better facilitate traffic flow and enhance pedestrian safety.

CIVIC CENTER MUSIC HALL

Built in 1937 as a part of the Works Progress Administration (WPA) program, the art-deco style Civic Center Music Hall is home to the Oklahoma City Philharmonic Orchestra, Ballet Oklahoma, Broadway Series, Canterbury Choral Society, and many other cultural and entertainment events. Remodeling of this facility will create an acoustically ideal 2,500-seat Main Hall that will better serve major performing arts groups. The current Main Hall area will be completely removed and a new auditorium built in its place. The new shoebox-shaped design will enhance stage viewing and improve acoustics dramatically. The stage will be lowered approximately 10 feet, and the back wall of the auditorium will be moved forward. Three levels of balconies will be put in place, along with box seats with donor suites.

Other renovations to the building include a new rehearsal hall, a west-end addition housing backstage support, and upgrades to the box office, lounge, dressing rooms, and restrooms. A new stage door entry will be added to the hall, as well as new star dressing rooms and a loading dock. The facility will also receive new heating, air conditioning, and lighting. The Civic Center Music Hall also houses the Hall of Mirrors for banquets and the Little Theatre for smaller productions.

METROPOLITAN LIBRARY/ LEARNING CENTER

A new modern library facility will replace the current downtown library. The new 113,000-square-foot Library/Learning Center will be inviting, flexible, efficient to operate, and responsive to the technological demands of the 21st century. It will be built just north of the Myriad Gardens on top of the Galleria parking garage facing Sheridan Avenue, placing it in a highly visible cultural district with ample parking.

The first two floors of the four-story facility will house book circulation operations, book collections, and a business information center. The third level will house library administration and conference rooms. The top floor will accommodate a learning center with conference rooms, classrooms, and a 120-seat auditorium.

The new library will originally house 500,000 materials and can accommodate an increased capacity to 800,000 materials. An oversized lobby on the first floor will provide a gathering space for literary readings, small musical performances, and art displays. Space will also be included for a ground-level cafe.

OKLAHOMA CITY FAIRGROUNDS

Many improvements to the fairgrounds area have been recently completed. A new $15 million, world-class horse and livestock arena and significant improvements to the barn areas are a part of the fairgrounds program. These improvements include lighting, sprinkler systems, electric power, doors, reinsulation of the roofs, and new horse stalls.

Most of the fairgrounds exhibit buildings are receiving improvements. The Kitchens of America Building, the Travel and Transportation Building, and the Made in Oklahoma Building will have new air conditioning, electrical systems, doors, restrooms, lighting, and sprinkler systems. Outdated interior insulation is being replaced with attractive new wall coverings in the Made in Oklahoma Building.

The 8500-seat Norick Arena at the State Fairgrounds also received a face-lift. An exterior addition allows space for more restrooms, larger corridors, and a new exhibit and ticketing area. New air conditioning, electrical power, a dimmer system, and sprinkler system add to the expanded facility. The arena itself includes new acoustic

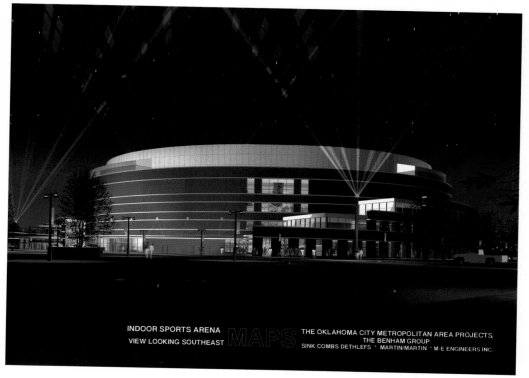

(right) The new multilevel arena will host major sporting events, concerts, shows, exhibits, and conventions. Courtesy of The Benham Group.

INDOOR SPORTS ARENA
VIEW LOOKING SOUTHEAST

THE OKLAHOMA CITY METROPOLITAN AREA PROJECTS
THE BENHAM GROUP
SINK COMBS DETHLEFS • MARTIN/MARTIN • M·E ENGINEERS INC.

panels that not only improve sound, but brighten the interior look of the arena bowl.

NORTH CANADIAN RIVERFRONT

A seven-mile expanse of the North Canadian Riverfront, from Meridian Avenue to Eastern Avenue, is involved in the Riverfront Development project. A series of low-water dams will raise the river's water level, creating attractive recreational areas along the north and south banks. The river will be bordered with landscaped areas, naturalistic trails, and recreational facilities.

TROLLEY SYSTEM

Connecting it all together will be a new mass transit system making it easy for residents and visitors to take advantage of all that Oklahoma City has to offer.

A trolley bus will connect the I-40 and Meridian hotel district to downtown and the fairgrounds area. Meanwhile, a new downtown trolley system will make a two-and-a-half-mile loop through the downtown/Bricktown district.

The Metropolitan Area Projects will bring the city into the 21st century with style, making it not only a perfect place to live, but a perfect place to visit as well. MAPS is made possible by the citizens themselves, those individuals who love their hometown and want to make it even better. With this community's combination of proud, hardworking individuals, leaders with ingenuity, and the space to make it happen, Oklahoma City will truly offer an even better living . . . an even better life. ✇

(above) The area surrounding Lake Hefner in northwest Oklahoma City has undergone many recent improvements. This artist's rendering depicts a new wharf now under construction. Courtesy of East Wharf Development LLC.

(right) Photo by Joe Ownbey.

The new 16,000-seat natural-turf baseball stadium will include first-class spectator accommodations, private luxury suites, restaurants, shops, and facilities for teams and media. The stadium will also host outdoor concerts and other entertainment events. (above left) Photo by Jack Hammett.

(left) Ballpark rendering designed by Chuck Jones, Architectural Design Group, Inc.

Major remodeling will turn this facility into a 2,500-seat performing arts center featuring state-of-the-art acoustics and enhanced stage viewing.

Photo by Jack Hammett.

(above and right) **Photos by Fred Marvel, Oklahoma Tourism and Recreation Department.**

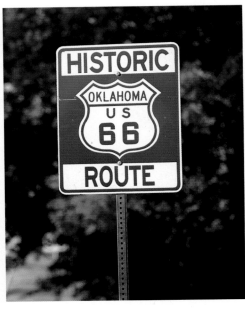

(on following page) **Photo by Joe Ownbey.**

Part Two

Oklahoma City's Enterprises

Leadership Square photo by Jack Hammett.

Networks

Southwestern Bell

In 1879, the first telephone conversation took place in what would officially become Oklahoma in 1907. In 1917, Southwestern Bell Telephone Co. was formed to service a five-state region, offering the first long-distance service in the Southwest. By 1954, operators in Oklahoma City were dialing long-distance calls direct nationally.

The telecommunications industry has drastically changed since 1954. Changes continue to bring new rules, challenges, laws, technology, and consumer needs. Riding this crest of change, Southwestern Bell Oklahoma Division, headquartered in Oklahoma City, has invested nearly $3 billion in the state. This figure translates to roughly $2,000 invested in every single telephone line the company serves.

This past investment ensured state-of-the-art telephone service for Oklahomans. But, Southwestern Bell's commitment to "connecting with the future" continues.

Southwestern Bell's advanced digital telecommunications network technologies eliminate the restraints of distance through voice, data, and image transmission systems. Additionally, recent developments in video technology help bring telemedicine into the realm of a cost-efficient, lifesaving tool.

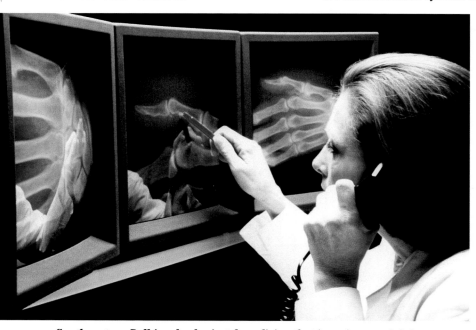

Southwestern Bell understands the value of a telecommunications infrastructure when it comes to bringing new business to Oklahoma City.

Southwestern Bell's decades-long partnership with Oklahoma educators continues with investments in Distance Learning which involves high-speed data and image transmission, plus interactive communications systems. Small schools that may find it difficult to provide teachers in areas such as calculus or foreign language are now able to offer course work through interactive video displays. Open to a wide range of subjects or events, this technique also introduces innovation and diversity in a classroom setting.

"Schools in rural areas now have interactive access to professors at some of our greatest universities," said David Lopez, president, Southwestern Bell—Oklahoma. "And we are just scratching the surface of possibilities."

Southwestern Bell and the Oklahoma Telephone Association worked as partners to create OneNet, Oklahoma's statewide telecommunications and information network for education

and government. OneNet's potential is two million connected users. With reduced rates for dedicated telephone circuits, OneNet's potential can be realized. Information can be more easily shared and state government armed with the capability of better serving Oklahomans.

Oklahoma City and Southwestern Bell have a record-setting relationship. Oklahoma City has the largest toll-free calling area in the nation, nearly 5,000 square miles. Oklahoma City was also one of the first areas in the nation to receive Caller ID and other services.

This high level of technological infrastructure has impacted the economic development of Oklahoma City. Just one of many examples is Hertz Rent-A-Car's Worldwide Reservations Center. Southwestern Bell outlines the scenario:

Hertz Rent-A-Car's Worldwide Reservations Center in Oklahoma City employs more than 2,000 people—more than the company's New Jersey headquarters—largely because of the city's sophisticated telecommunications capabilities. The center handles about 20 million calls annually through 800-number technology, enabling Hertz to serve customers quickly and stay competitive in the marketplace. Hertz customers can reserve a vehicle, report problems, and resolve billing questions more simply by dialing one phone number. Before Hertz activated 800 service, the company took calls at 14 regional centers across the United States.

As a catalyst for economic development and a dynamic corporate citizen, Southwestern Bell understands the value of a telecommunications infrastructure when it comes to bringing new businesses and jobs to Oklahoma City. The

Southwestern Bell is a leader in telemedicine that is saving crucial time in diagnosis.

One connection is telemedicine, a link between the health care provider and patient, who may be hundreds of miles apart. In the technological arena of health care, Southwestern Bell is a leader in the development and implementation of new technologies, such as telemedicine, that is saving crucial time in diagnosis, improving quality of rural health care, and, literally, saving lives.

Schools in rural areas now have interactive access to professors at some of our greatest universities.

company noted: "A study by Wharton Econometric Forecasting Associates estimates that for every $1 million telecommunication investment, 33.3 jobs will be created annually in our state."

Southwestern Bell contributes to economic development and the quality of life in Oklahoma City through various avenues. In fact, the company donates in excess of $2 million to the community in a year's time. The company's managers and associates are consistently involved in the Oklahoma City Chamber's efforts—as chairman of the board, committee members, or through funding of the Chamber's projects.

Southwestern Bell believes the investment of time and resources enhances the potential of citizens. The company supports the Metropolitan Area Projects plan (MAPS) and the city's leadership organizations. In an innovative business-public school partnership, Southwestern Bell's minority managers personally visit the Oklahoma City classroom for one-on-one mentoring sessions. This education and encouragement for students provides role models from the business arena.

The Southwestern Bell building, One Bell Central, served as headquarters for the rescue and relief operations following

the bombing of the Alfred P. Murrah Federal Building. The company also donated $1 million to the relief fund, which was distributed to several care-giving entities.

In sports, Southwestern Bell is a sponsor for the Oklahoma City 89ers baseball team, Cavalry basketball team, Blazers hockey team, and programs at area universities. The company is also heavily involved in the arts, education, literacy, youth organizations, and quality of life events.

One major gift to Oklahoma City from Southwestern Bell is the One Bell Central building, state headquarters for Southwestern Bell. This three-year, multimillion-dollar challenge refurbished the city's historic and dilapidated Central High School into a landmark business building.

Historian Dr. Bob Blackburn writes about One Bell Central in a publication called *New Life for a Landmark:*

"The rehabilitation of One Bell Central was a landmark project. For preservationists and graduates of Central High, it saved an important building from ultimate destruction or conversion. For Southwestern Bell, it provided a unique and beautiful building that reflected the corporate commitment to improving the quality of life in Oklahoma."

Financially, both Bell and the city benefited from the project. By renovating an existing building instead of erecting a new structure, Bell saved about $4 million. By demanding labor intensive craftsmanship, Bell created jobs for hundreds of people and companies. And by improving a central city building, Bell added another structure to the tax rolls and encouraged other rehabilitation projects."

From acquisition and design to completion and occupation, the rehabilitation of One Bell Central was a success with benefits for all. It preserved what was best of the past, improved the city of today, and held forth promise for what would come tomorrow."

Many can speculate what landmark dates of telecommunications history will be on the calendar for the year 2000 and beyond. Although the telecommunications industry is everchanging, one fact remains consistent: Southwestern Bell's commitment to simplicity, cutting-edge service, and to the people of Oklahoma City.

"At Southwestern Bell, our promise to you, our customers, is to make this brave new world of telecommunications as simple and convenient as it can be," said Lopez. "Many companies talk about service and quality. At Southwestern Bell, that is our legacy, our goal, and our reward." ✸

Southwestern Bell promises to make the world of telecommunications as simple and convenient as it can be.

Oklahoma Natural Gas Company

Nine Decades of Performance

In 1906, a year before statehood, a few industrious pioneers decided Oklahoma City was destined to become a center of commerce. They recognized the need for energy to fuel the future of this municipal upstart, and they set out to build a 100-mile pipeline to transport natural gas to the area.

Undaunted by detractors who said gas would never be a viable energy source, these entrepreneurs overcame many challenges and, as they had promised, completed the pipeline that jump-started an era of remarkable growth. They sensed this abundant fuel supply would one day play an enormous role in their town's life. Time proved them right.

THE COMPANY THEY FORMED WAS OKLAHOMA NATURAL GAS.

The contributions made to our quality of life by the fuel they brought us are indeed enormous. For today natural gas heats our homes, prepares our meals, dries our clothes, produces hot water on demand and even powers vehicles.

Gas fuels industry in our state with tremendous reliability and versatility. Its power is provided here with extraordinary efficiency and affordability. Independent studies show the rates paid for Oklahoma Natural Gas service are among the lowest in the nation.

The benefits of nature's miracle fuel are environmental as well, because gas is quite gentle on the earth from which it comes.

Now the dynamic business begun by those high-plains pioneers gathers, stores, transports and delivers natural gas to people and industries across most of Oklahoma. Oklahoma Natural now serves over 725,000 residential, commercial and industrial customers. It owns and operates over 18,500 miles of pipeline and supports its expansive system with five underground storage facilities.

Oklahoma City is the company's largest service area. Over 27% of the company's workforce is based in this district.

With a heritage spanning nine decades, Oklahoma Natural views its responsibility

Oklahoma City is the company's largest service area. The district headquarters are located downtown at Third Street and Harvey Avenue.

with great sincerity. Daily operations are guided by the desire to be "The Customer's Choice," a principle of paramount importance today.

Direct competition is approaching as Oklahoma Natural prepares to "unbundle" its services. As this occurs, many businesses will compete in earnest for share of market.

Oklahoma Natural welcomes this change. The company is an early leader in deregulation. A strong advocate of customer choice, it was one of the first utilities to give large volume industrial and commercial customers the opportunity to purchase their own gas on the open market. Plans are now being made to extend this option to residential customers as well.

The organization founded by pioneers has expanded with time to include the ownership and operation of other business units. The corporate parent of Oklahoma Natural Gas is ONEOK Inc. (pronounced "One-Oak"). ONEOK is an S&P 500 company trading on the New York Stock Exchange under the symbol OKE.

ONEOK (www.oneok.com) operates energy companies involved in natural gas processing,

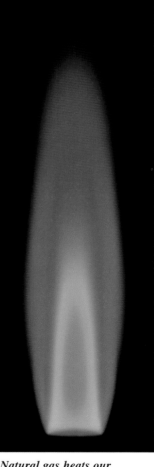

Natural gas heats our homes, prepares our meals, dries our clothes and produces hot water on demand.

oil and gas production, and natural gas marketing. It is emerging as one of the nation's premiere energy service providers. Through strategic ventures and acquisitions, the corporation is positioning itself to be a solid competitor in the energy service and supply industry of the 21st century.

Following the course set by its founders, Oklahoma Natural continues its contributions right here in Oklahoma City.

The company works closely with local officials and business leaders to ensure the community always will have the vital energy it needs to sustain its economic growth and maintain its position as one of America's most livable cities.

Recent efforts on behalf of Tinker Air Force Base, the state's largest employer, helped this vital military installation transform itself into one of the most energy-efficient bases in the country. Securing Tinker's long-term viability as one of the nation's core logistic depots remains a top priority.

But there's much more to Oklahoma Natural Gas than just fueling the economic engine that helps the city run. The company's history is deeply woven

Joint efforts have helped transform Tinker Air Force Base into one of the most energy-efficient military installations in the country.

executive program and its managers often fill key leadership roles on the United Way's fund-raising teams.

Oklahoma Natural Gas is very proud of the contributions its employees make to the community and the recognition the company receives as a result of their efforts.

From relatively modest roots seeded by our pioneer founders over nine decades ago has sprung a large and dynamic organization vitally connected to the city and the people it serves and supports.

Much opportunity and change lie ahead. We approach the new frontier holding onto the vision, courage, commitment and energy of those who walked the path before us. We know that no matter what the future holds, one thing is certain — Oklahoma Natural Gas always will be Pure Oklahoma. ✪

into the city's history. Whenever there's a vital need in the community — Oklahoma Natural is there to assist.

Like so many businesses, the company and its employees were there to lend support when the Murrah Federal Building was destroyed and the city faced its most profound tragedy.

In 1983, Oklahoma Natural Gas established its Share the Warmth program. Administered by The Salvation Army, Share the Warmth has raised millions of dollars to help thousands of Oklahomans experiencing temporary financial difficulties maintain essential energy services.

Oklahoma Natural Gas was a founding member of Oklahoma One Call System, Inc. This non-profit entity was formed in 1981 to increase public safety and reduce accidents related to striking underground pipes, wires and cables during excavations.

The Oklahoma Natural Gas Volunteers in Action are a formally organized group of employees who pour their personal energies into helping their community. Over 75 employees in Oklahoma City donate their time to projects that assist the needy, raise money for medical research and help children learn and grow in the Oklahoma education system. These volunteers also are active in the efforts of the Arts Council's annual Red Earth festival.

Oklahoma City employees serve on more than 30 civic, community service and professional boards. They provide active leadership to these diverse groups, which multiply their personal efforts exponentially.

Oklahoma Natural Gas Volunteers in Action pour their personal energies into helping the community. A float is created each year for the Martin Luther King, Jr. parade.

The annual United Way Campaign is the city's largest fund-raising program. Oklahoma Natural and its employees are deeply involved in this effort, which is the primary source of support for over 50 social service agencies in Central Oklahoma. The company perennially ranks in the top five in per capita corporate giving to the campaign. It participates actively in the loaned

Will Rogers World Airport

"Get yourself a flying field—even if you have to trade your Chamber of Commerce for it!" When Oklahoma humorist Will Rogers made that statement in the early 1930's, not everyone was convinced that the new form of travel would someday become a primary means of transportation. But for Will, there was no question that airplanes were the way of the future.

Many citizens of Oklahoma City shared Will's vision. In 1924, they acted on their faith in the fledgling industry by purchasing a section of land and building an airport. Today that airport has become the major business and industrial complex known as Will Rogers World Airport.

FLYING FIRST AND HIGHEST

Right from the start, Oklahoma City made aviation history with several notable firsts. Municipal Aviation Park, as it was then called, was home to the Tibbs Flying School, one of the first in the nation to train pilots and airplane mechanics. The inauguration of airmail service in 1926 put Oklahoma City on the nation's airmail route, a big coup for city leaders. A few years later, Paul Braniff took off for Tulsa in his five-passenger plane, the forerunner to Braniff Airlines. The first aerial photography company in the country got its start at the young airport, pioneering yet another area of aviation. By 1936, there was no question that the aviation industry would flourish in Oklahoma City.

Today, Will Rogers World Airport encompasses more than 8,000 acres, serves 3.6 million passengers a year, is home to 80 businesses and industries, and is a vital economic contributor to the community of Oklahoma City.

CONVENIENT AIR PASSENGER AND CARGO SERVICE

Will Rogers World Airport is the primary commercial service airport for Oklahoma City and the surrounding areas. The airport is served by ten commercial carriers that land an average of 85 flights a day—an unexpectedly high level of service for a community the size of Oklahoma City. Passengers can fly directly from Will Rogers to New York City, Los Angeles, and Detroit, or nonstop to Chicago (both O'Hare and Midway), Houston (Hobby and Intercontinental), Dallas

Over the past five years, freight movement increased more than 32 percent, a trend that is expected to continue at Will Rogers.

Easy access to departure gates allows travelers time to call the office, select a last-minute gift, grab a bite to eat, or just catch up on the news of the day.

(DFW and Love), St. Louis, Kansas City, Phoenix, Denver, Atlanta, Salt Lake City, Cincinnati, and Colorado Springs.

Mail and cargo activity have been cause for excitement at Will Rogers, with double-digit increases over the past five years. Air freight services are provided by all the major carriers as well as numerous freight charters.

PRIME LOCATION

Located on 8,000 acres of land, Will Rogers ranks among the ten largest airports in the United States in land area. Nearly 3,400 acres are available

for immediate development, much with prime airside access. These sites are ideal for manufacturing, fabricating, maintenance, shipping, cargo, and warehousing operations.

For many entrepreneurs, the close proximity of Will Rogers to the NAFTA corridor (I-35) and other major interstate highways (I-40, I-44 and I-240) is a distinct asset. Located in the heart of America, Will Rogers is a prime intermodal site for businesses and manufacturers. For those companies that rely on exports and imports of materials, components or products, Oklahoma City's Foreign Trade Zone is located on airport property.

Will Rogers World Airport has an average of 85 scheduled departures daily, with non-stop service to eleven major cities.

EXCEPTIONAL TENANTS

The family of tenants at the airport round out this impressive airport complex. The largest tenant is the Mike Monroney Aeronautical Center. This 1,050-acre complex of 50-plus buildings employs an average of 5,000 personnel and pumps an estimated $450 million annually into the Oklahoma economy. The center comprises a number of vital services and supports missions for the Federal Aviation Administration and the U.S. Department of Transportation, including the training of more than 30,000 students a year in air traffic control, electronics, inspection, and management.

Other prominent tenants include Southwest Airlines Reservations Center, AAR Oklahoma, Organon Teknika, the U.S. Bureau of Prisons Transfer Center, and Metro Tech Career Aviation Center. Will Rogers also provides space for seven major car rental agencies, five international cargo haulers, several corporate tenants, as well as the various retail vendors in the terminal.

AIRPORT PARTNERS

In addition to Will Rogers World Airport, the City of Oklahoma City and the Oklahoma City Airport Trust own and operate two other airports—Wiley Post and Clarence E. Page.

Wiley Post Airport, named after another Oklahoma aviation pioneer, is the designated reliever airport for Will Rogers. It handles much of the city's business and corporate travel. In addition, Wiley Post serves as a base for more than 300 aircraft and provides a thriving environment for aviation-related industry. Gulfstream Aerospace Technologies and Commander Aircraft are two notable tenants.

Clarence E. Page Airport, a small general aviation airport, is home to weekend aviators and several thriving aviation businesses. In 1996, the airport hosted the International Aerobatic Championships. A prestigious accomplishment for the airport and city, since only one other U.S. airport has ever been chosen in the event's history.

GROWTH AND IMPACT

What began as a facility to accommodate air passenger service and airmail transportation has evolved into a major business complex. The economic impact of the airports for Oklahoma City and the state of Oklahoma is significant. During the last year, over $10 million was committed to service, engineering, and construction contracts, much of it to state firms. In addition, the airports and their tenants provide jobs to more than 8,000 people with payrolls exceeding $300 million.

Future prospects for Will Rogers World Airport continue to be promising. The airport has an outstanding location with large tracts of land ready for expansion and an excellent airfield able to handle healthy growth in operations. Its economic impact is also expected to grow as airport staff, city, and state officials progressively seek to develop the airport system and to identify and recruit new businesses and tenants. "Will Rogers is securely positioned to accommodate continued growth," says Luther Trent, Director of Airports. "As we approach the new millennium, our attention will be focused on expanding terminal and airfield facilities, increasing air service, and attracting new industry, never losing sight of our primary goal of providing a safe, convenient, and enjoyable means of transportation."

Will Rogers confidently said, "Aviation is the greatest advancement of our times." There is no question that the spirit and vision of Will Rogers flourishes in Oklahoma City and in the airport community that proudly bears his name. ✪

Will Rogers World Airport is home to 80 businesses and more than 8,000 employees.

Waste Management of Oklahoma, Inc.

Waste Management of Oklahoma has grown from a division of fewer than 50 employees to almost 200 employees.

Service and education-oriented, dynamic, fun, and friendly may not be words often associated with a waste management company. Yet state-of-the-art waste hauling trucks with robotic arms and a "Bin to the Curb Lately?" recycling informational campaign from Waste Management of Oklahoma, Inc. may offer a new perspective.

Waste Management of Oklahoma is a wholly owned subsidiary of Waste Management, Inc., a company that reached its 25th year of operation in 1996, having entered the Oklahoma City market in September, 1984. Its motto is "Count on Us."

Waste Management of Oklahoma has grown from a division of fewer than 50 employees to almost 200 employees. Its burgundy trucks with white stripes provide solid waste service to approximately 4,600 commercial and industrial customers and more than 90,000 residential customers in the Oklahoma City area. Through its Waste Hauling and Recycle America divisions, Waste Management collects solid waste from 100 percent of Oklahoma City rural residents and the western half of Oklahoma City, which comprises approximately 60 percent of Oklahoma City residents. It also provides curbside recycling services to 100 percent of the urban Oklahoma City residents.

Waste Management of Oklahoma is the largest waste hauler in the state, and Waste Management, Inc. is the largest waste management company in the world. This magnitude of business proves the company's successful business philosophy of viewing the customer as number one.

By placing the customer's needs first, a waste management plan can be tailored to fit specific requirements. Clients also can be assured the company properly takes care of disposing of a product, whether it is general refuse or other industrial waste.

Commercial and residential recycling services also have been significant factors in the company's success. Recycle America of Oklahoma City provides curbside recycling pickup for 140,000 households in Oklahoma City. Six employees and a million-dollar, state-of-the-art machine can classify and sort 12 tons

Commercial and residential recycling services also have been significant factors in the company's success.

of commingled recyclables every hour.

Recycle America supplemented this full-scale curbside program with the educational campaign: "Bin to the Curb Lately?" Along with informational brochures, advertising, door hangers, neighborhood events, and refrigerator magnets, the program incorporated a poster and essay contest in area schools.

The campaign literature cites that each year Oklahoma Citians generate about 365,000 tons of residential and commercial solid waste, and fortunately, these Citians have a recycling program with the easiest pickup method: curbside pickup of a household's bin.

From paper and glass, to plastic and tin cans, the curbside recycling is easy, quick, and makes for a healthier planet. Taking only two minutes per household, there is no sorting or transporting of the material past the curb required of customers.

"Recycling is an opportunity to invest in the future of our natural resources," says Phil Smith, division president. "To prove this commitment to recycling in Oklahoma City, Waste Management's corporate office made a sizable donation to the Omniplex Science Museum for a recycling exhibit scheduled to open in late 1997."

Waste Management of Oklahoma, Inc. hauls more than 30,000 tons of recyclables a year for Oklahoma City, providing premier solid waste collection with service and attitude that can make it a pleasure to take out the trash. ✺

East Oak Recycling & Disposal Facility

East Oak Recycling & Disposal Facility (RDF), a division of Waste Management of Oklahoma, Inc., is not the usual "dump." It is environmentally friendly, landscaped with grass, flowers, and trees, and is Oklahoma's most technologically advanced landfill.

Increased knowledge, better engineering, and strong environmental concerns have changed the local landfill of a generation ago. Today's landfills are stringently governed by state and federal agencies, and government regulations get tougher every year.

East Oak RDF has always performed one step beyond government standards. Its design and operation lead the way in the safe disposal of solid waste. Internal and external controls, in addition to doing it properly the first time around, have ensured growth for East Oak RDF.

"We're on top of the heap," said James Meinholz, division president/general manager. The company is proud of its accomplishments and stands by an open-door policy. A viewing stand with bleachers is set up for students, general public, and environmental groups for landfill education tours. In fact, the company hosts approximately 1,500 students per year.

There are many examples of how East Oak RDF offers a safer, more environmentally sound facility to protect the environment as well as area citizens and their property.

East Oak RDF is the only landfill in central Oklahoma with a composite (synthetic) liner that meets government requirements and is permitted by the Oklahoma Department of Environmental Quality. The liner protects ground water by collecting the wastewater, called leachate, in specific areas from where it is pumped to holding ponds, then tested and pumped to a wastewater treatment plant.

This facility pioneered the use of waste tire chips in its leachate collection system, a method that not only recycles old tires, but increases environmental safety. It is the only facility in central Oklahoma that offers solidification of liquid and semiliquid, making disposal easier and safer.

The Mosley Road landfill is also managed by East Oak RDF. This remediated Superfund site received Environmental Protection Agency approval in October, 1995, to accept construction and demolition waste as fill to improve its environmental quality. The work to make Mosley Road a safe, viable disposal facility is a national success story for Superfund sites.

Employee relationships and loyalty are a high priority at East Oak RDF. Eighty percent of personnel have been at the East Oak RDF from 5 to 8 years, employee longevity for a facility established only 10 years ago. The full-service staff takes pride in their work and dealing with customers.

The safety specialist, licensed professional engineer, maintenance personnel, mechanics, general office staff, and equipment operators are among the best trained in the industry. They have completed 24 hours of Occupational Safety and Health Administration and 40 hours of hazardous materials training, in addition to monthly seminars and additional training.

"Being in the service industry, we achieve our goals by addressing customer service issues and critiquing our operation on a customer-needs basis," says Meinholz. "As a leader in the landfill industry, we've positioned ourselves to maintain environmentally sound disposal procedures that reduce potential liabilities and protect the environment for our children and the future." ✪

Oklahoma's most technologically advanced landfill.

East Oak's design and operation lead the way in the safe disposal of solid waste.

The Oklahoma Publishing Company

The company headquarters and newspaper offices are at 9000 North Broadway.

The flagship of the company is The Daily and Sunday Oklahoman.

From a newspaper page to the on-line page, from statehood to the turn of the century, from real estate to nonprofit support, The Oklahoma Publishing Company (OPUBCO) has strongly established its presence and commitment to Oklahoma City and the state. Through leadership in publishing, broadcasting, dedication to economic development, and growth in diversity, this company has been part of Oklahoma City's heritage for almost a century.

A bronze plaque in the lobby of the publishing company's office building reflects the conviction of this company with a quotation from Edward L. Gaylord, chairman of The Oklahoma Publishing Company and editor and publisher of *The Daily Oklahoman.*

"The newspaper is much more than just publishing the news and advertisements. *The Oklahoman* is mightily involved in building a city, a state, and a better America."

LEGACIES OF GROWTH

The Oklahoma Publishing Company was founded in 1903 by E.K. Gaylord when he acquired an interest in *The Daily Oklahoman*, one of four frontier newspapers dating back to 1894. Since before statehood, OPUBCO has stood as a voice of leadership. *The Oklahoman* was a staunch advocate of Oklahoma Territory and Indian Territory merging into one state, and the newspaper published an "Extra" edition in 1907 for the declaration of statehood.

This leadership legacy continues through the Gaylord family. Building upon the foundation set by his father and progressively positioning the company for the future, Edward L. Gaylord now serves as chairman and CEO of The Oklahoma Publishing Company. E.K. Gaylord II is the president. They continue to build a company that looks to the future through tradition, technology, diversity, and communications.

THE DAILY AND SUNDAY OKLAHOMAN

While OPUBCO is a diversified company that has achieved exceptional growth throughout its history, the flagship business is publishing *The Daily and Sunday Oklahoman. The Oklahoman* serves as the state newspaper and touches the lives of more than one million readers.

For forty years, *The Oklahoman* was the circulation leader in the Southwest. Now Dallas and Houston with larger markets are ahead in circulation, but *The Oklahoman* is still ahead in penetration of its market.

In becoming a leading newspaper in the Southwest, *The Oklahoman* has provided the best equipment and facilities for its 1,100 employees and 3,000 agents and carriers to produce the best newspaper possible. Along with being noted for quality journalism, *The Oklahoman* has always been an industry leader in embracing new technology.

The Oklahoman was the first newspaper in the world to set type by computer. In the early 1980s, the newspaper became the first in the United States to have an on-line, searchable database of its news content. In 1991, *The Oklahoman* was the first newspaper to use a fiber-optic linkage for a computer system integrating word processing, spreadsheet computation, graphics, and pagination.

One of the first major newspapers in the United States to take its product on-line was *The Oklahoman.* To facilitate this advancement, Connect Oklahoma, Inc. was formed in 1996. This new media subsidiary creates a presence for *The Daily and Sunday Oklahoman* on the Internet and provides additional information and entertainment services.

The Oklahoma Publishing Company's leadership reflects a company strong in its industry and strong in its community. OPUBCO's commitment to community growth and economic development is evident not only in its editorial advocacy of important issues, but also in key leadership roles and civic support.

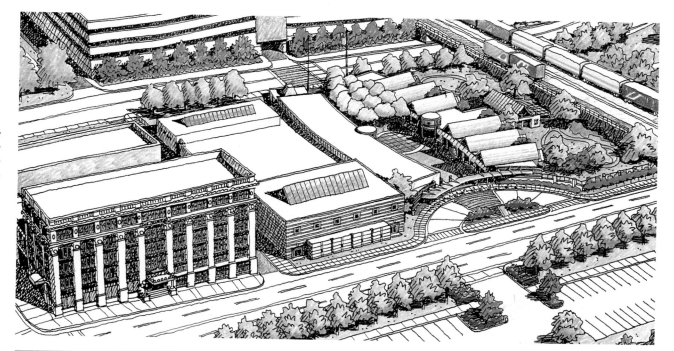

The historical office building and property at 4th and Broadway were donated for a new YMCA complex.

OPUBCO Development Company's Gaillardia Country Club clubhouse.

STRONG TRADITION OF LEADERSHIP

OPUBCO maintains a strong tradition of support and leadership in the Greater Oklahoma City Chamber of Commerce with chairman roles fulfilled by E.K. Gaylord and Edward L. Gaylord. Other company executives have served as chairmen, including two within the past seven years.

Public service and responsibility are a high priority for this company. OPUBCO has long been a supporter of the State Fair of Oklahoma and has sustained this support since being one of the first corporate sponsors in 1906. OPUBCO and the Gaylord family have been deeply dedicated to the National Cowboy Hall of Fame and Western Heritage Center since its inception through both leadership and financial contributions. Involvement in the development of Oklahoma Christian University has been extensive over the years.

Quiet, but significant philanthropy helps a myriad of organizations such as the Oklahoma Medical Research Foundation and the Boy Scouts. Employees can be found throughout the community serving as board members and volunteers of civic organizations. OPUBCO is always a pacesetter in United Way giving.

The downtown OPUBCO historical office building was built in 1909 and served as this company's home for 81 years. Following the move to the new, state-of-the-art location on Broadway Extension and the YMCA's loss of its home due to the bombing of the Alfred P. Murrah Federal Building, OPUBCO's historic building and adjoining property were donated for a new YMCA complex.

This type of consistent and gracious giving by OPUBCO ties back to the quotation etched in bronze in the Oklahoma Publishing Company building. Yet the community involvement moves beyond the newspaper and community leadership.

DIVERSE BUSINESSES

OPUBCO has invested heavily in the Oklahoma economy over the years. Its diverse growth has included properties in broadcasting, trucking, oil exploration, commercial printing, real estate, and professional sports.

OPUBCO's radio station, WKY, was the first commercial station west of the Mississippi River. The company also founded WKY-TV, the first TV station in Oklahoma and one of the first television stations to broadcast color in the nation. Mistletoe Express Service was a revolutionary trucking enterprise.

OPUBCO Development Company, a subsidiary of The Oklahoma Publishing Company, continues a company tradition of improving the quality of life in Oklahoma with Gaillardia Golf & Country Club and Residential Estates. The 600-acre mixed-use master planned community in far northwest Oklahoma City features a stately clubhouse, now under construction, and championship golf course surrounded by luxury residential enclaves.

Another one of OPUBCO's prime properties is the Broadmoor Hotel in Colorado Springs.

The Oklahoma Publishing Company for more than 95 years has been instrumental in building a better city and state. A special centennial supplement section to *The Sunday Oklahoman* titled "Those We Touch," is filled with 22 pages of how OPUBCO accepts challenges and makes an impact on the community. The opening lines read; "A century full of words has passed between *The Oklahoman* and its readers. Many of those words were used to speak out against corruption and expose stumbles in the standards the public adopts. Many have spoken for public service and responsibility. At *The Oklahoman*, it hasn't stopped with words. . . ."

At *The Oklahoman*, it starts with those it touches each day. ✪

OG&E Electric Services

Providing "Power at the Speed of Life" is not just a slogan for OG&E Electric Services. It is a corporate philosophy that translates to low-cost and reliable electric service, economic development initiatives, a sound infrastructure for long-term dependability, and a sense of dedication to customers.

Headquartered in Oklahoma City, OG&E supplies electric power to 684,000 retail customers in Oklahoma and western Arkansas, with additional wholesale customers throughout the region. OG&E's electricity comes from eight company-owned power plants (coal-fired and natural gas-fired), plus purchased power. Its natural gas business is conducted through Enogex, a non-utility subsidiary involved in gas transportation, gathering, marketing, processing, and production.

The OG&E record speaks for itself:

• OG&E rates are about 30 percent lower than the national average.

• OG&E has steadily decreased customers' electric rates over the past decade.

• From 1994 to 1996, OG&E lowered rates twice and lowered fuel costs resulting in customer savings in excess of $100 million per year.

• In 1996, OG&E paid its 200th consecutive dividend to shareowners.

OG&E, which began service in 1902, has a long history of supporting Oklahoma's growth with a consistent presence in the economic development arena. For decades, the company has successfully helped businesses to locate or expand in the OG&E service area.

When a company is considering a move or expansion, the site search can begin with OG&E's Business Resource Center (BRC). Without charge, the BRC offers comprehensive services to companies looking to relocate or expand in Oklahoma.

The industrial properties database system continually updates state-of-the-art inventory of prime industrial properties in Oklahoma and western Arkansas. The Site Efficiency Evaluation (SEE) System, through the latest multimedia technology, allows viewers to visit available buildings and sites, get a feel for communities, compare tax and utility rates, and learn about labor availability and lifestyles.

OG&E's Business Resource Center offers the latest multimedia technology.

Providing reliable service and rates that are among the lowest in the country make OG&E the power of choice.

Extensive demographic and market information assembled from national and on-line data services and other respected sources is ready for analysis to make a site assessment quicker and more accurate. The BRC's Community Database and SEE System can give an instant snapshot of communities fitting a search profile. All work at the BRC is strictly confidential.

Another facet of the BRC's zeal toward economic development is a series of advertising supplements in *The Wall Street Journal*. OG&E is responsible for producing and underwriting the distribution of this piece, which skillfully bundles all assets of the state into a dynamic personality that shines to business executives in various regions of the nation. Five supplements, issued through 1995 and 1996, were considered one of the singly most substantial economic development efforts made on the state's behalf by a private business.

"At OG&E, we know the significance of electricity in the daily lives of our customers. As we position ourselves for the future, we will continue to provide reliable service at the lowest possible price," said Steve Moore, chairman of the board, president and CEO. "We will continue our efforts to make Oklahoma an even better place to live through ongoing partnerships with the communities we serve." ✵

Cox Communications

Cox Communications is your telecommunications link to the future. Offering video, voice, and data services with a variety of customer-focused options, there is no one company in Oklahoma City better positioned to give the customer what they want, when they want it, all with ease and convenience never before imagined.

In today's fast-paced world, telecommunications means change—rapid change and a rapid flow of information. Over the past few years, Cox has prepared for this change by investing heavily in technology and customer care. The company's philosophy of diversification welcomed the challenge of offering new services in the vast and quickly expanding technology industry.

Now, Cox Cable and Cox Fibernet, two robust businesses in the Cox Communication's family, are powerfully positioned to deliver Oklahoma City the best in telecommunications services and customer care. With 450 miles of fiber optics and 2,200 miles of high-capacity coaxial cable, Cox offers its customers advantages like crystal-clear reception, greater reliability and unmatched capacity. This capacity means Cox can offer unparalleled choices for entertainment, video-on-demand, high-speed internet access through cable modems, long-distance telephone service, commercial and residential telephone service, and wireless communications.

No other telecommunications provider will be able to offer this kind of one-stop shopping. This powerful mix of technological strengths and strategies translates to greater benefits for every customer—whether residential or commercial.

Cox Fibernet is the premier provider of high-end telecommunications services for Oklahoma City's business community. Offering services to commercial customers that range from basic telephone service to high-speed private data networks, Fibernet brings the highest standards to its state-of-the-art technology. For any business with critical communications needs, Fibernet's network provides the opportunity for choice and flexibility demanded by today's businesses.

Cox Communications is one of the five largest cable operators in the country with more than 3.2 million customers nationwide. Cox systems are the largest, on average, in the industry. Each cable operation responds to the unique challenges and needs faced in the local marketplace with local management, which means a greater ability to provide customers with what they need, when they need it.

In the midst of a rapidly changing industry, Cox remains true to its core values—quality customer service and commitment to the community, education, and its people. At Cox Communications, customers are the number one priority. With locally staffed and operated 24-hour telephone service, Cox responds to customer inquiries any time, any day, and with its unique On-Time Guarantee program, the customer is guaranteed service will be provided as promised.

Cox Communications believes in taking an active and leading role in the community, anchoring that belief in educational outreach. Cox is a member of Cable in the Classroom, the cable industry's education initiative. Cox provides free cable service to 170 Oklahoma City area schools and airs 525 hours of commercial-free educational programming each month. Through its adopted school program, Cox provides significant financial support and ongoing encouragement to two schools in high-risk areas.

Cox makes tremendous contributions to community organizations each year, with close to a million dollars of in-kind contributions annually. Cox also makes a substantial number of cash donations to community organizations, and Cox employees donate countless volunteer hours to various organizations throughout Oklahoma City.

Through its focus on customer service, technology, and service to the community, Cox offers a unique mix of telecommunications products over a powerful network to both residential and commercial customers. Cox Communications is progressive. Fast-paced. Upbeat. Caring. Positioning Oklahoma City for the future, Cox is leading the way to tomorrow through the world of telecommunications. ✪

Multimedia Cablevision

Fiber-optic thread connects people to the world, and Multimedia Cablevision is leading the way on this powerful highway of the future. Once simply a basic television programming provider, Multimedia has catapulted into a full-service telecommunications company. It is a driving force that is connecting people. . . and sending data at the speed of light.

Multimedia has invested $50 million to upgrade its existing systems using state-of-the-art fiber-optic technology. With an infrastructure that makes the Oklahoma City metropolitan area competitive on a global basis, and an Intranet that links at least 30 Oklahoma communities, businesses can be assured that sophisticated high-speed electronic commerce is a very real part of the metropolitan Oklahoma City landscape.

Thanks to this effort, businesses communicating with each other, and linked to a broad spectrum of customers, and potential customers, experience the benefits of freedom, convenience, and the competitive edge.

Multimedia is part of a diversified media company that operates newspapers, television and radio stations, syndicated programs, security systems, cable television, and telecommunications systems. Its parent company is Gannett, one of the largest media groups in the United States. Multimedia serves approximately 460,000 cable subscribers in the five states of Oklahoma, Kansas, Illinois, Indiana, and North Carolina.

The Oklahoma region provides cable service with more than 70 channels of television networks to more than 110,000 customers. It also provides telecommunications solutions and cutting-edge opportunities for schools, municipalities, businesses, and individuals. From Guthrie to Norman, El Reno to Midwest City, the mastery of Multimedia both encircles and connects the metropolitan Oklahoma City area.

"Multimedia Cablevision is taking a leadership role in making sure the Oklahoma City metropolitan area is in the forefront of the telecommunications revolution," says Terry Gorsuch, vice president/regional manager. "We are a vibrant part of the communications world. Most cities are not at this stage."

Multimedia's advanced telecommunications services and state-of-the-art fiber-optic system supports Wide Area Networks (WAN),

Multimedia Cablevision ran 470 miles of fiber optic cable as part of the $50,000,000 upgrade to the Oklahoma Region. Photo by Jack Hammett.

Multimedia Cablevision's commitment to Cable in the Classroom and Distance Learning Technology plays a critical role in preparing our young people for the 21st century. Photo by Jack Hammett.

Metropolitan Area Network (MAN), distance learning, cable in the classroom, two-way interactive video-conferencing, Internet access, and telemedicine. This translates to electronic field trips, a high school senior taking a college level course, face-to-face video conferencing, and much more.

Multimedia Cablevision has connected a multitude of schools and businesses in many area communities, and more are being connected every day. In Norman, 15 schools are connected to a fiber-optic network, which is used for e-mail, Internet access, and administrative functions.

Example: Eighth-grade students are touring the Louvre Museum in Paris via computers at school. Within minutes, they have studied the Mona Lisa and other works of art by famous painters spanning many centuries.

In Edmond, Oklahoma, the company

connected the Renaissance Center, which specializes in hospital care for women, to a radiology clinic about seven miles away.

Example: While giving a mammogram at the center, medical images are simultaneously being transmitted to the radiology clinic where physicians and other colleagues can view the results and immediately confer about other tests.

Using data, video, and voice, Multimedia is simply facilitating information flow and improving the quality of life in our communities.

"Our long-range goal is to build a metropolitan area network that connects schools, municipalities, businesses, and consumers," says Gorsuch. "With the use of fiber optics and the new developing cable modem technology, we are building the electronic commerce network of the future." ✪

Whinery's Off Airport Parking

A service-oriented family, who cared for people through their funeral home business in western Oklahoma, found a special market niche when they expanded their service to car rental and parking in the Oklahoma City airport area. This business is known as Whinery's Off Airport Parking, currently the largest off-airport valet parking lot in the United States.

In 1982, A.L. and Dala Whinery purchased a rent-a-car franchise and saw an opportunity to fill the real estate when cars were out on rental. They allowed patrons to park their car on the lot for $2 a day and even took these customers to the airport in their own car for no charge.

A door has never closed on this business since these humble beginnings. No longer in the car rental business, Whinery's has evolved into a 50-employee, off-airport valet and self-parking business encompassing 22 acres. It is owned and operated by the Whinery's son, Don, who is president of the company.

Service and security are the keys to success at Whinery's. When customers drive onto the valet parking lot, a courteous employee immediately meets them at the entrance. While the customer is checking in at the service desk, his or her luggage is carefully placed in one of the seven Whinery's vans, ready to leave for the airport. These vans run on demand, not time-table

When customers return to the airport after their trip, a quick call from the Whinery Customer Courtesy telephone in the baggage claim area brings a Whinery's vehicle in minutes, ready to deliver customers back to Whinery's and their car. Claim check and photo identification are several security procedures Whinery's has in place for car pick-up.

Whinery's takes each car personally. The company works around the clock, 365 days a year to deliver thousands of customers to and from the airport and keep their vehicle safe during the trip.

Whinery's is the only off-airport parking service in the nation that provides 100 percent insurance against auto theft. An electronic fence completely encircles the property, and on-site security personnel patrol the premises 24 hours a day. Additionally, the premises are videotaped at all times.

Other customer comforts include services such as hand-washing a vehicle on the day of return, oil and filter change, a complete tune-up by a certified mechanic, and body repair work by a certified body repair technician. Covered parking is also available at Whinery's to protect vehicles from Mother Nature's mood swings.

"Anyone can park a car," said Whinery. "It's the service that makes a difference."

The Whinery's are pioneers in the valet off-airport parking business. "We've been breaking new ground since we've opened," said Whinery. There was not a blueprint for implementing this type of business, and all aspects have been custom designed. In addition, the company was changing very rapidly. Whinery's grew 600 percent in 18 months.

As the first off-airport parking facility in Oklahoma City, Whinery's plays its part in helping the airport with clean air standards. Four of Whinery's valet vans run on Compressed Natural Gas (CNG), and the company is in the process of converting all vehicles to CNG. Whinery's Off Airport Parking is family-oriented, service-oriented, pioneering, and believes in the business of making it easier for the flying public. In describing his business, Whinery said, "We care." ✪

When customers drive onto the valet parking lot, a courteous employee immediately meets them at the entrance.

Whinery's is the only off-airport parking service in the nation that provides 100 percent insurance against auto theft.

AT&T Wireless Services

On June 25, 1876, at the Centennial Exposition in Philadelphia, Alexander Graham Bell—a little-known teacher of the deaf—began demonstrating his amazing new invention, the telephone. Bell's invention marked the beginning of a communications revolution, that continues with more intensity than ever before. It also laid the foundation for a new company that would become an American institution and the leader of the communications revolution, AT&T.

AT&T went through its adolescence in what was called the "Progressive Era" in American politics and social policy. It served Americans through two World Wars and the great Depression. As the 20th century marched on, AT&T provided the communications technology that enabled the business growth that has made America the world's biggest economy.

AT&T connected the big cities, small towns, and family farms of America together with universal telephone service and integrated them into the global village of the information age.

AT&T Wireless Services, Inc. began in the 1970's with the vision of cellular pioneer Craig O. McCaw, who built the world's largest wireless service provider. In 1994, McCaw Cellular Communications, Inc. merged with AT&T to form AT&T Wireless Services, Inc.

Today, AT&T Wireless is the leading wireless communications provider in the country. A unique collection of licenses allows it to offer a full range of wireless services including voice, data, paging, advanced messaging and aviation communications:

• The Cellular/PCS division—the largest in the United States—includes license areas covering nearly 80 percent of the country and more than 200 million potential customers. Operationally, the company includes five geographical regions responsible for network development and management, sales and marketing, and all aspects of customer service.

• The Wireless Data division provides cellular digital packet data (CDPD) and circuit-switched data services. The division's applications include mobile computing, personal data assistant, and telemetric, and are expected to be among the fastest-growing segments of the wireless communications industry.

• The Messaging (Paging) Division currently is the fifth largest in the United States.

• The International Development and Operations Group is a team of partners around the world. These partnerships license, construct, market, and provide services to a broad range of international markets. The international wireless market is the fastest-growing segment of the industry.

• The Aviation Communications division operates a digital air-to-ground phone service on commercial airlines and corporate aircraft. This technically unique system has contracts to provide services for Southwest, Alaska, Northwest, American, Air France, Lufthansa Airlines, SAS Airlines, Austrian, Kuwait, KLM, Lauda, Swiss Air, and Delta Airlines.

Phenomenal growth experienced by the wireless industry has had a positive economic impact in many cities. AT&T's need for full-time employment of salespeople, cell site technicians, and executives has added more that 10,000 jobs in communities across the country, with 376 employed in Oklahoma City. Tens of millions of dollars go directly into communities we serve, large or small, when we build our networks and provide our services.

Without a doubt, wireless technology will continue to shape the way we live—giving us the tools we need to keep our families closer, make our communities safer and our lives easier. AT&T Wireless is there for people to take along for personal safety, to send e-mails, fax a message to the office, or simply to call home. AT&T Wireless Services make wireless communications work for Oklahomans and the Oklahoma City community. ❂

AT&T Wireless Services keeps families closer by covering more of North America than any other cellular provider in the industry.

AT&T Wireless Data division allows packets of data to transmit across our wireless network to deliver e-mail messages, internet service and data base information, remotely, to your fingertips.

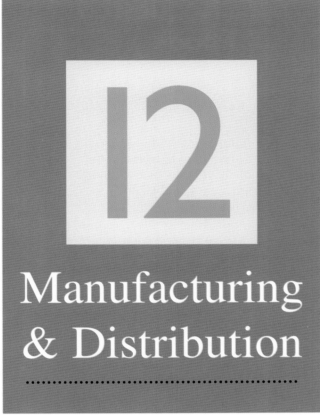

12

Manufacturing & Distribution

••••••••••••••••••••••••••••••••••••

Photo by Jack Hammett.

Chef's Requested Foods

When it comes to food preparation, a chef's skill and quality are held in highest regard. That is the philosophy behind Chef's Requested Foods, Inc. and why this company has grown to become one of the Southwest's premier providers of portion-controlled meats. By doing many things exceedingly well, Chef's Requested provides maximum quality products and a high degree of service to meet the requests of all types of chefs.

EARLY REQUESTS

Chef's Requested Foods, Inc. began as a local meat purveyor in Oklahoma City in 1979. Modest beginnings taught one important concept: the unique needs of each customer. This customized approach is still the foundation of Chef's Requested business—a business on the grow.

Processing capacity was expanded in 1982 to allow Chef's Requested to enter the food service and retail distributor markets. Then as the company's portion-controlled red meat products became more "requested" in these segments of the food industry, it became necessary again to expand operations.

In 1992, a major corporate expansion project was completed that involved plant relocation and an increase to six times the original processing capacity.

Growth has not hindered the company's ability to make quality count. Chef's Requested is driven and equipped to provide high-quality meat products and an unsurpassed level of service.

SATISFACTION GUARANTEED

Chef's Requested Foods is meticulous in meeting the requirements of each customer, whether the concern is grade, size, shape, trim, thickness, aging, or packaging. It takes pride in providing products that meet the most discriminating specifications with unparalleled consistency. No request is too great. No order is insignificant.

This commitment to perfection and living by the basic standards of the golden rule: do unto others, is the reason a high level of service exists. "If we say we'll do something, we will do it. If we make a commitment, we will live by that commitment," said John H. Williams, president. "Our company's growth is proof of our commitment."

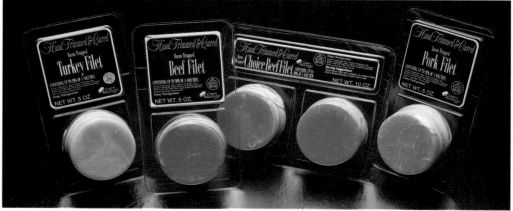

SERVING TWO MARKETS

Chef's Requested supports two basic divisions: food service and retail. It is rapidly becoming a most requested custom processor for the food service industry and a recognized brand name for the retail grocery market.

Through wholesale food service distributors, the food service component reaches all types of food preparation environments, including the health care industry, colleges and universities, restaurants, and hotel markets. Chef's Requested knows its ultimate customer is the chef that serves food to his or her patrons.

The retail division sells the product line in grocery fresh meat cases across the nation. In addition to the high-quality beef, pork, and poultry products, Chef's Requested is also characterized by value, ease of preparation, variety, and convenient packaging.

Williams explained Chef's Requested retails high-quality products that fit a need and are difficult for the market manager to prepare in the store. "We keep looking for ways to enhance satisfaction," said Williams. "Whatever the category, we want to be at the top."

There are many reasons Chef's Requested products bring satisfied customers back to meat cases. The company focuses on three areas of excellence: product, shelf life, and presentation.

First, Chef's Requested foods have outstanding value and good flavor, plus are tender and well trimmed.

Secondly, the consistent freshness of the products and the stable shelf-life are important. Chef's Requested uses special packaging processes to

west as Arizona. Retail is even more widespread; New England to Florida in the east and the Rocky Mountains in the west. The company has also begun initial entry into exporting.

ABOUT THE PRODUCT AND ITS FUTURE

The best value for every taste is found in Chef's Requested Foods.

Chef's Requested offers a variety of quality beef products—from more than 10 selections of beef steaks and ground beef patties to cubed beef steaks and other beef items. There is also a choice of 15 pork products. Any of these products can be customized to meet specific requests and some products are available marinated.

Hand-trimmed beef, pork, and turkey bacon-wrapped filets are made from top grade raw materials and are lightly marinated to ensure the freshest and most flavorful product available. Chefs save time at the cutting board with Chef's Requested thin-sliced beef and pork. The chicken breast is also cut into bite-sized pieces for convenience.

Chef's Request Foods, Inc. looks to the future of its products through its work force and the youth of America. Williams indicated he is impressed with the state's vo-tech system and its ability to train people who already possess a pride in ownership regarding their work. He is pleased to recruit for his work force from Oklahoma State University's food and agricultural students and to participate in summer internship programs.

This company also works with local Future Farmers of America (FFA) chapters to develop fund-raisers. Chef's Requested assembled a package of six different steak and chicken product items for the students to sell. Williams said this type of fund-raiser allows for general purchase of a quality item and is a good opportunity for the youth and for the company.

The future for Chef's Requested Foods, Inc. includes continued growth through expansion of the physical plant, product line, and distribution network. "I believe we are at the beginning of our life cycle," said Williams.

Through maximum quality products, consistency, and a high degree of service, the mastery of portion-controlled meats is met with pride in precision and emphasis on each customer's needs. At Chef's Requested Foods, Inc., "At Your Request" is not just a slogan, it is a commitment to a way of life. ✸

ensure quality in the meat case. The fresh meat has a 30-day shelf life from the day it is packed, and the product is guaranteed for 25 days.

All products have attractive packaging with bright, informational labels. Chef's Requested believes appearance is part of the decision making process and making a good first impression is imperative. The development of an attractive label and unique packaging is accompanied by nutritional labeling.

Distribution in food service spans as far east as Kentucky, Tennessee, and Mississippi, and as far

Gulfstream Aerospace Technologies

A HISTORY OF DISTINCTION

The company was formed in 1944 by a small group of engineers with plans to develop a twin engine airplane to serve executive business travel needs. The company organized in Culver City, California, as Aero Design & Engineering Company, and the prototype Aero Commander made its first flight in April of 1948. That airplane is now on display at the Oklahoma City Fairgrounds.

During 1949, several investors in Oklahoma City provided capital to get the Commander through flight testing and FAA Certification. Construction of the factory to build the airplane was completed in 1950, with assistance from the Oklahoma City Chamber of Commerce.

The year of 1957 brought a fire that destroyed the production facility at Wiley Post Airport. Fortunately, a new factory was already under construction, and the completion schedule of the new building was accelerated, putting the Commander back in production with minimal delays.

In 1958, Aero Design was acquired by Rockwell-Standard and operated as a wholly owned subsidiary. In 1967, Rockwell-Standard merged with North American Aviation, and the Oklahoma City facility continued to grow and develop new, innovative products. In 1981, the Commander factory was purchased by Gulfstream American Corporation located in Savannah, Georgia. The company name was later changed to Gulfstream Aerospace Corporation in recognition of its prominent role in the aviation community. Gulfstream Aerospace is a publicly owned corporation, with stock listed on the New York Stock Exchange.

A CHANGING VISION

The twin engine Aero Commander remains a fine airplane today, having set several world records for speed, altitude, and endurance during the 1970s. However, beginning in 1982, the general aviation market declined dramatically for several years. Sales of new twin engine turboprop aircraft dropped from 918 in 1982 to only 129, for the entire industry, worldwide in 1986. This change in market dynamics forced Gulfstream to stop production of the Commander aircraft manufactured and sold from Oklahoma City. Since its inception, the Commander management team has been guided by a spirit that has challenged the conventional concepts limiting business opportunities, so it was an easy decision for Gulfstream Aerospace to give its Oklahoma facility major responsibilities in providing parts shipped to Savannah, Georgia, and used to assemble the Gulfstream IV-SP and Gulfstream V business jets, recognized around the world as the finest aircraft available.

GULFSTREAM AEROSPACE AND OKLAHOMA CITY: NAMES FOR EXCELLENCE

Two-thirds of all the large-cabin general aviation jet aircraft operated by Fortune 500 companies bear the Gulfstream name. Balancing exceptional performance with comfort and reliability, Gulfstream has a prominent role in helping companies conduct business competitively on a global scale. Also, the President of the United States travels in a Gulfstream, as do the leaders of more than 30

Original Factory

Present Facilities

other governments around the world. Gulfstream aircraft have been extending the boundaries of executive transportation for decades. On a single day in 1995, the Gulfstream IV-SP broke 31 national and world records for performance. The Gulfstream IV-SP and Gulfstream V jointly hold 81 national and world records. The Gulfstream V engines were developed in close partnership with BMW Rolls-Royce. Considering all the factors, it was no surprise when aviation directors voted the Gulfstream family of aircraft the best turbine powered aircraft in the world.

The Gulfstream V aircraft was certified by the FAA during late 1996, and the first demonstrator made a whirlwind tour of the United States and abroad, including several long-distance flights. The Gulfstream V is pioneering a new era of worldwide business travel. With a range in excess of 6,500 miles and cruising at altitudes up to 51,000 feet, the Gulfstream V flies high above adverse winds and inclement weather and makes nonstop travel possible from cities such as New York to Tokyo, London to Singapore, Beijing to New York, and from Los Angeles to any city in Europe.

Gulfstream Aerospace maintains a highly visible presence in every area of the world and recently opened an office in Hong Kong in order to support the expansion of business aircraft into China, Hong Kong, and Macau—all emerging markets for corporate aviation in a region traditionally noted for its restricted airspace and congested airports.

Gulfstream's Oklahoma City employment level is at 650 and will grow, due to a decision to increase the production rate of the Gulfstream IV-SP and Gulfstream V over the next two years. With more than 500,000 square feet of space, the Oklahoma City facility can accommodate the additional work. A wide variety of parts are supplied by the Oklahoma facility, ranging from a small 1 inch-by-4-inch aluminum clip to a large 8-by-20-foot outer fuselage skin panel. The assembly work is also varied. Many assemblies are small and contain few parts, while the largest assembly is the vertical tail structure, which has several thousand parts.

In addition to the fabrication and assembly work, the Oklahoma City facility has a general aviation maintenance and modification center. This service center provides maintenance and repair capabilities for the twin engine Commander aircraft, as well as all other small and medium-sized turbine powered aircraft. Clients come from across the United States,

New Gulfstream V

Service Center Operations

Canada, Latin America, and Europe, and work has also been accomplished for customers as far away as Japan and Thailand.

As Gulfstream Aerospace Technologies approaches 50 years of business in Oklahoma City, it is well positioned to share many opportunities and benefits of the growth and success achieved by its parent company, Gulfstream Aerospace, in Savannah, Georgia. ✪

Fleming

If a product is in a U.S. supermarket, chances are Fleming distributes it. Fleming is one of the leading food distributors and marketers in the world, a top 20 food retailer, and a leading service provider in the food retailing industry.

Fleming employs about 40,000 dedicated associates across the country, including nearly 2,000 in Fleming's headquarters community of Oklahoma City.

DYNAMIC LOCAL IMPACT

Acknowledging the company's global status, it is no wonder that Fleming is the largest publicly held firm in Oklahoma. This Fortune 100 company serves about 150 stores and several military bases in Oklahoma, making it an integral part of the economy.

The company's corporate headquarters at the Waterford office complex and two product supply centers in Oklahoma City encompass more than 1.6 million square feet. Because of Fleming's importance in the industry and its location in Oklahoma City, a number of national manufacturers and vendors have established facilities and staffs in the area.

"Fleming's corporate staff has been headquartered in Oklahoma since 1972. This is a great location for us, first because of the state's centralized geography. Another attraction is Oklahoma's low cost of living," said Chairman and CEO Robert Stauth. "Perhaps most importantly, we continue to be impressed with the excellent work ethic of people in this state."

FOOD DISTRIBUTION AND RETAIL CHAIN OPERATIONS

Fleming's food distribution network serves about 7,000 stores around the country with a broad base of more than 3,000 supermarkets in 42 states. This includes hundreds of company-owned stores, operating in regional chains. These retail operations are a growing component of Fleming's business.

The company supplies these stores with approximately 100,000 different products consisting of frozen foods, meat, produce, dairy, dry grocery, health and beauty, general merchandise, and specialty items. Store brands offered by Fleming include the popular BestYet®, Piggly Wiggly®, Rainbow®, IGA®, and Marquee® lines. These brands offer consumers a quality alternative

Fleming delivers customer-focus solutions, designed to help Fleming-serviced stores around the country win at retail.

to national brands, at an appealing price.

Each year, the company delivers more than 630,000 truckloads of grocery, frozen, and diary products—enough trucks to wrap around the earth six-and-a-half times. Fleming's annual French bread volume is 30 million loaves, which laid end to end would reach from Oklahoma City to Paris. The number of bananas Fleming sells each year would fill Chicago's Sears Tower 5,000 times!

To meet specific needs of each customer, Fleming concentrates its efforts to offer consistent quality, wide product selections, competitive

pricing, and efficient service. Fleming also exports to retailers as far away as Australia.

LEADING INDUSTRY SERVICE PROVIDER

Along with products, Fleming provides an array of innovative retail support services. The American grocery industry has a long history of providing value to customers. Fleming's services add to that value and include expertise in virtually every discipline a food retailer needs to succeed in an increasingly complex and competitive retail environment.

Fleming believes its customers' retail success is closely tied to the quality of its support services. Its 12 major service areas address competitive challenges, changing consumer behaviors, and the application of rapidly advancing technology to provide the right package of solutions.

Many of these service areas involve leading-edge computer systems that help manage inventory, pricing, and human resources. Others focus on education, financial support, and advertising. They are often customized and combined to meet unique needs of individual stores. Each service is developed and supported by the company's professionals, finding solutions to help retailers compete more effectively and increase profit potential.

Fleming's VISIONET™ service is an interactive electronic communications network. This advanced technology gives retailers a dynamic way to make more timely and effective business decisions.

Fleming retail advertising service, which functions as a full-service advertising agency to clients throughout the United States, is one of the largest advertising agencies in Oklahoma and the

Besides distributing national brands, store brands and perishables, Fleming also offers a wide range of professional consulting services, offering retailers leading-edge ideas and solutions.

Fleming monitors emerging consumer trends, such as the rapid growth in demand for prepared meals, then develops innovative retail concepts to respond.

largest supermarket advertising agency in the country. Fleming is one of the top lenders in Oklahoma, providing financing for the development and improvement of supermarkets.

Fleming's training programs, with headquarters in specially constructed facilities in Oklahoma City, are formally known as the Supermarket Management University. Here retailers and their employees learn every aspect of the business, from shelf management to cake decorating. Many of these retail students travel across America, or from foreign countries, to study with the Fleming experts.

The Fleming team also offers market research services to help retailers identify and respond to industry and consumer trends. Through market research—tracking and monitoring customer trends by surveying approximately 25,000 consumers a year—important statistics can help answer the questions that will affect the design and operations of supermarkets of the future.

COMMUNITY SUPPORT

Fleming has a history of corporate citizenship in every community and state where it operates. The company has provided support for a wide array of programs from social services to the fine arts, from education to the environment, from historical preservation to medical research.

Company associates are encouraged to participate in community life. They volunteer thousands of hours each year, offering their time and leadership to worthy community projects across the nation. Fleming-served retailers also are active citizens of the communities where they operate. Often with support and coordination supplied by Fleming, these retailers participate in and sponsor a multitude of projects that make their cities better places to live.

The company also has worked with many of its national vendors to provide equipment and dollars for educational programs across the

nation. One of the best known of these is IGA's Hometown Trees project, which has planted more than 70,000 seedlings in the state of Oklahoma.

As a food distributor, retail chain operator, and service provider, Fleming has grown to be one of the biggest and the best in its industry. For more than 80 years, Fleming has been working behind the scenes to help food retailers—both independents and chain customers. This food industry leader uses the dynamics of its diversity to bring innovative and responsive strategies, products, and success to supermarkets around the country. ✪

The Fleming-developed Visionet™ computer network gives retailers access to inventory information, financial data, retail support services and on-line promotion ordering, along with rapid communications with Fleming.

Macklanburg-Duncan Company

Home and family are a natural tie for this dynamic, family-based business that offers more than 2,500 home improvement products. Along with the Macklanburg family heritage of corporate ownership and leadership, Macklanburg-Duncan (M-D) employees are a part of the family circle. They are the heart of the company and make the organization what it is today.

What started in a 25' x 130' building on Main Street in downtown Oklahoma City has grown to over 1,400,000 square feet of plant and distribution sites, in four other states and one foreign country. Nearly 1,750 employees across the country and more than 700 employees in the Oklahoma City location look to the history of this company as a strong part of the present and the future.

COLORFUL HISTORY

The history of Macklanburg-Duncan began in 1920, with L.A. Macklanburg. He was an entrepreneur who responded to his wife's request to keep the cold wind out of their home in the winter and the dust out in the summer. Pansy Macklanburg got her wish, and as a result, weatherstrip was patented under the name Numetal.

Numetal is the familiar strip of metal with spring brass that fits between a window frame and sash, or doorframe and door, that acts as a seal from the weather. Business was good for M-D as the demand for Numetal continued to grow.

Over nearly 80 years, many things have changed at Macklanburg-Duncan: the buildings, the faces of management, and the products—all adapting to remain successful in a competitive environment. One thing has not changed and that is M-D's appreciation of its customers.

The words written by L.A. Macklanburg back in the early years are as timely today as they were decades ago: "A man can do as much as he wants if he's willing to work. If you make a better product, give better service, gain the confidence of your customers and always treat them right, they'll stay with you always."

L.A. Macklanburg was a giant of his time in Oklahoma City business, starting with the invention of Numetal weatherstrip and then building Macklanburg-Duncan into a national manufacturer of a variety of home improvement products.

THOUSANDS OF PRODUCTS

Macklanburg-Duncan offers a high-quality product for the customer's value. The company sells its products directly to retail outlets, which include more than 20,000 hardware and home improvement stores across the United States, Canada, Mexico, and a variety of foreign countries.

The six main product categories are weatherproofing; caulks, adhesives and coatings; levels and tools; floor and carpet trim; steel and aluminum shapes and moldings (thresholds); and builders hardware.

M-D is currently the number one manufacturer of carpet metals, weatherstrip, thresholds, and overall weatherproofing products. The company is number two in the caulking industry.

The company's number one selling product is the 35-year acrylic latex with silicone caulk. In honor of this product and the company's 75th anniversary, M-D converted a large silo tank to a yellow and black caulk tube. This 50-foot caulk tube stands as an icon for Macklanburg-Duncan in the Oklahoma City community.

Of the 2,500 different home building and improvement products, M-D manufactures approximately 90 percent. The additional 10 percent are purchased outside and offered to complete a product line. Sixty-five percent of the manufactured products are made in Oklahoma City.

PEOPLE-FOCUSED COMPANY

M-D is privately owned by descendants of L.A. Macklanburg and by employees through a stock ownership plan started in 1985. This stock program, illustrates the company's investment in its employees as an integral part of the corporate structure.

"The strength behind Macklanburg-Duncan's nearly 80-year reputation for consistent quality and dependable services is the M-D employee," says Mike Samis, president and chief executive officer. "Each of our employees is a stakeholder in this company and everyone, from the machine operator to the chief executive officer, takes extreme pride in their work."

Samis, who is married to the founder's granddaughter, also speaks with passion about the Macklanburg-Duncan business concept. Introduced in 1996, the business concept offers a fresh, new vision for the future. It brings into focus how M-D conducts itself in its external and internal affairs.

The business concept has very definite goals, strategic plans, and specific steps of action that include a commitment and input from employees. Through this concept, employees are involved in the goals and direction of the company, with a mission that everyone will move forward— together.

"In reviewing the past success of M-D and viewing the future of our industry and business, it

Mike Samis, President and CEO. In honor of the company's number one selling product, M-D converted a large silo tank to a yellow and black caulk tube.

was clear for M-D to be successful in the future, we had to have employees who were able to merge their personal goals with those of the company," says Samis. "Macklanburg-Duncan's business concept is committed to providing employees with a safe environment of mutual trust and respect, rich with opportunities for personal and professional growth."

"Today, the success of most companies can usually be measured by the individual success of its employees," Samis added. "At Macklanburg-Duncan, our future is brighter than ever."

With a strategic goal of earning a market share position of one or two in all product categories, Macklanburg-Duncan works through every employee and 2,500 home building and remodeling products to reach this success. It is because of this inclusive corporate environment that the combination of family and home build the framework for a spectacular company with a strong foundation for the future. ✺

M-D manufactures approximately 90 percent of the 2,500 different home building and improvement products.

Tech Pack, Inc.

VISION

Tech Pack, an Oklahoma-owned corporation, has slated customer service as its top priority. Tech Pack, Inc. manufactures custom corrugated boxes and also distributes specialized packaging materials such as bubble wrap, foam cushioning, tape, and plastic bags for all types of commercial and retail applications throughout Oklahoma and surrounding states.

"We are a full-service company," says Gary Clonts, president and CEO. "We can supply every item needed in packaging a customer's product, which has allowed us a niche in the marketplace. A company can get all its packaging needs met with us."

TAKING CARE OF BUSINESS

Since Tech Pack was established as a distributor in 1985 by Gary Clonts; his son, Randy Clonts, vice president of operations; and partner Dan Meyer, vice president of marketing, the enterprise has focused on customer service, market niche, and growing the company to meet client needs.

By taking care of all the manufacturer's packaging requirements, Tech Pack makes the manufacturing and delivery process much easier for its customers. "We meet every packaging challenge," says Randy Clonts. "We have a wide range of equipment to handle all packaging requirements."

Tech Pack, Inc. frequently updates its computer technology to accommodate an array of different packaging needs. As a result of its focus on state-of-the-art design, the company was chosen as a pilot program for a software prototype developed specifically for the corrugated box industry. After its development, Tech Pack will be used as an industry model for the software's design capabilities.

Oklahoma City businesses appreciate Tech Pack's full service and technologically advanced approach. The majority of Tech Pack's customer base is in the Oklahoma City area. However, Tech Pack's Tulsa facility coordinates with the Oklahoma City plant to service all major cities in Oklahoma, as well as in Texas and Kansas.

By taking care of all packaging requirements, Tech Pack makes doing business much easier for its customers. Despite operating in a highly competitive field, Tech Pack continues to enjoy steady growth because of its unique company philosophy.

Founder Gary Clonts (center) along with Tech Pack's management and sales teams. (l to r) J.R. Clonts, Dan Meyer, JoAnn Clonts, Gary Clonts, Randy Clonts, Chuck Friedlander, and Steve Maness.

An example of one of the modern flexographic presses at the Tech Pack facility in Oklahoma City.

GROWTH AND EXPANSION

After its first three years in business, Tech Pack expanded from distributing packaging materials into manufacturing corrugated boxes. Manufacturing now comprises 95 percent of its gross sales. Broadening its services encouraged the company to repeatedly multiply its production space to house its growing number of flexographic presses and manufacturing equipment. In 1996, Tech Pack tripled its floor space when it moved to its new 113,000-square-foot facility in Oklahoma City. The expansion was part of a 10-year growth program. "This will ensure that we maintain our goal of increasing revenues by an additional $1 million per year. So far, we have met that goal," says Dan Meyer.

Along with its production growth, Tech Pack has evolved from an initial staff of two people to approximately 60 employees. Gary's wife, JoAnn, and younger son, J.R., have also joined the Tech Pack team. At Tech Pack, every employee and customer is considered family. "We respect the jobs our employees do," says Gary. "Since my family has worked in every area of production, we clearly understand the process. We operate on an open-door policy for the staff and assist with whatever needs to be done."

With more than 140,000 square feet of warehouse throughout Oklahoma, Tech Pack offers a unique J.I.T. delivery program to customers. Photo by Keith Ball.

Tech Pack's manufacturing facility located in Oklahoma City. Prompt delivery is ensured by using a fleet of company owned trucks.

Deliveries are made each day to customers throughout Oklahoma on Tech Pack's own fleet of trucks. Frequently, several trips may be made to one location every day.

Recognizing the value of the Internet, Tech Pack also developed a website to display its products and services to make it easier for clients to review and order products. "Customers can always count on us to fulfill their needs. We offer quality service to assure our clients operate more productively. It all weaves together in a coordinated effort between our company and theirs," says J.R. Clonts. The executive staff credits this reciprocal relationship for its continued growth.

The Tech Pack "customer first" approach has been frequently recognized by the Oklahoma City business community. In 1990, the Oklahoma Private Enterprise Forum and the Oklahoma Venture Forum applauded Tech Pack with a joint award for the company's outstanding contribution to Oklahoma's economic growth, productivity, and innovation. In 1992, the Greater Oklahoma City Chamber of Commerce awarded Tech Pack "The Best in Business."

TECH PACK GROWS OKLAHOMA

Gary Clonts is very optimistic about business opportunities in Oklahoma. "The bright side of the story is what Oklahoma is doing. We are very excited about what is happening in the state and about the people who live and work here."

Future plans for the company include adding new employees to the sales force to provide better coverage for the estimated 6,500 Oklahoma manufacturers. The company also plans to continue upgrading equipment as it addresses the fast-paced demands of its manufacturing clientele.

"We thrive on solving our clients' packaging problems," comments Gary. "When we do, we not only gain a valued customer, but more importantly, a friend. To sum up our company, we are Tech Pack, the Packaging People—with a strong emphasis on people." ✺

PACKAGING CUSTOMER SERVICE

Developing long-term relationships with customers and employees has been the guiding force for Tech Pack. Solving challenges, Just-In-Time (JIT) delivery programs, and quality products are the Tech Pack standard. Interdepartmental cooperation is essential for Tech Pack's consultative sales group, which brings clients' packaging challenges back to the design team for solutions. In turn, the design team coordinates closely with the marketing department to assist Tech Pack customers with

an increase in their sales volume through the creation of attractive packaging with strong shelf appeal. Monitoring the market and listening to customers' needs, Tech Pack began manufacturing custom-designed pallets to assist clients in meeting strict due dates and specifications as another service to their customers. The Just-In-Time (JIT) delivery program is another unique aspect of Tech Pack, Inc.'s "customer first" focus. This service eliminates clients' need to assemble or warehouse packaging stocks, allowing them to focus more of their efforts on primary products.

Great Plains Coca-Cola Bottling Company

The Oklahoma City Coca-Cola franchise was incorporated in 1903.

Behind each can, 12-pack, plastic bottle, or case of Coca-Cola products shine the faces of more than 900 associates that make up the Great Plains Coca-Cola Bottling Company team. This team reflects the company's grassroots organizational structure—organic, alive, and interdependent.

The collaboration and relationships among these associates, other innovative techniques, and progressive use of technology serve as the key to Coca-Cola's success in Oklahoma City. This commitment of knowledge and resources ensures there is "Always Coca-Cola."

The history of Coca-Cola and Oklahoma City is intertwined through generations of business leaders. In 1886, Coca-Cola was invented by Dr. John S. Pemberton, a pharmacist in Atlanta, Georgia. Pemberton is the great-grandfather of Art Pemberton who owns and operates Oklahoma

director and officer of the Oklahoma City Chamber of Commerce for 26 consecutive years. Seventy-five years later, Browne's descendants—Henry Browne, Sr., now Chairman Emeritus and his sons, Bob and Henry, Jr.,—and an expert management team are in Oklahoma City carrying on the company and community leadership under the revised name of the Great Plains Coca-Cola Bottling Company.

Just as history of the Browne family brings continuity and strength to this bottling company, the vitality of continuous improvement also positions the company for the next millennium.

Behind each case of Coca-Cola products shine the faces of more than 900 associates that make up the Great Plains Coca-Cola Bottling Company team.

City's oldest grocery store, Crescent Market.

The Oklahoma City Coca-Cola franchise was incorporated in 1903 and purchased in 1922 by a group of business partners headed by Virgil Browne. At the age of 44, Browne moved to Oklahoma City from New Orleans for a short-term stint to establish the business.

This short-term business venture turned into a lifetime. Browne is acknowledged as one of Oklahoma City's most prominent civic leaders, most notably recognized for his service as a

The Just-in-Time Inventory procedure is a result of this continuous improvement path. The Great Plains Coca-Cola Bottling Company no longer stocks inventory in its sales centers. The product is produced and shipped direct, ensuring freshness—the world's freshest according to audits and surveys. No wonder the company's product list encompasses six of Oklahoma City's top eight sellers: Coca-Cola Classic #1, Dr Pepper #2, diet Coke #3 Sprite #5, caffeine free diet Coke #7, and diet Dr Pepper #8.

Through respect for people and the desire for progress, the internal organizational structure has become a Team Cell Network. This network serves to bond individual plant functions and associates into a seamless flow of development. The ideologies of empowerment, teamwork, fact-based improvement, shared responsibility, trust, customer focus, flexibility, and cross-functional teams describe the components of this work environment.

Enhancements in all aspects of the company evolved to facilitate this just-in-time procedure, which has proven successful in a number of areas.

An educational video for the company explains: "Great Plains will always be a better place than the way it was just a few years ago. On our journey of continuous improvement, we listen, we learn, we share. As a result, our process becomes more precise. We become even better."

The Great Plains Coca-Cola Bottling Company also continues to better the community by being a generous corporate citizen. The company is a major sponsor of the Oklahoma City Blazers hockey team, Redbud Classic, Lazy E. Arena, Red Earth Native American Festival, Aerospace America, the Children's Miracle Network, to mention just a few.

Coca-Cola has been a supporter of the Olympics since 1928. Last year, the Great Plains franchise acted as a host for the 1996 torch relay run through the Oklahoma City area. This company is also the exclusive soft drink distributor for sporting events at both the University of Oklahoma and Oklahoma State University.

The spirit of Great Plains Coca-Cola Bottling Company can be seen in the faces of maintenance, sales, distribution, production, and service department workers. It is their commitment to learning and working together that guarantees it's "Always Coca-Cola." ✪

Unit Parts Company

Unit Parts is recognized in the automotive aftermarket as the premier electrical remanufacturer for starters and alternators. By aggressively seeking avenues for improvement, demonstrating unmatched quality and service in the industry, and making a commitment to success, Unit Parts is dedicated to being the absolute best in the industry.

Since its beginning in 1981, Unit Parts' strategy has been to build copies of original equipment and serve a narrow market niche with only two products: premium starters and alternators for domestic models of General Motors, Ford, and Chrysler automobiles. As a consequence of this focus and pledge to excellence, the company has doubled its employment within the past five years.

The 1,500-plus employees in the Unit Parts' organization act on the belief that the customers' expectations must be achieved or surpassed. These actions are based on the company's mission to achieve 100 percent customer satisfaction. Unit Parts is committed to continuous improvement, doing it right the first time, and maintaining the 100 percent order fill guarantee.

These values are demonstrated repeatedly through unrivaled customer service. Unit Parts utilizes a 50,000-square-foot facility to ensure on-time product delivery to more than 30 customers in North America. The company also guarantees service quality through on-time shipping performance featuring "speed of delivery," demonstrated problem resolution performance, direct ship to retailer service, and Electronic Data Input (EDI) capabilities for placing orders.

Along with service, product quality is paramount. Unit Parts offers a lifetime warranty on all of its products. Unit Parts' engineering team, organized in a variety of disciplines, is established to direct specific continuous improvement activities targeted at warranty and efficiency improvements. To improve return on a customer's investment, the company routinely

The distribution center is equipped with the latest computer-controlled material handling, storage, and retrieval system and a carousel system for rapid picking of small order quantities.

tests, disassembles, and analyzes samples of warranty returns to determine the root cause of return.

Resulting analyses are reviewed for corrective action. Does this philosophy work? Unit Parts has received numerous quality and service awards from its large and small customers, certifying its excellence in supplier quality and

service. The company, however, is not content with past accomplishments and actively seeks to add value in return for its customers' investment.

By meeting stringent quality standards, Unit Parts received ISO 9000 status in April, 1997. Other initiatives that will continue the company's preeminence in the remanufacturing industry include the expansion and upgrading of manufacturing facilities and equipment, along with manufacturing/distribution control system upgrade and integration. State-of-the-art cataloging services keep customers up to date on changes and additions to the company line through the Internet web page that features technical bulletins, detailed application pictures, and service procedures.

Striving to be on the leading edge of technology, Unit Parts recognizes the need to give employees the most current tools to enable them to do their best. Personnel work two shifts at four facilities in Oklahoma City and Edmond, which include a main plant, distribution center, and warehouse.

Through its ongoing strategy of aggressively appraising products and service for improvements, while integrating innovation and leadership, Unit Parts remains true to its mission to achieve unmatched 100 percent customer satisfaction in remanufacturing premium quality generator and starting motors. Working from this foundation, the company believes it will continue to grow and bring new jobs to Oklahoma. ✪

Crescent Market

Upon entering the Crescent Market in Nichols Hills Plaza, it is obvious that there is something different about this grocery store.

The Pemberton family has been operating the Crescent Market since 1942.

The Crescent Market shares a birthday with Oklahoma City. In 1889, a tent with a dirt floor served as the predecessor for what would later earn the prestigious Crescent Market name. Through several owners, name and location changes, the continuous tradition of excellence at the Crescent Market has flourished through generations.

The Pemberton family has been operating the Crescent Market since 1942 when Art L. Pemberton became a half-interest owner, then full owner in 1948. Art E. Pemberton joined his father in the business. Robert A. Pemberton followed in his father's footsteps. This family-oriented culture ensures the same quality and service from day to day and decade to decade.

Upon entering the Crescent Market in Nichols Hills Plaza, it is obvious there is something different about this grocery store.

The Crescent Market is still a mom and pop grocery. There is always an owner on premise and an open door to their office—Monday through Saturday from 9:00 a.m. to 6:30 p.m. The exceptional service and individual care of the store starts from this office. If an item is not in stock, customers can make special request orders.

The dirt floors have long been replaced by red carpet, which covers the entire shopping area. A hand-carved, 18th century couch is placed in front of a large fireplace with logs ablaze when it is cold outside. This couch is available for customers who may want to stop and enjoy a complimentary cup of coffee and cookie while shopping.

While walking along the large aisles, customers can choose from a selection of standard and hard-to-find items. The produce section's specialty is baby zucchini, squash, and spinach along with white asparagus. Domestic and imported caviar are intermingled on the shelves with the tuna fish, sardines, and smoked oysters. Fresh caviar requests can be filled overnight.

A trademark of Crescent Market is the wide variety of exotic meats. In addition to the usual meat selections, ostrich and buffalo are on hand at all times. Other exotic meat opportunities include alligator and bear. Crescent Market is also the only store in Oklahoma City with fresh, frozen homemade stock.

Eight meat cutters are employed by the Crescent Market, which touts a personal-service meat counter. There are no prewrapped meats at this grocery store. The market takes pride in its

top-choice beef that is aged in the store's meat cooler and large deli offering an array of sliced lunch meats.

Approved charge accounts are a special perk when shopping at this store. When leaving, courtesy clerks carry groceries in the multiuse, handled bags to the customer's car.

The Crescent Market's future plans include the same goals that have been passed on from father to son and father to son: individualized service and quality products.

The Pemberton Crescent Market tradition stands a good chance of continuing through four generations. Alexis, the daughter of Robert Pemberton, indicated in a first-grade class assignment that she wanted to run the Crescent Market when she grew up.

The Crescent Market and Oklahoma City are growing together. Both have a bright and prosperous future. ✸

Corrugated Packaging & Design

Box it, ship it, display it—Corrugated Packaging & Design (CPD) can do it. From containers to displays and packaging for shipment of manufactured goods, CPD understands that product quality is preserved and enhanced by proper packing and shipping. Where there is a need, CPD designs a way.

Founded in 1979 by Tom Roe with only two employees, CPD now operates out of almost 100,000 square feet of manufacturing and warehouse space with approximately 50 employees. Roe credits the success of the company to "lots of hard work and commitment from all our staff."

The CPD staff works with clients on conception and design, then manufactures and tests each new corrugated product. "We are always doing something new because our clients are growing and changing their products to something new," says Roe. Roe explains that everyone in the client's office must be happy with a finished product—from president of the company to the personnel on the shipping dock.

CPD manufactures products for companies based in Oklahoma City and the surrounding areas. These customers include Fortune 500 corporations as well as those just starting new businesses and developing new products. CPD's packages ship products throughout the world for these national and international corporations.

Because products must be packaged and displayed properly for maximum protection and appeal, one graphic artist and two structural engineers on staff design the packaging and develop the art or graphics for the corrugated product. An automated sample table allows designers to cut a sample product from a computerized drawing or pattern, and then to refine the product as needed. Additionally, using a new four-color press with varnish capabilities. CPD is one of the first companies in the Southwest to print directly onto corrugated material.

Colorful boxes fill CPD's sample room. These samples represent thousands of displays, boxes, tests, and colors. Hot pink-and-purple displays are interspersed with white-and-black boxes. A 1 1/2-by-1 1/2 inch box sits across from the giant navy- and-gray floor display made from die-cut, corrugated plastic. Products from dog bones to flowers, gourmet food to vegetables, health care to novelty items are sold in these corrugated products.

Along with packaging, shipping, and display purposes, CPD has found other important uses for its corrugated materials. An elementary school class used corrugated pads donated by

Roe credits the success of the company to the hard work and commitment of the staff.

The CPD staff works with clients on conception and design.

CPD to back a writing assignment about what the children wanted to be when they grow up. A binder full of creatively written and illustrated thank-you letters from these second graders lies open for the enjoyment of visitors to CPD's inviting reception area.

CPD's plans for the future include plant expansion and diversity into other areas of packaging. "As Oklahoma City grows, we will grow and create more employment for our city," says Roe.

"Corrugated Packaging & Design is proud to be a part of the Oklahoma City business community," Roe adds. "Our professional sales, design, and service departments save our customers time and money every day with efficiently designed packaging being delivered when they want it, the way they want it," Roe added. "Our customers are our first priority." ✪

CMI Corporation

Since the 1960s, CMI Corporation has literally been moving the world to new levels with its road-building equipment and technology. Starting with the $500 he borrowed to incorporate the company, Bill Swisher has led the charge to usher in a line of more than 100 major road-building products that have increased the quality and productivity of road construction projects across the world.

"From the beginning, CMI embraced change as the one dependable constant," Swisher says. "We have been driven by the philosophy any machine or process could be improved. Eventually, they will all become obsolete."

Applying his CB credentials as an expert "blademan" to early interstate highway development, Swisher identified the need for reliable, automated grading equipment. Because of his vision, the CMI Corporation was born in Oklahoma City with the company's manufacturing of the AUTOGRADE trimming and paving machines. This durable equipment increased productivity of grading contractors over 500 percent and provided a quantum leap in fine-grade accuracy.

Success with the AUTOGRADE led to product innovations in multiple areas of paving technology. By the 1970s, CMI had branched into related construction equipment markets with the acquisition of companies manufacturing asphalt hot mix production plants and bridge deck concrete finishing machines. A complete line of trailers was also included to meet the heavy hauling requirements of road contractors and materials producers.

CMI's history of industry firsts is among the most impressive of any company serving the pavement construction and maintenance industries. With an ongoing commitment to the natural environment, CMI's design team pioneered ROTO-MILL processes of pavement mining, leading to hot mix recycling and cold in-place recycling. Old asphalt is ground up with virgin material to create the new pavement. These innovations have significantly reduced the costs of maintaining highway and airport surfaces.

Swisher credits a strong customer base as a key component for his company's success. "We have been blessed with wonderful customers, who have caught the CMI vision for quality and productivity. We owe much of our success to these companies," he comments.

CMI's new ROTO-MIXER is a giant 650 horsepower pavement roto-tiller. It breaks up old, worn pavement and re-mixes it with new bonding agents to produce a brand new, long life roadway surface.

CMI's 750,000 square foot facility, located on the western edge of Oklahoma City, is the world's largest paving products manufacturing plant.

The hard work from a team of creative people at CMI is another major ingredient Swisher attributes to the company's continued expansion. Of the 1,100 CMI employees, over 900 work at the Oklahoma City plant.

"I have traveled the world over many times, and Oklahoma City remains one of my very favorite cities," Swisher said.

His community pride reflects a notable record of civic accomplishments, which includes service as past president of the Oklahoma City Chamber of Commerce and as a member of numerous state and federal advisory boards. All employees are encouraged to be civic-minded. The company spearheads a yearly United Way campaign and invites soccer teams to enjoy play on the spacious grounds.

Future goals for CMI include continued international expansion through worldwide marketing. Eighty foreign countries currently build their roads, pave parking lots, and create aircraft runways with CMI equipment.

"We plan to address tomorrow's changes by providing road-building machinery to transform the world's transportation infrastructures," Swisher concludes. "Innovation will always be a part of our future." ✪

Lopez Foods, Inc.

The Lopez family: John P., John C., David and (seated) Pat.

Nestled on 32 acres in western Oklahoma City resides the third-largest Hispanic-owned manufacturing company in the United States. Although the company manufactures products most Americans consume frequently, it doesn't even advertise.

Lopez Foods' secret is one steady customer that always wants its hamburgers, pork sausage patties, and Canadian-style bacon made the same quality way. The client is McDonald's Corporation, which depends on Lopez Foods to supply thousands of its restaurants.

In 1971, McDonald's selected Oklahoma City's Wilson Foods to build a state-of-the-art processing facility to manufacture beef patties. As the breakfast menu expanded during the 1970s, the company also began producing fully cooked pork sausage patties and the sliced Canadian-style bacon used on the famous Egg McMuffin.

Wilson spun this division off in 1989 as an independent company, Normac Foods, Inc. In 1992, John C. Lopez sold his four Los Angeles-based McDonald's restaurants to move to Oklahoma and purchase controlling interest in Normac. The company name was changed to Lopez Foods, Inc. in 1995.

Lopez credits McDonald's diversity program for making it possible for him to become the majority owner of a major food production company. As a 19-year veteran of the McDonald's system, Lopez has used his experience to make Lopez Foods more compatible with McDonald's operations and management philosophies.

"By implementing the same "total quality management" standards that established McDonald's as the leader in its field, we have

State-of-the-art technology, preventive maintenance programs, and an on-site analytical laboratory ensure continuous benchmark performance for this industry leader.

Service does not stop when the product leaves the plant. The company employs trained field representatives who work closely with McDonald's distribution centers, regional offices, and restaurants to maintain quality from the grinder to the grill.

Lopez Foods employs over 300 people who operate around the clock, making it one of Oklahoma City's major employers and sales leaders. Approximately three million pounds of meat are produced every week in the 150,000-square-foot facility.

"We are accomplishing our goals for Lopez Foods to grow financially and as a family-owned business with the help of our associates and support from the community," Lopez concluded. "We are extremely proud of the success the company has enjoyed and look forward to increasing our civic charge here in Oklahoma City." ❂

President Charlie Barajas, Chief Financial Officer Jim English, Senior Vice President Frank McKee and (seated) Chairman and Chief Executive Officer John C. Lopez, not pictured, Vice President Marvin Smith.

become a top supplier of the world's most successful chain of quality restaurants," said Lopez, chairman and chief executive officer.

Lopez Foods' first priority is to maintain the highest standards for quality food production, employee safety, and environmental protection.

13

Business & Finance

Greater Oklahoma City Chamber of Commerce

Downtown Oklahoma City skyline.

The year was 1889. In a little wooden shack at the corner of Broadway and Sheridan, a group of businessmen organized the Board of Trade. More than 100 years later, the Greater Oklahoma City Chamber of Commerce perpetuates the same dreams of its founding group—unifying business vision, working to accomplish shared goals, and planning for the future.

PRINCIPAL ACCOMPLISHMENTS

Since its inception, the Chamber has played a leading role in the major developments which improve the economy and quality of life in Oklahoma City. In 1907, the Chamber's predecessor organization promoted statehood and organized the state fair.

Other projects which became reality because of the Chamber's aggressive efforts included the railroad and U.S. Highway 66, state capitol relocation to Oklahoma City, packing plants and stockyards, and the Great Southwest Distribution Center, which is now revitalized as Bricktown.

This organization was also an initiator and major player in the drilling ordinance to develop the Oklahoma City oil field discovered in 1928 and bringing the air depot and aircraft companies to the city.

"If you look at the history of Oklahoma City's industrial and economic development you will find that the Chamber's role has been paramount in the location of almost every major industry we have here," said Paul Strasbaugh, general manager, Oklahoma Industries Authority.

The Chamber's post-World War II economic and aviation development activities originated premier facilities with a nationwide impact: Tinker Air Force Base, Federal Aviation Administration Aeronautic Center, and Will Rogers World Airport. The Turner Turnpike, Interstate System, and Inter-City Connector System were also a part of the Chamber's post-World War II highway development projects.

Principle industrial projects secured as a direct result of the Chamber's pursuit include Lucent Technologies, Firestone Dayton Tire Plant, General Motors, Seagate Technologies, Xerox, Altec Industries, and Macklanburg Duncan. Other notable industrial projects include Autocraft Industries, Unit Parts Company, Hertz Corp., Southwest Airlines, American Paging and America Online.

Oklahoma City's strength in the areas of health care and education is by design—and hard work of the Chamber. The Oklahoma Health Center, Baptist Medical Center, Oklahoma State University (OSU) Technical Institute, Oklahoma City University, the vocational-technical system and a host of development organizations were fostered by the dedication and commitment of resources from the Greater Oklahoma City Chamber of Commerce and its leaders.

The important resource of water is abundant in Oklahoma City because of Lake Hefner, the Atoka pipeline and storage at Canton Lake. The Chamber's handprint is on each of these projects.

Results of the Chamber's efforts also include important quality of life components. Cultural, sports and entertainment facilities such as the National Cowboy Hall of Fame and Western Heritage Center and Remington Park highlight the heritage and assets of Oklahoma City, the state and region.

Oklahoma City's designation as the "Horse Show Capital of the World" is possible because of the Chamber's efforts in securing and keeping national horse shows in Oklahoma City. The Chamber further develops the city as a key tourist and convention destination through the Oklahoma City Convention & Visitors Bureau.

The Greater Oklahoma City Chamber of Commerce continues to grow its list of exceptional accomplishments. Recent examples consist of continually adding new and expanding businesses to the Oklahoma City landscape, successfully passing bond referendums for improved city infrastructure in the mid-1980s and in 1995, bringing the Metropolitan Area Projects (MAPS) to life, and creating *Forward Oklahoma City—The New Agenda.*

POSITIONED FOR PROGRESS

The Greater Oklahoma City Chamber of Commerce's leadership and continuity of vision for Oklahoma City continues with the Chamber's bold, $10 million economic development plan for the metropolitan Oklahoma City area.

Forward Oklahoma City—The New Agenda is a five-year, aggressive, economic development strategy which positions the city for continued

(left) Members of the Chamber board of directors with artwork from the Forward Oklahoma City-The New Agenda advertising campaign.

(left) Members of the Total Resource Development Campaign Steering Committee. The TRDC Campaign raised more than $1.1 million.

(above right) Employees of The Hanover Company at a ground breaking. The Chamber supports more than 40 ground breakings and announcements a year.

progress. This carefully structured action plan was adopted by the Chamber's board of directors in November of 1994. Since that time, more than $10 million has been raised from private companies for underwriting this program, exceeding the initial fund-raising goal.

Forward Oklahoma City—The New Agenda has seven initiatives including:

* Aggressively market Oklahoma City to the world.
* Promote a probusiness environment through right-to-work and business regulations.
* Expand existing business through the Oklahoma City Business Network.
* Develop infrastructure for technology transfer through the Advanced Technology Corporation.
* Expand international trade and seek foreign investment through the World Trade Council.
* Promote private sector development with MAPS through Oklahoma City Partners, Inc.
* Expand the role of Tinker Air Force Base and preserve the mission of the Air Logistics Center.

The goals and objectives of this aggressive strategy are designed to increase jobs, capital investment, office space occupancy rates, retail sales, per capita income, convention visitation, companies that export, and expand the employment of high-tech companies. *Forward Oklahoma City* goals more specifically include

creating 28,000 new jobs and increasing capital investments by $450 million by the year 2000.

The Chamber's ambitious *Forward Oklahoma City* plan can easily be described through an early theme used for building enthusiasm toward growth and public betterment: "Greater Oklahoma City Chamber of Commerce Makes Things Happen."

BENEFITS TO MEMBERS

When companies join the Chamber—the state's largest coalition of businesses—they are making an investment into a group initiative. These businesspeople believe more can be accomplished together than one person or business alone.

Along with alliance of purpose, there are many other benefits of membership in the Greater Oklahoma City Chamber of Commerce. New business contacts can be made through a variety of the Chamber's events, including Business After Hours, committee and task force meetings, Chamber luncheons, orientations, and MegaMarket Place.

Governmental advocacy, referrals, guidance and counseling, and access to business statistics all help point a business in the right direction. Publicity and exposure from the newsletter and other publications is a premier way to get a company name before the business community.

Numerous seminars and programs throughout the year are dedicated to professional development opportunities.

Overall, the Greater Oklahoma City Chamber of Commerce provides tangible resources for business success.

FRAMEWORK FOR SUCCESS

Ongoing efforts on behalf of business are formulated through a program of work and implemented through eight divisions within the Chamber's organizational structure. These divisions each have a purpose and goals for a given year.

Current Greater Oklahoma City Chamber of Commerce divisions include economic development, small business, metropolitan events, government relations, community improvement/education, communications, membership/marketing, finance and the Oklahoma City Convention and Visitors Bureau. Of the nine committees appointed by the first Board of Trade president, five are still divisions of the Chamber and one is a task force.

Henry Overholser served as the first Board of Trade president. Past volunteer chairmen include familiar names like John A. Brown, Buttram, Classen, Colcord, Gaylord, Nichols, and Kilpatrick. The same quality of leaders are guiding the Greater Oklahoma City Chamber of Commerce into 2000 and beyond.

Along with the volunteer leadership, Chamber staff works with a large volunteer force to execute and achieve the immediate and long-term goals. These willing volunteers commit their time and resources to carry out the Chamber's mission, which in part reads: "The Greater Oklahoma City Chamber of Commerce is the voice of business and the visionary organization in this community!".

Although the street is Park Avenue, the building is a contemporary, multi-level structure and the name is different, the same vision of early Board of Trade members lives on in the Greater Oklahoma City Chamber of Commerce. Early founders would agree that "A Better Living. A Better Life." is a just and noble cause for the Greater Oklahoma City Chamber of Commerce—now and for the future. www.okcchamber.com ❀

Bank of Oklahoma

Bank of Oklahoma (BOk), founded in 1910, has earned the reputation of setting the standards in meeting the needs of businesses, consumers, and other users of banking services. BOk entered the Oklahoma City market by acquiring Fidelity Bank in 1984. Fidelity was founded in 1921. In 1991, George Kaiser, a native Oklahoman and Tulsa oil man, recapitalized the bank to preserve local ownership of significant banking assets and to maintain a large Oklahoma headquartered bank. Additional acquisitions included purchasing assests and locations of Continental Savings and Loan and Sooner Federal Savings and Loan. This expanded the number of offices to over 20 in the Oklahoma City area. With subsequent acquisitions in nine other communities throughout the state, in Arkansas and the Dallas areas, locations increased to over 70 and BOk became a regional bank with total assets exceeding $5 billion.

Bank of Oklahoma is the dominant market share provider of many services including Transfund, which operates hundreds of ATMs throughout Oklahoma. BOk is also the largest residential mortgage provider in the state. BOk introduced the first supermarket bank and now offers supermarket banking services in over 15 locations. Consumers can use the supermarket locations to conduct their banking seven days a week. Consumers can also use the ExpressBank (via telephone), which allows customers to open accounts, apply for loans, make transfers and complete a variety of other transactions 24 hours a day using a live Express Bank operator.

BOk has a strong history as an active commercial lender for both large and small businesses. It provides superior business banking services and is the largest commercial lender in the state. Commercial lending is complemented by a unique offering of cash management, international and investment banking services. BOk's Commercial Client Service Group provides a responsive, daily contact for its many business customers.

Bank of Oklahoma has extensively grown its services to small business customers and is a Preferred Lender as designated by the Small Business Administration (SBA). All business customers have interest in the high-

TransFund Electronic Funds Transfer Network, owned by BOK Financial Corp., has more than 600 machines installed in a five-state area and services approximately 750,000 cardholders.

Bank of Oklahoma offices in Albertson's supermarkets provide 7-day a week, 365-day a year banking convenience.

technology services available. OnlineExpress and BusinessExpress offer account information electronically and allows customers to initiate contact with the bank whenever it is convenient to them.

BOk has donated millions of dollars to dozens of local charities and community events and hundreds of volunteer hours to give back to those communities where it does business.

Bank of Oklahoma, due to its Oklahoma roots, is dedicated to the community, anticipating at all times the changing needs of business banking and setting trends in consumer banking products. Bank of Oklahoma is providing the leadership and financial services—convenience, quality, responsiveness, and sophistication—that Oklahoma City's consumers and corporations deserve. BOk has helped build this state for 85 years and will be investing in Oklahoma for years to come. Bank of Oklahoma is truly the Home Advantage.

Located in the heart of downtown Oklahoma City, Bank of Oklahoma Plaza is a distinctive Oklahoma City landmark.

The interior of Bank of Oklahoma's downtown lobby displays unique product and service merchandising.

MISSION STATEMENT

BOK Financial Corporation's goal is to be the financial institution of first choice in its chosen markets. By delivering its best to its customers, employees, and communities, BOk will maximize long term value for its shareholders.

CORPORATE VALUES
BOK VALUES ITS PEOPLE

BOk believes that its people are the heart of its success. Therefore, it strives to provide a workplace that:

* Gives employees the same concern and respect that they are expected to give its customers.
* Rewards exceptional initiative, performance, and commitment.
* Promotes a vigorous work ethic that is characterized by trust, enthusiasm, creativity, entrepreneurial thinking, and new ideas.
* Encourages listening and interaction with others that is based on honesty, courtesy, and respect.
* Provides opportunity for training and career development.
* Provides a supportive/professional environment where objectives are clear and adequate resources are available to perform the tasks.

BOK VALUES ITS CUSTOMERS

Customers, both internal and external, are the focus of everything BOk does. Therefore, it embraces the following:

* Highly responsive employees working together to understand and anticipate customers' needs and to provide workable solutions.
* Customer service better than that expected as customers themselves.
* Customer service delivered with courtesy, competence, concern, and individual pride.
* Products of the highest quality that satisfy customers' needs for timeliness, convenience, and functionality.

BOK IS COMMITTED TO THE COMMUNITIES IT SERVES

BOk will demonstrate this commitment by:
* Filling a leadership role in the community on important local and regional causes.
* Financially supporting key civic activities critical to the vitality of the community.
* Encouraging and supporting employee participation as volunteers. ✦

Hilb, Rogal and Hamilton Company of Oklahoma

William L. Chaufty, president.

Hilb, Rogal, and Hamilton (HRH) is the 18th-largest insurance intermediary in the world and the 8th largest in the United States, and is among the leaders nationwide and worldwide with deep roots in Oklahoma. HRH is an aggressive, customer-oriented insurance agency with competitive products, a national presence, and strength to move swiftly in the rapidly changing industry.

HISTORIC PRINCIPLES

HRH Company of Oklahoma, a continuation of McEldowney, McWilliams, Deardeuff and Journey, Inc., finds strength in its heritage and the philosophy of its founders. It is based on a legacy to operate the agency with "forthright honesty and integrity." It is a name that represents generations of clients and employees who have made this company a success since 1911.

In 1989, the merger with Hilb, Rogal and Hamilton Company changed the name, yet the people remained the same. "This merger was a way for the company to become stronger and perpetuate history," says William L. Chaufty, President, Hilb, Rogal and Hamilton Company of Oklahoma.

Now a part of Hilb, Rogal and Hamilton's agency system, the stability and negotiating power blended from both companies create an agency with a product and commitment to people that makes it one of the best in the business. The firm's presence remains very much a part of the local landscape with it's building located at 125 West Park Avenue in downtown Oklahoma City next door to the Greater Oklahoma City Chamber of Commerce building.

PEOPLE MAKE THE DIFFERENCE

Many HRH employees have multiple designations that confirm their allegiance to additional training and continuing industry education. The wide range of commercial and industrial client's (HRH insures businesses from a storefront retail outlet to one of the largest steel fabricators in the world) makes such training a necessity. Employees develop finely-tuned expertise in various special aspects of the insurance and risk management industry. Local HRH employees, many of them native to the Oklahoma City area, become heavily involved in civic and cultural pursuits and community programs.

HRH's giving of resources and employee time impacts a variety of areas—from individual events to specific organizations. Its presence is strong in civic and non-profit organizations, client trade associations, and the insurance industry. Whether underwriting a cultural event or supporting the downtown Oklahoma City efforts, HRH considers the avenue of corporate citizenship another positive way to invest in its people and community.

CLIENT SERVICE FORMULA

"We have agency-client relationships that date back 50 years and beyond," says Chaufty. "We are blessed with very good clients, stable and successful in their own right." Commitment to service and a strongly held belief that clients represent the firm's first priority is the cornerstone of the business philosophy of Hilb, Rogal and Hamilton Company of Oklahoma.

The firm employs the team approach in serving clients and solving problems. Team work is viewed as a key to success in mastering a complex and client-sensitive business such as insurance.

COVERAGE SOLUTIONS

Traditional insurance challenges faced by clients are met with a fresh approach by HRH professionals. Automation technology, changes in coverages and insurance methods are continually monitored and mastered. When a client signs up with HRH, they operate secure in the knowledge that through counseling and an intimate knowledge of the client's needs, HRH will place a competitively priced comprehensive insurance program on their behalf. Any insurance need can be met by HRH. "You name it: we write it," says Chaufty. Construction, oil and gas, health care institutions, financial services, property management, manufacturing, retail grocers, private prisons, and professional

Back row (left to right) Billye Keister, Kent Taylor, CIC; Dwight Journey, CIC; Larry Mitchell, CPCU, CIC; Keith Shideler, CIC; Gary Liles, CIC; Frank Shadid; and Dave Deardeuff, CIC. Front row: Bill McWilliams, CPCU; William Chaufty, CIC; Stan Deardeuff and Richard Ross.

businesses find their insurance needs covered by HRH, whose loss-control experts can also evaluate operations, safety, and security for the client's company. Hilb, Rogal and Hamilton Company of Oklahoma offers a new way to look at insurance. With knowledge, technical expertise, high levels of service and carriers, the company offers a range of traditional insurance placement and alternatives for risk transfer.

FUTURE TRENDS

"Our industry is changing and the nature of risk transfer for our clients in changing. HRH intends to be ahead of these changes through new products, services, and skills needed to compete in the year 2000 and beyond," says Chaufty.

Through loss control, risk management services, self-insured retention programs, and alternative risk financing, HRH is aggressively preparing to meet the future and the ever-changing needs of its clients. The company continues to invest heavily in technology and its people in order to handle the service needs of its clients. ✪

Hilb, Rogal and Hamilton State Headquarters located at 125 W. Park Avenue in Oklahoma City.

Mutual Assurance Administrators, Inc.

MAA works with businesses consisting of more than 100 employees and a stable benefit history, to custom design and administer health insurance packages for their employees. Photo by Tami Isaac.

With the same visionary spirit of Oklahoma's forefathers, Mutual Assurance Administrators, Inc. (MAA) has pioneered the way for companies to develop self-funded benefit programs for their employees. MAA works with businesses consisting of more than 100 employees and a stable benefit history, to custom design and administer health insurance packages for their employees.

In 1975, following the enactment of the Employee Retirement Income Security Act (ERISA), Richard E. Carllson and his wife, Jacki, founded their original company as the Mutual Assurance Agency, Inc. Their first client was Ashland Oil Company who hired MAA to administer claims for their service station lessees and commission agents.

Recent surveys point toward the growing trend the Carllsons initiated over two decades ago with MAA. Almost half of all employers in the United States utilize partial or full self-funded programs to provide their employee's benefits.

LEADING THE EMPLOYEE BENEFITS ADMINISTRATION INDUSTRY

The idea of self-funding was so new when MAA was developing in the 1970s, other corporations struggled to follow Ashland's lead to self-funding benefits. Consequently, MAA focused much of its attention on educating companies about the substantial savings they could expect with a self-funded program.

Several financial advantages become increasingly apparent to employers who begin self-funding programs with MAA. Because claim reserves are managed by the employer, greater financial controls result. Funds may be contributed to an employer-established account and transferred when needed to pay claims. Excess funds remain in the account to collect investment earnings. By using a 501 C(9) trust, these earnings may accrue to the trust, tax free.

When companies develop a self-funded benefit plan, most premium tax is eliminated, automatically saving the company two to three percent.

The profit margin of an insurance carrier is also removed from the bulk of the plan.

"One of the aspects our clients like most is the option they have to redesign their benefit plans," says Todd Archer, executive vice president. "Doing so eliminates any unwanted or unneeded aspects and often saves the client money."

By consistently demonstrating the cost control available with self-funding, MAA now serves over 40 corporate clients, processing claims for approximately 30,000 employees. Oklahoma City businesses comprise the majority of MAA's clientele. Additional customers outside the metro area are found in Ohio, Kansas, Mississippi, Arkansas, and Texas.

SHIFT OF EMPLOYEE BENEFIT PROGRAMS

Traditionally, employee protection was the top priority as employers met the needs of a look-alike work force with conventionally insured group health care plans. After two decades of unprecedented inflation in medical care costs, companies now need the flexibility MAA offers when planning health care coverage.

Over the years, service on medical claims has moved from a simple payment turnaround to an emphasis on quality medical care, including competitive pricing, expert claims analysis, state-of- the-art technology, and financial resiliency. MAA continues to establish and expand its services as the health care industry moves in this new direction.

SELF-FUNDING FUNDAMENTALS

With the assistance of MAA, moving into a self-funded benefits program is not difficult for a business. The employer decides on an employee benefit plan. The new benefit package is often similar to the one currently provided on an insured basis.

MAA arranges Stop-Loss insurance to protect the self-funded plan against catastrophic losses. The amount of risk to be insured is adjusted according to the employer's size, nature of business, and location. MAA also considers financial resources, prior experience, plan of benefits, and tolerance for risk.

A master plan is created which contains all provisions of the plan. MAA also provides its clients with employee benefit descriptions, identification cards, and other materials necessary to operate the program.

When MAA begins administering the new self-funded plan for the customer, it establishes eligibility, banking arrangements, risk transfer, benefit parameters, provider network arrangements, and system programmers in preparation for paying claims. It prepares required reports, maintains data for the plan, and handles government compliance issues if needed. MAA bills and collects premiums and other administrative fees for its clients.

MAA continues to establish and expand its services as the health care industry moves in new directions. Photo by Tami Isaac.

claims submission from providers, electronic screening of claims for billing irregularities, automated coding and adjudication of claims, and electronic funds transfer. All offer the opportunity to improve service while at the same time assisting in controlling costs. Archer predicts this trend will eventually result in revolutionary changes in both the level and scope of service that can be provided.

The benchmark MAA uses to determine its achievements is the success of its clients' benefit plans. The company's commitment to customer service drives every corporate decision, from marketing to operational policy. The customer's interests always receive primary consideration.

"As the longest practicing third-party administrator in Oklahoma, Mutual Assurance offers a pronounced competitive edge," says Archer. "It has demonstrated consistent financial integrity with the capability to interact with brokers, insurance carriers, and health care providers, to give clients a variety of options to include in their employee benefits." ✪

SERVICES EXPANDING TO MEET CLIENT NEEDS

In 1988, the Oklahoma State Insurance Commission ruled no company could be issued both an insurance agency license and a third-party administrator's license. As a result, MAA divided into two corporations.

Mutual Assurance Administrators, Inc. focuses on third-party administration of self-funded employee benefits. Its sister company, Mutual Assurance Agency, Inc., is a licensed insurance agency, which places life insurance and/or re-insurance for employee benefit plans administered by Mutual Assurance Administrators, Inc.

At the request of many clients, the Carllsons formed a third company in 1990. Mutual Assurance Advantage, Inc. provides utilization review services, which consist of precertification for hospital admission, discharge planning, and large case management.

In 1995, MAA formed a workers' compensation unit designed to administer "own risk" and insured workers' compensation programs for stand-alone or combined medical and workers' compensation clients.

MAA's staff has multiplied as the three companies have expanded. Currently, more than 70 employees provide service to accounts. In 1990, Richard and Jacki Carllson invited Todd Archer to join the executive team. Together they have

The benchmark MAA uses to determine its achievements is the success of its clients' benefit plans. Photo by Tami Isaac.

developed a management structure that will assure continued success for the company.

TECHNOLOGY EFFICIENCY

MAA is committed to implementing all proven technological advances available in the industry to improve services to their clients. Innovations such as electronic verification of benefits, electronic

Phillips Securities

Thompson S. Phillips, president of Phillips Securities, describes the company as, "simply a reflection of the quality of the brokers that work here. Each broker is a veteran with a proven track record and an average of 10 years or more experience."

Phillips Securities opened its doors in January of 1990. It is what has come to be known in the industry as an Independent Brokerage Operation. It has a contractual relationship with Correspondence Services Corporation, a subsidiary of Paine Webber. This arrangement allows client accounts to enjoy the custody and security of a major firm while allowing individual brokers the independence of making local investment decisions that are in the client's best interest. It's the best of both worlds, all the support of a major New York firm without the bureaucracy.

As a result, Phillips Securities is truly client driven. It has neatly bundled all the positive aspects of the brokerage business 30 years ago and tied them into advanced technology that sets the company apart from its competition. It's full range of services includes: portfolio planning; estate planning; business valuations; portfolio management; stocks and bonds, both foreign and domestic; mutual funds; commodities; insurance and annuities; investment research, both foreign and domestic; and investment management consulting for institutions and individuals.

Our state-of-the-art technology provides our clients with global execution and trading services and up-to-the-minute information and research via several on-line databases, the Internet and the World Wide Web.

The company has grown substantially since 1990, both in clients and brokers. The client base has grown to represent hundreds of millions of dollars under management. The other important asset, of course, is the quality of the brokers and staff of the company itself. Along with Mr. Phillips, Dr. Robert Black serves as a Managing Partner and part owner. Other brokers who serve as vice presidents include Carol Anne Rakosky, Charles E. Oliver, Terry R. Holcomb, and Homer C. Myers.

The growth of the company lies in the people. "We want to expand, but at a very controlled pace," said Phillips. "We will grow as a result of

(standing L-R) Robert H. Black, Terry R. Holcomb, Thompson S. Phillips, Charles E. Oliver, Carol Anne Rakosky (seated L-R) Sharon Allman, Kim Allman, Calista Saunders.

(seated L-R) Robert H. Black, Thompson S. Phillips.

He explains that his military background and that of his father have given him a broad view of different locations. "I have a perspective that makes me aware of what a unique and wonderful place this is."

Trust is the foundation for what we do. At Phillips Securities, trust is built on developing professional relationships, a focus on service, and a friendly staff with a commitment to putting our clients first. ✪

being able to find the kind of brokers that meet our standards of professionalism and dedication to the client." Phillips explained that his yardstick for hiring is based on the answer to the following question. Would I let this broker manage my mother's money?

Phillips' tie to Oklahoma City came with a transfer from E.F. Hutton, where, at one time, he served as Hutton's youngest branch manager. Upon starting his business, Phillips decided to remain in Oklahoma City because he liked the people, the excellent standard of living, and the quality of life.

Roger Hicks & Associates Group Insurance, Inc.

He may not have been "Sooner born and Sooner bred," but Michigan native Roger Hicks is filled with the Oklahoma spirit as much as any native son of the nation's 46th state. He loves his adopted state and considers any disparaging remarks about Oklahoma to be "fightin' words!"

Hicks has a passion to help Oklahoma be even better than it ever has been as the turn of the century approaches, and to share its heritage with its friends and neighbors. His company, Roger Hicks & Associates Group Insurance, Inc., celebrated its 13th anniversary this past June as the Southwest's oldest and largest group-only benefits consulting firm.

A veteran of 30 years in the group insurance industry, Hicks has been the president of two successful corporations, as well as a marketer, manager, and performance contractor. He is recognized nationally as one of the Southwest's leading authorities on group health insurance programs. He is a past president of the Health Underwriters Association of Central Oklahoma and is active in the community, at the state legislature, and nationally, serving on numerous boards and commissions.

His active participation and leadership with the Stockyards City Main Street program is just one way in which he has given himself back to Oklahoma City and Oklahoma.

Meanwhile, a well-prepared and dedicated staff helps keep the Roger Hicks & Associates standard flying high in a very competitive market. "Each member of my staff has been recruited based upon performance and has been properly credentialed and required to commit to statutory and optional continuing education training on an annual basis," Hicks said.

Group health insurance providers know that when a representative of Roger Hicks & Associates Group Insurance, Inc. comes through their door, they're dealing with an educated, experienced professional who's ready to go the extra mile for his or her client.

One of its services is to review and evaluate prospective insurance companies, third-party administrators, and reinsurance companies on behalf of clients and to apply stringent standards that will ensure the best performance and coverage for the client and its most valuable asset—its employees. That's why the company has grown into the largest group coverage counseling, marketing, and servicing firm in the Southwest.

"We're committed to traditional service that ensures quality, value, and professionalism," he added. "That's our philosophy, and we know our clients appreciate the difference. They see it reflected in our services, in our products, and in our tenacious spirit on their behalf!"

"Our agency works for the client, not for the insurance company or third-party administrator," Hicks concluded. "We go that extra mile for the insured. Our service policy is `whatever you want when you want it,' and the energy and resources of our team are committed to that concept. That total dedication to the client is what enables us to stand out from our competitors."

"Satisfaction keeps our clients, and we are proud to include many of the Southwest's largest and most prestigious public and private employers on our client list."

Roger Hicks & Associates: experienced, committed Oklahoma professionals dedicated to providing solid advice and strong advocacy when you need it! ✪

A well-prepared and dedicated staff helps keep the Roger Hicks & Associates standard flying high in a very competitive market.

North American Group

North American Group's determination to offer clients a tangible difference—through solutions and service—has made it the largest independent insurance agency in Oklahoma. Its innovative, full-service approach is deliberately segmented to fit the varying needs of small businesses, large businesses, and individuals, giving clients the attention and products that match their business requirements and style.

"We've spent a lot of money and time training and developing products and services to specifically serve segmented customers," said Jack McMahan, president and CEO.

North American Group is dedicated to making a tangible difference in clients' businesses and lives. It has been providing quality property/casualty, risk management, and employee benefits products and services to its clients for more than 35 years. Three successful niche marketing areas for the company are schools, trailer manufacturers, and independent grocers.

Partnership with North American Group comes with the value of knowledge and the benefit of specifically addressing solutions, not just dealing with a company selling products. Approximately 12,000 clients worldwide are served by 70 employees based in Oklahoma City. Four other locations in Tulsa, Norman, Memphis, and Houston service this client base.

A solutions approach to problems and challenges is paramount in the North American Group's dealings with large complex businesses. Its deep and comprehensive involvement in the company's business helps build an integrated financial system focused on solving a problem and finding better ways for the company to achieve its objectives.

"The value may be in something we can do with insurance or non-insurance, moving to self insurance, loss prevention or loss control in lieu of insurance. These options help manage what the customer is trying to achieve and can directly assist in improving a particular situation," McMahan said. "It's an integrated solution approach that requires listening, financial analysis, and stretching ourselves creatively to look at all the ways we can help our customers win."

Small businesses receive personal attention for what are normally simple transactions handled in

(above) North American Group is committed to its clients and growth in Oklahoma.

Jack McMahan, president and CEO of North American Group. Photos by J.D. Merryweather

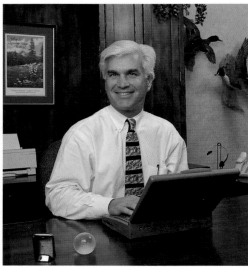

an expedient manner. "We deliver personal service, different products, and technology with speed and cost efficiency to the small commercial customer," McMahan said.

Another example of the company's ingenuity and commitment to clients is an extra element of value for the public school clients. Because safety is the number one concern with schools, North American Group distributed a comic book series program with arch villains and heroes throughout the schools. Characters Red Alert and The Extinguisher, along with stickers and posters, were created to sustain a fire safety message with children.

North American Group is committed to its clients and growth in Oklahoma. It takes a leadership role in the Oklahoma Independent Agent's Association and is an active corporate citizen and annual sponsor of the Redbud Classic, Ballet Oklahoma, Lyric Theatre, and more. The agency is also a sponsor of the Oklahoma State University Executive Management Briefing Series, which features such world leaders as George Bush, Colin Powell, and Henry Kissinger.

Effective solutions to the most difficult and involved commercial insurance problems or simple transactions handled quickly and efficiently, are met with strategic confidence within the full-service, segmented approach of the North American Group. Through innovation, knowledge, and a committed partnership, the North American Group's value extends beyond insurance into the realm of true service and solutions. ✵

American Fidelity Assurance Company

American Fidelity Assurance Company (AFA) is your financial security network. With emphasis on customers, colleagues, and community, AFA has been ranked among the nation's leading insurance companies by A.M. Best since 1981. It has set itself apart from the competition by taking a leadership role in implementing customer-friendly programs and finding a market niche.

The modest beginnings of this company reflect a true American success story, dating back to the early 1930s when C.W. Cameron entered the insurance business in rural Oklahoma. Cameron's work successfully created the largest life and health insurance firm owned and operated in Oklahoma, with headquarters in Oklahoma City.

Incorporated as American Fidelity Assurance Company in 1960 by Cameron and his son C.B. Cameron, AFA is a subsidiary of the American Fidelity Corporation, a financial services organization with more than $1 billion in assets and 1,700 employees.

American Fidelity Assurance Company specializes in marketing annuities, life insurance, disability income insurance, comprehensive and specialized medical insurance, and group fringe benefit programs to trade association members and educational system personnel. Through 900 employees, it services more than 650,000 insured nationwide and is licensed in 48 states and the District of Columbia.

William E. Durrett is chairman, John Rex is president and COO, and Bill Cameron serves as vice chairman and CEO. The company focuses on meeting or exceeding the needs of customers.

The Educational Services Division is solely dedicated to providing employees, administrators, and other members of the educational field with financial solutions in today's complex world.

The Association Group Division strives to furnish each association with customized programs and personal service. The Brokerage Division markets to traditional insurance brokers.

The products offered by this company are supported by a commitment to provide the best possible service at competitive rates.

AFA was the first insurance company in the United States to start the payroll deduction method of payment for its groups. It was also one of the first companies to utilize the salaried account manager for marketing and distribution so the focus could be shifted to a secure, customer-service environment.

Customer service was a part of AFA before it was the industry trend. Currently, these niche-focused

The modest beginnings of this company reflect a true American success story, dating back to the early 1930s when C.W. Cameron entered the insurance business in rural Oklahoma.

customer service divisions include a field representative and home office service team dedicated to a particular industry and part of the country.

Meeting the ongoing needs of people is not limited to customers at American Fidelity Assurance. The company provides its employees one of two accredited wellness programs in the state and acknowledges employees for their contributions to the company in a variety of ways. It also maintains a family-oriented atmosphere by sponsoring summer family picnics and holiday gatherings.

As a community-spirited corporation, the company commits money and manpower to

support its deep commitment to the community, especially from an educational standpoint. The goal is to make a difference in improving the quality of education to strengthen the society.

AFA's strict adherence to ethical market conduct and business practices guides the company's march toward a world-class environment. Focus on customers, colleagues, and community are unwavering, and allow the "assurance" of this American Fidelity company to insure the current and future needs of Oklahomans and Americans across the country. ✪

Dorchester Capital

(l to r) G. Rainey Williams, Jr.; G. Jeffrey Records, Jr.; E. Carey Joullian, IV; and Clayton I. Bennett.

Dorchester Capital is a diversified private investment company formed to make equity investments in private, small to midsize companies. Dorchester Capital targets profitable companies having an established market share and a sustainable competitive advantage within their industry. Dorchester Capital seeks to provide capital in partnership with management to facilitate later stage internal growth, strategic acquisitions, and buyouts, all with the objective of achieving superior long-term capital appreciation.

Clayton I. Bennett, director of real estate and investments for The Oklahoma Publishing Company, founded Dorchester Capital in 1991. Dorchester Capital was formed to respond to the difficulties faced by small to midsize companies — those with market capitalization of less than $50 million — in obtaining financing from traditional Wall Street sources. The company is owned by Bennett; E. Carey Joullian IV, president, Mustang Fuel Corporation; G. Jeffrey Records, Jr., president, MidFirst Bank; and G. Rainey Williams, Jr., president, Pinnacle Asset Management, Inc. This ownership group possesses a long-term commitment to the economic growth and development of both Oklahoma City and the state of Oklahoma.

Dorchester Capital's philosophy precludes its involvement in the management of portfolio companies. Rather, Dorchester Capital provides management with the financial resources to pursue a specific strategic plan or acquisition while assisting management where needed through active board participation. Dorchester Capital focuses on historically successful companies that have long-term growth opportunities. Dorchester Capital bases its investment decisions largely on the integrity, experience, and qualifications of existing ownership and management. An intimate understanding of the company's goals, strategies, and philosophies is critical to the decision process.

A key component of Dorchester's success is the development of a long-term, mutually beneficial partnership with company owners and management. The ultimate goal is to increase shareholder value through the resources of the partnership by:

1. identifying growth prospects via new markets, expansion opportunities, and strategic acquisitions;
2. rewarding management and employee performance through incentive-based compensation and ownership opportunities;
3. enhancing company financial structure through opportunities in the capital markets.

Dorchester Capital targets companies that have experienced profitability for three to five years, report annual revenues of at least $5 million, possess strong name recognition for their products or services, and possess opportunities to grow internally or through acquisition. Investment transaction sizes range from $2 million to $25 million. Industries served by Dorchester extend from real estate, manufacturing, and sports and entertainment to health care and technology.

Dorchester Capital views value creation as a long-term process and anticipates extended relationships with its portfolio companies. Dorchester Capital will work in conjunction with management to explore timely opportunities to realize investment returns. ❂

Tinker Federal Credit Union

For more than 50 years, Tinker Federal Credit Union (TFCU) has been helping its members achieve their goals and realize their dreams. Building a strong future from its rich heritage, TFCU has grown into one of the nation's largest, most dynamic, cooperative financial institutions.

With assets closing in on a billion dollars, the credit union now provides its membership with choices from three Tinker Air Force Base locations, five additional branches in the Oklahoma City metro area, and sites in Shawnee, Enid and Vance Air Force Base.

Members are civilian and military employees of Tinker Air Force Base and Vance Air Force Base, a variety of military-related organizations, and an extensive list of small and medium-sized companies that wish to offer credit union services to employees.

The credit union boasts services extending from checking, savings, loans, credit cards, and electronic services to a number of convenience services.

TFCU's checking accounts are easy to open and operate. A variety of savings opportunities include share accounts, money market accounts, share certificates and IRAs, youth savings club accounts, and Christmas club accounts.

Loan options certain to fit almost any borrowing need include new and used vehicle, mortgage, home equity, secured, unsecured, and student loans. Great rates and competitive terms make these loans one of the most valuable credit union benefits.

Visa®, MasterCard®, and Gold MasterCard® are the credit card options. Rapidly growing electronic services encompass ATM cards, Visa check card, electronic bill paying, automated telephone access, quick-cash dispensers, and personal account teller machines.

The Specialized Investments Division located at TFCU offers the services of professionally trained, fully licensed investment representatives available for personal and business goals. Mutual funds, annuities, stocks, bonds, IRAs, and 401(k) plans are a part of this package.

Additional benefits extend across the nation. Borrowing from the basic credit union philosophy of pooling resources, Tinker Federal Credit

Sonia Shelton assists a member in the Area A branch on Tinker Air Force Base. Tinker Federal Credit Union serves civilian and military employees with three branches on TAFB and a branch on Vance Air Force Base in Enid.

Union and other credit unions across the country, as well as a number of state credit union leagues, are sharing the expense of operating generic branches called Shared Service Centers.

All members of participating credit unions may use any of the centers, located in 16 different states and expanding all the time.

TFCU prides itself on making all decisions on behalf of the membership. The board of directors,

supervisory committee, and credit committee members are volunteers. Board members and credit committee members are elected from the membership base.

This sense of community is apparent in the unique and collective mission of the credit union and through its impact on the people it serves. The credit union partners with the community in many ways. Individually and collectively, TFCU employees support many nonprofit organizations by donations of both time and funds.

Each year, TFCU sponsors numerous community activities through donations of thousands of dollars to approximately 125 organizations. TFCU employees also pledge annually to the Combined Federal Campaign, which supports dozens of charitable organizations.

After 50 years of building a legacy, Tinker Federal Credit Union intends to spend the next half-century serving the growing membership with expanded services, advanced technologies, new locations, and rewarding dividends. The dynamics of this mutual prosperity will play an important role in helping members achieve their goals and realize their dreams. ❂

TFCU's newly built branch at 4626 NW 39th Street serves members living and working in northwest Oklahoma City, Bethany and as far away as Yukon, Mustang, El Reno and Piedmont.

Union Bank & Trust Company

Then...Union Bank and Trust Company had its beginning as May Avenue Bank in 1952 at the corner of 49th and North May Avenue in Oklahoma City.

Union Bank's foundation and continued strength are manifested in its mission statement: "the delivery of superior service to our customers by providing high-quality, competitive, and need-driven products in a professional, timely, and profitable manner." The foundations of stability, relationships, flexibility, and solutions represent the bank's commitment to remain a customer-sensitive, locally managed banking institution.

Established in 1952 as the May Avenue Bank and Trust Company, Union Bank began its long history of service to the people of Oklahoma City. In 1976, as the bank experienced steady growth and profits, the bank's name changed to Union Bank and Trust Company. MidCity

the plant upon completion. The project was completed in 1910, and by the end of 1911, some 3,000 people were working in "Packingtown."

Under the dedicated leadership of E.M. Bakwin, Union Bank moves into the 21st century as one of the soundest financial institutions in Oklahoma. It successfully blends the favorable qualities of community banking with the vast

services, and commercial loans structured to meet customer needs. The bank's strong legal lending limit allows it to comfortably accommodate the borrowing needs of most any individual or company.

Matching a customer's needs with the right products and then servicing the customer is what "relationship banking" is all about at Union Bank. Local people who understand individuals' unique financial needs manage and operate the bank, and are committed to giving them the utmost in customer service. Union Bank's lending staff makes hundreds of personal visits to customers' and prospective customers' places of business to understand their needs and structure loans and other accounts to meet those needs. That, after all, is the very essence of community banking.

In 1997, Union Bank is celebrating 45 years of service and the relationship between the bank and Oklahoma City. While marking this anniversary, the bank recognizes the loyalty and trust of its customers, the commitment of its employees, and contributions to the community.

Continuing a tradition of community investment, Union Bank annually donates to more than 120 groups and organizations in the Oklahoma City area. Union Bank officers and employees personally serve on many boards and are members of many community organizations. As an example, employees and customers teamed to raise more than $25,000 or an equivalent of more than 100,000 pounds of food for the 1996 Harvest Food Drive. Allied Arts and the United Way are other primary recipients of the bank's giving.

Union Bank is proud to be a foundation in the Oklahoma City banking community and takes great pride in providing customers—corporate and individual—with the highest quality of financial services. ✪

Now...Union Bank and Trust Company's main bank sits in the same location today where it started 45 years ago. With six branches located throughout Oklahoma City and Edmond, Union Bank and Trust Company continues to be one of Oklahoma City's leading banks.

Financial Corporation, a multibank holding company, and its chairman, E.M. Bakwin, acquired Union Bank in 1988.

Like Union Bank, Bakwin has a strong foundation in Oklahoma City through his grandfather, Edward Morris. Historians have credited Mr. Morris for being the first large industrial developer in Oklahoma City, founding Stockyards City or "Packingtown" in 1910. Morris entered a contract with the Oklahoma City Chamber of Commerce in 1908 that called for him to invest $3 million in a packing plant and the Chamber of Commerce to give him 10 percent of the cost of

resources available through MidCity Financial Corporation.

Through its six branches located as far north as Edmond and Memorial Road to Capitol Hill and South Pennsylvania, Union Bank is well positioned to offer its full array of products and services. These include such consumer products as free checking accounts, high-yield investment accounts, special feature Certificates of Deposit, and a variety of competitively priced consumer loans.

Commercial products include corporate cash management services, sweep accounts, lock box

Express Personnel Services

Introducing Express Personnel Services—a company with sales growth anticipated to reach a billion dollars in revenues before the end of the century. Founded in 1983, this privately held franchisor of staffing services boasts 1997 figures of over 350 franchised offices and employment of more than 250,000 people annually.

Express opened its doors in the temporary industry when downsizing was becoming a familiar word in the corporate world. This impeccable timing, creation of three divisions sensitive to customer needs, the decision to franchise, and cultivation of quality employees, comprise the Express formula for success.

The company employs more than 150 team members at International Headquarters in Oklahoma City. It touts more than 25 offices in Oklahoma and more than 5 in the Oklahoma City metropolitan area. The company's divisions include Express Personnel Services—temporary and direct-hire placement; Express Human Resources—professional employer services; and Robert William James & Associates—professional search/contract staffing.

The executive management team consists of Robert A. Funk, founder and chairman of the board; William H. Stoller, vice chairman; and David B. Gillogly, president. As a renowned businessman and entrepreneur, Funk owns more than 30 Express offices in Oklahoma, Missouri, Texas, and Kansas.

Recognition for Express has surpassed 350 awards, including *Entrepreneur* magazine's Franchise 500 List for five consecutive years and *SUCCESS* magazine's Gold 100 List for three consecutive years. Express also captured *Inc.* magazine's 500 List, an annual listing of the 500 fastest- growing, privately held companies in the United States, for three consecutive years.

In 1996, Express was the top designee in Oklahoma for the Blue Chip Enterprise Award, established by Massachusetts Mutual and *Nation's Business* magazine to recognize the highest standards of small business excellence.

Robert A. Funk, William H. Stoller and David B. Gillogly discuss site plans for the new corporate headquarters of Express Personnel Services on Northwest Highway.

Providing qualified temporary and direct-hire employees for its client companies is an important part of the Express success story.

Express also received the Oklahoma Venture of the Year Award from Oklahoma Venture Forum.

The future for Express Personnel Services is to offer all the services related to human resources within one entity—from temporary help to management of a company's employees, and everything between. As another step toward expansion, Express stepped into the international arena with affiliations in Canada, Great Britain, Russia, and the Ukraine.

Because of the company's growth, construction is scheduled to begin for a new corporate headquarters on Northwest Highway.

Express offers training and ongoing support to its increasing employee base. "At Express we spend a great deal of time locating, retaining, and further developing skills of our staff, franchisees, and staffing associates," Funk said.

The company promotes continuing education for its team through an active scholarship program and is a strong believer in corporate citizenship and community involvement. As a national sponsor of the Children's Miracle Network, Express offices nationwide are also involved in local civic and charity work.

"Our decision to provide a full range of staffing options and to franchise has brought us exponential growth," Funk said. "In just a few short years, Express has reached the forefront of the booming staffing industry; add the work ethic, loyalty, and dedication of the people here in Oklahoma, and you have a guaranteed formula for success." ✿

MidFirst Bank

"An entrepreneurial, tenacious spirit is not likely to be altered from its course." The strength of MidFirst Bank is evidenced by its ability to excel. For almost five decades, MidFirst and its predecessors have encouraged the growth of Oklahoma's businesses and enhanced the financial success of its citizens. With an average annual growth rate of 65 percent, MidFirst Bank has advanced to become the largest Oklahoma City-based and owned financial institution. A fitting accomplishment for a bank whose heritage is grounded in Oklahoma.

MidFirst has deep roots in Oklahoma City, dating back to 1950 when Midland Mortgage Co. was organized. For 31 years, the mortgage company operated as a leading mortgage originator. In 1981, the executive team of Midland Mortgage Co. developed Midland Financial Co. and acquired MidFirst, a newly-formed financial institution. From a humble beginning of five full-time employees, Midland Financial Co., Midland Mortgage Co., and MidFirst Bank have enjoyed continuous growth and now employ more than 900 Oklahomans.

MidFirst Bank, now the dominant company of the group, has grown to $2.8 billion in assets and 25 locations. By providing exemplary customer service and highly competitive products, MidFirst has become a top performing financial institution in Oklahoma.

"Credit for MidFirst's success belongs to our dedicated employees and loyal customers," says G. Jeffrey Records, president and CEO. "Thanks to the focus, integrity, and vision of our staff, MidFirst stands out as a solid and respected corporate citizen in every community we serve."

MidFirst offers a full array of traditional banking services, as well as business loans and brokerage services. Also, with mergers and acquisitions being a mainstay in the banking industry, MidFirst has remained under a single family's private ownership since its inception.

With a path distinctly carved, MidFirst Bank will continue on its course to become the premier financial services provider in Oklahoma. The bank's destination will surely be reached by an unyielding spirit of determination. ✺

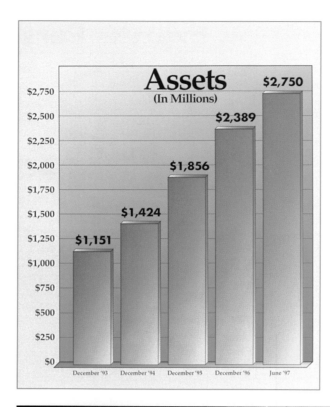

Assets
(In Millions)

	December '93	December '94	December '95	December '96	June '97
	$1,151	$1,424	$1,856	$2,389	$2,750

MidFirst Bank is the largest Oklahoma City-based and owned financial institution.

MidFirst Bank, along with its affiliates, employs more than 900 Oklahomans.

UMB Oklahoma Bank

Stockyards City—Agnew and Exchange

Downtown—Robinson and Park Avenue

U MB Oklahoma Bank continues to leave its mark on Oklahoma City's business community. For over 70 years, the bank has shared in the hard work that has built one of the nation's great cities. It has been accomplished with leadership, determination, and resourcefulness.

The financial institution that would become UMB Oklahoma Bank was established in 1925, when a group of pioneer businessmen received a state charter to open the doors of the Stock Yards Bank at the corner of Agnew and Exchange. The Stock Yards itself was only 15 years old, and trolleys each day brought thousands of workers to and from the then rural setting of the Stock Yards.

During World War II, brave Oklahoma men joined the fight for freedom. Here at home, the bank was one of the first financial institutions in the state to employ women in vital bank functions. With the post war boom in automobile production, the bank was the first financial institution in Oklahoma to offer modern conveniences such as drive-through banking windows to its customers.

The bank's name was changed to UMB Oklahoma Bank in 1995 when UMB Financial Corporation acquired the bank. Led by R. Crosby Kemper, chairman and chief executive officer, UMB Financial Corporation is ranked among the larger banking companies in the country with over $6.5 billion in assets. UMB's Trust Department ranks among the top 20 in the United States.

"The affiliation with UMB Financial Corporation has provided the strong resources of a diversified financial services company combined with local pride and commitment to the Oklahoma City community," stated UMB Oklahoma Bank Chairman and President Royce Hammons. "It is a great combination."

Known for years as "America's Strongest Banks," UMB Financial Corp. offers complete banking and related financial services to both individual and business customers.

Headquartered in Kansas City, Missouri, the company currently owns and operates 16 affiliate banks with 140 locations from Oklahoma City to Omaha and from Denver to St. Louis. Other subsidiaries include a corporate trust and securities processing company in New York, a trust employee benefit office in San Francisco, and companies that deal with brokerage services, leasing, venture capital, and insurance.

In 1997, UMB Oklahoma Bank opened a new downtown branch and banking offices at Park Avenue and Robinson. This location houses corporate banking, private banking, and trust. Middle market and small business customers continue to be served primarily at the Agnew and Exchange office. The Broadway Extension office, opened in 1989, serves individuals and businesses in north Oklahoma City.

Under the 10-year leadership of Chief Executive Officer Royce Hammons and his management team, assets have grown from $100 million to $150 million and stockholders' equity has increased from $6 million to $16 million. This dynamic growth has occurred through a recognition that customer relationships are the bank's greatest assets.

A recognized leader in business banking, UMB Oklahoma Bank also provides a complete line of consumer banking, mortgage banking, trust, and investment products. UMB bankers are career professionals who distinguish themselves by their readiness to serve their customers and their community. ✺

North Office—Broadway and 63rd Street

Local Federal Bank

A s the end of the 20th century witnessed intense upheaval in the banking industry, one Oklahoma institution remained stable and financially solvent. Local Federal Bank had the strength to address the national savings and loan crisis of the 1980s with many long-term solutions for the state's extensive financial challenges.

"Of the Savings and Loans in Oklahoma during 1980, with assets over $500 million, Local Federal was the only one to survive. We were one of the first banks to purchase other financial institutions and to acquire loans throughout the country," says Alan Pollock, senior vice president and general counsel.

The century of success Local Federal Bank has enjoyed is due to a very strict underwriting philosophy. Selecting markets advantageous to growth has moved the bank from $900 million in assets to $2.7 billion in the last 10 years.

Local Federal's long history in Oklahoma began in 1908, when it incorporated as the Local Building and Loan Association. Early headquarters were on the second floor in the heart of Oklahoma City's downtown area at 205 1/2 W. Main Street.

In 1935, with the invention of deposit insurance, the bank initiated a federal charter. Since 1990, it has been known as Local Federal Bank.

Local Federal created an impressive history of firsts as it established deep financial roots in Oklahoma. After World War II, it issued the first Veteran's Administration (VA) loan in the state, which was also one of the first two VA loans in the nation. In 1964, this financial vanguard was the first savings and loan institution in Oklahoma to open a branch office. Today, the majority of its 540 employees work at one of the 41 Local Federal Bank branches spread throughout the state.

Local Federal has initiated mortgage practices designed for low and moderate income levels. Other financial innovations the bank introduced include commercial real estate lending and alternative investment products. It was one of the first financial institutions to offer annuities and mutual funds.

From the beginning, the focus of lending practices has been on home mortgages. The same mission remains true today.

"Local Federal continues to provide the funding to give people the opportunity to be home-

Local Federal's long history in Oklahoma began in 1908, when it incorporated as the Local Building and Loan Association.

Oklahomans can depend on the long-term stability of Local Federal to meet their expanding financial needs. Photo by J.D. Merryweather.

owners. This goal remains the lifeblood for much of our current business," Pollock says.

However, as he points out, loans have changed dramatically over the years. "In the early 1900s, citizens borrowed up to the amount they had in their accounts. They had three options for monthly repayment: 75 cents, one dollar, or one dollar and a quarter toward the principal," he says.

"Currently, we offer a dynamic array of financial services from the $1,000 personal loan to the $15 million commercial real estate loan," comments Robert Vanden, president and chief executive officer. "Customers remain our focus. We continue in our commitment to provide them with the best financial services and products in Oklahoma. They can depend on the long-term stability of Local Federal to meet their expanding financial needs." ❂

Metro Journal

"Our mission at the *Metro Journal* magazine is to promote the Oklahoma City business community by focusing our news and editorial commentary on commercial trends and stories," says James Durocher, publisher. "What sets us apart is our research team. We do all original research ourselves."

The initial vision for the publication began with Tim Bales, who developed the *Metro Journal* in 1993 to report commercial real estate transactions in the Oklahoma City metro area. When Durocher moved to Oklahoma in 1995 and saw the publication, he knew it had the potential to become a premier business magazine.

"I saw the opportunity for the *Metro Journal* to be a publication the local community could be very proud to claim. Our intention is to spotlight Oklahoma City as the land of opportunity. I want the *Metro Journal* to be a selling tool for commercial interests in Oklahoma City," Durocher continues.

Before buying the publication, Durocher had successfully developed comparable publications in different areas of the country. After relocating to Oklahoma, he began researching Oklahoma City's publishing history. He discovered multiple publishing giants had originated from the area.

Among these giants was the highly acclaimed *Guffy's Journal*, one of the first business publications in the United States. Durocher tracked down publisher Chan Guffy, who became his publishing mentor.

As a result of his intense and thorough research into publishing and the Oklahoma City business community, Durocher has been able to move the *Metro Journal* to exceptional levels very quickly. The magazine continues to address commercial real estate interests and has expanded to include in-depth reporting about Oklahoma City's thriving business community. With a longer shelf life than a tabloid, it is the only publication to offer readers rankings of annual business transactions.

Publishing innovations which have an impacting and recognize the local business community include the *Metro Journal's* Metro 50, which recognizes Oklahoma City's 50 fastest growing companies. To celebrate these entrepreneurial success stories, the *Metro Journal* hosts an annual awards banquet, which is co-sponsored by the Greater Oklahoma City Chamber of Commerce and individual corporations.

Fourth from left, James Durocher, publisher of the **Metro Journal**. *Photo by J.D. Merryweather.*

Forty under 40, another *Metro Journal* listing of Oklahoma City's *Who's Who of Business*, identifies the 40 most influential local businesspeople under the age of 40. Durocher applauds these leaders infusing ambition and vision into the commercial arena.

"It's important to take a long-term view for the city," Durocher says. "Our up-and-coming business leaders need to be held responsible for the future business climate in Oklahoma." Durocher believes strongly in connecting commercial interests with the community's well-being. As a result, the *Metro Journal* champions Big Brothers and Big Sisters. An advertising campaign is being planned by the magazine to attract more big brothers and big sisters in the metro area.

Future publishing plans include expanding circulation to a statewide readership of 50,000. Through an aggressive marketing strategy, Durocher plans to position Oklahoma City as a major contender for innovative, far-reaching business practices in the coming years.

"Oklahoma City has a different bottom line than other metropolitan areas," he concludes. "It has a broader, more genuine emphasis, which can demonstrate to the world what the next generation of business in this country can be." ✪

14

Professions

McAfee & Taft

McAfee & Taft is a law firm that works to facilitate the business of its clients.

Many consider McAfee & Taft one of the premier business firms in the state. The firm continually maintains the highest standards of legal excellence, which is reflected in its status as one of the principal firms that businesses look to for legal representation.

McAfee & Taft is one of the largest law firms in the state. The attorneys of McAfee & Taft are graduates of highly respected law schools and ranked among the top students in their graduating classes. They carry on the early vision of

McAfee & Taft serves a broad range of clients. Clients include diverse local, state, and regional companies; many publicly held companies based in Oklahoma; smaller start-up companies and individuals. The firm's clients engage in business on a local, national, and international scale.

AREAS OF PRACTICE

Areas of practice for the organization highlight the variety and breadth of the firm's

Kenneth E. McAfee and Richard G. Taft, who founded the firm in 1952.

Well known as one of Oklahoma's oldest law firms, McAfee & Taft continues the tradition of providing high quality legal services to clients. The attorneys of this full-service firm are readily accessible to clients and attentive to their affairs. They are adept at developing creative legal and business planning strategies and keep abreast of current legal developments that might affect the firm's clients.

capabilities. Collaboration and interdisciplinary work ensure clients are receiving a comprehensive range of legal services.

Business and Transactional

McAfee & Taft represents public and private companies in business, corporate, mergers and acquisitions, and securities matters, and in extraordinary transactions.

Other business and transactional work falls in the broad categories of oil, gas and energy law and real estate.

Trial

McAfee & Taft's trial lawyers have extensive trial experience in state and federal courts in Oklahoma, as well as across the nation. The firm also represents clients before state and federal administrative agencies and various governmental boards, agencies, and commissions.

Trial work includes business litigation, appellate, general commercial, products liability, oil and gas, employment, and employee benefits.

Tax

McAfee and Taft's tax lawyers, many with advanced degrees in tax law, are experts in analyzing and providing strategies in a broad range of tax transactions. Firm attorneys routinely address local, state, federal, and foreign tax matters for corporations, tax-exempt organizations, limited liability companies, partnerships, estates, trusts and individuals.

Corporate and business tax, estate planning, and employee benefits are major areas of the firm's tax practice.

Specialized Areas

Specialization in specific areas of law allows the firm to provide representation to clients in diverse industries and transactions. Areas of focus include aviation, business reorganizations and bankruptcy, environmental, financial institutions, franchise, international and health law.

COMMITMENT

McAfee & Taft's commitment to its clients is reflected in its strong commitment to the community. The firm's attorneys provide *pro bono* legal services to a number of charitable and civic organizations. They also provide legal representation through Legal Aid of Western Oklahoma and frequently speak at professional meetings and seminars.

A wide range of foundations, arts agencies, health care institutions, educational programs, schools and universities, and other nonprofit organizations benefit from the volunteer time and support of the firm.

The firm's attorneys also take their roles in professional associations very seriously and have assumed leadership positions in the Bar Association affairs. They assist in the courts by serving as court-appointed judges, arbitrators, or mediators. Each takes pride in his or her contributions and serving the community and their profession.

Prominent community leaders have emerged from the McAfee & Taft firm over the past five decades. The firm established the McAfee Centennial Chair at the University of Oklahoma College of Law in honor of founder Kenneth E. McAfee. Eugene O. Kuntz, an original lawyer with the firm, served as Dean of the University of Oklahoma College of Law. Kuntz is also author of the authoritative treatise on Oil and Gas law.

The firm subscribes to numerous periodicals and law-related reference materials, many of which can be accessed from every attorney's computer terminal. Attorneys are supported by a highly skilled staff of legal assistants, secretaries, clerical personnel, and litigation case support personnel. McAfee & Taft also houses one of the state's largest private law libraries.

The collection of attorneys at McAfee & Taft represents years of experience and highly developed legal talent for the Oklahoma business community. The work by this firm embraces tradition, ethical principles, and excellence in the practice of law. ✪

Ackerman McQueen, Inc.

Creativity. It is found in the more than 40-year-old B.C. Clark Christmas Anniversary Sale jingle and current Oklahoma Native America campaign. These icons reflect the ingenuity, award-winning, and long-term effectiveness of the work by Ackerman McQueen (AM)—the largest advertising agency based in Oklahoma.

From its humble beginnings in 1939, AM has moved from an agency with $250,000 to its present $110 million in billings. This growth propelled AM into the top 150 of the 8,000 advertising agencies in the United States.

Much has changed with this full-service communications agency since George W. Knox, Jr. founded the company. In 1952, a young navy pilot, Ray B. Ackerman, joined the agency and for the next 20 years grew the agency from $250,000 annual billings to $5 million. Marvin and Angus McQueen, father and son, joined Ackerman in the early 1970s, and together they built the company to its present status.

The technology-driven, one-office concept now links the five AM offices located across the United States. This unique model for managing a multi-office business puts AM on the cutting-edge of client service and computer sharing from city to city.

ONE-OFFICE CONCEPT

Technology. Several technological avenues connect AM offices in Oklahoma City, Tulsa, Dallas, Washington, D.C., and Colorado Springs. This equipment makes the distance between cities irrelevant and avoids duplication of business functions in each office. Fiber-optic videoconferencing, rapid-response paging networks, desk-to-desk direct telephone lines, and a wide-area network that provides a 24-hour link between the business and graphics computers are technological connectors for staff and client use.

Angus McQueen, chairman and chief executive officer, and William F. Winkler, Jr., chief operating officer, are dedicated to enhancing the company's technological edge, which translates into consistently improving the abilities to serve AM clients and to control costs. Six full-time staff are dedicated to the company's technological maintenance and growth. In fact, an entire room in each office is especially designed for

AM Chairman Emeritus Ray Ackerman, Chairman and CEO Angus McQueen, Vice Chairman Lee Allan Smith, and President Tom Millweard.

AM's Golden Voice in-house studio recording facility is a state-of-the-art computerized recording studio with full-time musicians, composers and producers who create original scores.

state-of-the-art, high-tech equipment that enhances interactive capabilities and speed of service.

VIDEOCONFERENCE CENTER

Interactive. State-of-the-art videoconferencing facilities are an integral part of each AM office. The similarly-designed videoconference centers function as a visual and interactive link for the agency's offices and clients.

The million-dollar Oklahoma City videoconference center features plush, theater-style seating and acoustical excellence.

Presentation capabilities include multimedia, multi-image, and video projection. A full 24-channel theater lighting system and black stage curtains add drama to presentations.

Many client benefits are derived from the videoconference facilities, including immediate review of works-in-progress and face-to-face interactions with agency staff across the miles.

OTHER HIGH-TECH FACILITIES

Quality. Ackerman McQueen touts other in-house facilities that enhance production capabilities. For video production, advanced in-house

Sophisticated videoconference facilities link all AM offices nationwide with multimedia, multi-image and video projection capabilities.

computerized editing systems are used to preview and edit video footage before final production takes place at a postproduction facility. For many clients, broadcast quality post-production can be handled in house. AM television producers, audio producers, and art directors have extensive production experience in all electronic and film formats.

Golden Voice is Ackerman McQueen's in-house audio recording facility with digital-linked capabilities to studios worldwide, including Los Angeles and New York. This state-of-the-art, computerized 32-track analog/digital recording studio is operated by four full-time staff composers, producers, and musicians who produce original lyrics, and compose and arrange music. Original recordings are edited and mixed in two adjoining production suites equipped with digital audio computers.

Additional facilities for multi-image audio-visual productions and photography provide comprehensive operations for assurance of quality.

A PHILOSOPHY OF GIVING

Community. The quality products Ackerman McQueen produces for its clients also equate to quality contributions to the community. AM gives back to its communities through executive time, agency products, and dedication to projects.

Following the Alfred P. Murrah Federal Building bombing in Oklahoma City, a full-page "Bless You America" ad, highlighting the purple, white, and yellow ribbons, was found in national newspapers and magazines. AM designed and produced the thank you ad, then negotiated free placement in the national publications.

Ray Ackerman, chairman emeritus, and Lee Allan Smith, vice chairman, are both distinguished for their long-standing, vast, and resolute leadership positions that strengthen all facets of Oklahoma City life. Smith served three terms and Ackerman served one term as chairman of the Greater Oklahoma City Chamber of Commerce. Both have been inducted into the Oklahoma Hall of Fame.

Ackerman is credited with spearheading the Oklahoma City promotional campaign: "It's a Wonderful Life!" Smith was president of Oklahoma Centennial Sports, which was the local organizing committee for the 1989 U.S. Olympic Festival in Oklahoma City.

AM is dedicated to dozens of causes and agencies in Oklahoma City. Involvement in United Way, Allied Arts, Omniplex, Oklahoma City Zoo, Boy Scouts of America, and National Cowboy Hall of Fame touch on only a few of the organizations served.

CLIENTS, PEOPLE, AND THE FUTURE

Continuity. Ackerman McQueen has developed an impressive list of clients over the years. Current clients include Bank of Oklahoma, Brunswick, *The Daily Oklahoman*, Dallas Cowboys, DataTimes, Foodbrands America, Homeland Stores, Kerr-McGee Corporation, and the Greater Oklahoma City Chamber of Commerce.

Other clients include OG&E Electric Services, Oklahoma Tourism and Recreation Department, Pizza Hut, Premier Parks, Southwestern Bell Telephone, Thrifty Car Rental, and Urocor.

AM won Madison Avenue's most coveted award, a Gold EFFIE, for its Oklahoma Tourism and Recreation Department's Native America campaign.

"As I look to the next decade, I see a transition to a new group of young people, many who have spent their careers with this company," said McQueen. "Longevity is not typical in this business, but we have been able to attract, motivate, and reward good people and they have stayed. Of all the things I have been a part of in this company over the last 23 years, I am most proud of this."

Strengthened by the fourth generation of leadership, this long-term commitment to clients is branded with the Ackerman McQueen legacy of creativity, technology, quality, community service, and positioning for the future. ✪

Frankfurt-Short-Bruza Associates, PC

Festival Place site improvements in downtown Oklahoma City.

Since its founding over 50 years ago, FSB has become one of the leading architecture, engineering, and planning firms in the state of Oklahoma. Through a consistent theme of excellence in architecture, trust, and project leadership, the firm has received many local and national awards. Perhaps, though, the awards that are most significant are the repeat business of satisfied clients.

A full-service team of architects, engineers, and planners has established a reputation for the firm based on traditional values, hard work, integrity, and performance. Since the company's beginning in 1945, these traits have guided the firm through steady growth.

CHOICE BASED ON REPUTATION AND ABILITY

Clients from New Jersey to Washington, Alaska to Hawaii consistently select FSB for a broad range of projects including world-class aviation facilities, sophisticated research and development laboratories, state-of-the-art health care complexes, and a variety of military, educational, and industrial facilities.

The growing national recognition of FSB's leadership role is based on the firm's architectural and engineering capabilities: a unique interdisciplinary approach that offers each client and project a spectrum of services, allowing clients to move from concept to completed project in a timely and efficient manner, fully informed and participating in each step of the process.

The resulting customer satisfaction is based on the firm's talent and integrity. This attention to client needs is reflected in an 89 percent repeat client base and testimonials from FSB's clients such as:
• "You and your people always respond with 'How can I help?' A real 'can do' attitude."
United Airlines
• "FSB has provided excellent professional services with an eye on quality. More important, this has been accomplished while meeting both time and fiscal constraints."
Oklahoma State University, College of Engineering, Architecture and Technology

Oklahoma City Metropolitan Area Projects Plan(MAPS) rendering.

• "Because of their superior management skills, expertise, and knowledge, FSB was able to maintain their aggressive schedule . . . and deliver the design for this technically complex building and site, complete and on time."
New York District Corps of Engineers

METROPOLITAN AREA PROJECTS (MAPS)

Through a qualifications-based interview process, Frankfurt-Short-Bruza was awarded the Metropolitan Area Projects (MAPS) contract. MAPS is one of the largest multiproject, urban-development programs in the country. As program coordinator, the firm's broad technical expertise is applied toward goals that will successfully develop the $300 million complex, nine-project master plan and conceptual design.

The nine MAPS components are the Bricktown Ballpark, Indoor Sports Arena, Metropolitan Learning Center, Downtown Canal, Myriad Convention Center renovation, Civic Center Music Hall renovation, Oklahoma State Fairgrounds improvements, the North Canadian Riverfront Development, and a transportation link between downtown and the I-40 Meridian business arc.

PROJECT SUCCESSES

Organization and management of complex projects are the cornerstone of FSB's success. It is based on a process that includes listening to—and understanding—client needs, interpreting those needs successfully, and taking action while providing innovative, cost-effective solutions.

FSB's clients trust this full-circle process. Local Oklahoma City clients working with FSB include Baptist Medical Center, the Federal Aviation Administration at Will Rogers World

United Airlines Indianapolis maintenance center.

Airport, Fife Manufacturing Company, Fleming Companies, Francis Tuttle, Hertz Corporation, Kerr McGee Corporation, OG&E Electric Services, Oklahoma Natural Gas Company, The Oklahoma Publishing Company, Second Century and the City of Oklahoma City, Southwestern Bell Telephone, Tinker Air Force Base, YMCA, and York International.

Nationally based clients include corporate names such as American Airlines, Conoco, Continental Airlines, Federal Express, Lawrence Livermore National Laboratory, Sandia National Laboratory, Uniroyal-Goodrich, United Airlines, USAir, and Whirlpool.

RECOGNITION, PEOPLE, AND TECHNOLOGY

FSB has been widely recognized for excellence in design. Numerous awards, including "Firm of the Year" and awards of excellence and honor awards from the American Institute of Architects (AIA) Central Oklahoma Chapter and the American Concrete Institute, are a tribute to the people of FSB: talented, experienced professionals working as a team through a variety of disciplines.

William Frankfurt, Glenn Short, and Jim Bruza, principals of the firm, understand FSB's people are the company's "inventory." They believe the proper mix of highly motivated people within a supportive work environment, able to take advantage of technically advanced tools, are the major components of the firm's success. For over half this century, FSB's consistent qualities of trust, leadership, and credibility have built more than just a thriving business—they've built relationships. ✪

Corporate office tower, Kerr-McGee Center, in downtown Oklahoma City.

Noble Research Center at Oklahoma State University in Stillwater.

Gable Gotwals Mock Schwabe Kihle Gaberino

Fortune 500 corporations, entrepreneurs, privately-owned companies, foundations, and individuals provide the attorneys of Gable Gotwals Mock Schwabe Kihle Gaberino with daily opportunities to address diverse legal challenges. The 69 attorneys of our firm respond by providing efficient, professional, and ethical representation—traits which have long been the firm's hallmark.

Gable Gotwals Mock Schwabe Kihle Gaberino is the result of the merger of three of Oklahoma's most respected law firms to provide law offices with significant size, experience and capability in both Tulsa and Oklahoma City. Gable & Gotwals, originally a Tulsa-based firm, was founded in 1944, and has experienced steady growth throughout its history. In July 1996, Gable & Gotwals merged with Mock, Schwabe, Waldo, Elder, Reeves & Bryant, an Oklahoma City-based firm founded in 1982. Most recently, in March 1997, the firm merged with Arrington Kihle Gaberino & Dunn, established 1919, in Tulsa, to create the present constituent firm.

Gable Gotwals Mock Schwabe Kihle Gaberino is now one of Oklahoma's largest firms with three locations. These locations enable its attorneys to place the greatest value on producing quality legal work, bringing a creative approach to problem solving, and rendering the finest possible legal services to its clients in a cost-effective manner. The combined talents of more than 120 shareholders, associates, and support personnel enable Gable Gotwals Mock Schwabe Kihle Gaberino to maintain a proud reputation as one of the Southwest's most respected law firms.

REACHING INTO OUR COMMUNITIES

Beyond the firm's reputation for employing distinguished graduates from law schools throughout the United States, the attorneys of Gable Gotwals Mock Schwabe Kihle Gaberino are noted for their presence in professional organizations and pro bono work. Our professionals have long had a responsibility to participate in and provide leadership and assistance to the organized bar and its programs.

Having been elected by their peers, one firm member is the current (1997) president of the

Gable & Gotwals, originally a Tulsa-based firm, was founded in 1944, and has experienced steady growth throughout its history.

The firm's attorneys provide efficient, professional, and ethical representation.

Oklahoma County Bar Association, one firm member is the current (1997) president of the Tulsa County Bar Association, and another of our members is president-elect (1998) of the Oklahoma Bar Association. The firm has been the recipient of numerous county and state bar association awards recognizing its commitment to the legal profession and pro bono activities.

Gable Gotwals Mock Schwabe Kihle Gaberino attorneys also are active on the boards or serve as officers of social welfare agencies, fine arts organizations, local education foundations, schools, churches, and other organizations.

Continuing legal education is strongly promoted. Additionally, the firm's attorneys are frequently requested to make presentations to or write materials for statewide and national semi-

nars. The firm believes that all of these activities enable our attorneys to maintain the integrity, knowledge, and experience that our clients demand and the legal profession expects.

LITIGATION, COMMERCIAL, AND REAL ESTATE PRACTICES

Gable Gotwals Mock Schwabe Kihle Gaberino provides a full complement of legal services to its clients throughout the county and abroad. The firm's trial attorneys represent and counsel major corporations, individuals, businesses and governmental agencies in both domestic and international controversies. They have extensive experience in all aspects of state and federal trial work and alternative dispute resolution processes, including arbitration, mediation, mini-trials, mock trials,

Our professionals have long had a responsibility to participate in and provide leadership and assistance to the organized bar and its programs.

The individual attorneys of Gable Gotwals Mock Schwabe Kihle Gaberino have achieved enviable reputations throughout Oklahoma and the Southwest.

have significant experience using the most sophisticated estate and gift planning techniques. Our attorneys also are active in will contests and guardianship litigation, as well as probate administration and guardianship.

LABOR, ENERGY, AND TAX

The firm's labor and employment law practice provides a full-range of legal service to management groups of various sizes and in a variety of industries. Our attorneys provide expertise at both the administrative and judicial levels in the specialized areas of employment discrimination, sexual harassment, equal pay, and wage and hour disputes. We also regularly represent clients before the Equal Employment Opportunity Commission, the Oklahoma Human Rights Commission, the Office of Federal Contract Compliance, the Oklahoma Employment Security Commission, and the Department of Labor, Wage & Hour Division.

The specialists in the firm's energy section offer comprehensive legal services to our oil and gas industry clients including majors, large independents, pipeline, gathering, marketing, and storage concerns, as well as public utility companies. These lawyers appear daily before the commissioners and the administrative law judges of the Oklahoma Corporation Commission and are widely acknowledged experts in their area of practice.

Our attorneys who practice in the tax and employee benefits area provide expertise in individual, corporate, partnership, trust and estates, tax-exempt organizations, and state and local taxation. They often appear in contested administrative and judicial proceedings before the Oklahoma Tax Commission, Oklahoma district and appellate courts, United States Tax Court, and federal district and appellate courts.

OUR FUTURE

The individual attorneys of Gable Gotwals Mock Schwabe Kihle Gaberino have achieved enviable reputations throughout Oklahoma and the Southwest. Our growing and dynamic firm is committed to continuing and enhancing this reputation. As our professionals resolve diverse legal challenges for our clients, we will continue the firm's legacy of efficient, professional, and ethical representation of our clients. ✪

and settlement conferences. Two former judges bring their unique perspective to litigation and alternative dispute resolution matters.

The firm's commercial practice attorneys are involved with transactions concerning business and personal property acquisitions, commercial loan documentation, banking, consumer law issues, debtor-creditor relations, including complex bankruptcy cases, and other commercial matters.

All areas of real estate law including acquisition, sale and development, real estate financing, municipal, and state regulation and environmental regulation are a part of the real estate practice. The firm also frequently renders title opinions, handles zoning disputes, and represents clients involved in condemnation/eminent domain litigation.

ENVIRONMENTAL, BUSINESS, AND ESTATE PLANNING

Environmental concerns are of paramount importance in today's business environment.

Whether it is CERCLA or Superfund claims or litigation, environmental issues in real estate transactions, or negotiation with the EPA or Oklahoma's Department of Environmental Quality, the firm can provide experienced attorneys to resolve these matters.

Our attorneys represent clients in a wide array of business planning opportunities and transactions. Each of these attorneys has significant expertise in transactional work. Several of the firm's attorneys are nationally-recognized, published specialists in the various legal disciplines within the business practice, including corporations and partnerships, securities, and contract law.

The firm also can assist clients with their estate planning needs, whether the estate is large or small. The firm provides counsel to young families just beginning to accumulate wealth, as well as to retirees with substantial estates. Gable Gotwals Mock Schwabe Kihle Gaberino attorneys

McKinney, Stringer & Webster, P.C.

Firm Co-founders: (left) N. Martin Stringer, (right) Kenneth N. McKinney.

McKinney, Stringer & Webster has built its reputation on the pursuit of innovative, but practical solutions to client's legal and business problems. As one of the largest law firms in Oklahoma, it has grown into a multi-specialty firm serving clients in a wide range of businesses, industries, and interests that stretch around the globe.

Consistent with its role as a leading firm in the Southwest, McKinney, Stringer & Webster is committed to the highest-quality, result-oriented practice of law. It seeks to develop a fine-tuned sensitivity to the needs and goals of clients, and accomplish these goals in a dynamic, enthusiastic, professional, and cost-effective manner.

SERVING DIVERSE CLIENTS

From large multinational companies to individual entrepreneurs, McKinney, Stringer & Webster seeks to represent clientele who are diversified and committed to excellence in their products and services. The firm actively develops a clientele who are on the cutting edge of their industries and markets and maintains client confidence through integrity, dependability, and high ethical standards.

McKinney, Stringer & Webster takes its ethical obligations seriously. "We believe our ethical duties are as important to the attorney-client relationship as the quality of our legal services," says Kenneth McKinney, Chairman of the Board.

The firm is progressive, service-oriented, and visionary in its service to clients in multi-specialty practice areas. "We attempt to anticipate clients' needs rather than just react to clients' problems," comments N. Martin Stringer, President.

McKinney, Stringer & Webster provides clients with a law firm that combines up-to-date technological resources and a professional support staff with dedicated, skillful attorneys who are experienced in virtually every dimension of the legal profession.

LITIGATION

The cornerstone of McKinney, Stringer & Webster's practice is the extensive experience in diverse litigation in every type of forum. Since

As one of the largest law firms in Oklahoma, McKinney, Stringer & Webster has grown into a multi-specialty firm, serving clients in a wide range of businesses. Photo by Dwayne Moore.

its founding in 1971, McKinney, Stringer & Webster has concentrated on providing clients the highest level of courtroom advocacy—experienced trial counsel who vigorously represent client interests and resolve a dispute as quickly and economically as possible.

The firm's attorneys regularly appear in state and federal courts, locally and nationally, at both the trial and appellate levels, before administrative agencies, and in arbitration and mediation proceedings.

BUSINESS PRACTICES

Corporate; securities; and other related areas, such as banking, health care, mergers and acquisitions, tax, and real estate; are areas of expertise that combine to establish the firm's business practice. A diverse business community is served through this practice area encompassing sole proprietors, partnerships, limited liability companies, closely held corporations, and large, publicly-held companies.

Corporate clients are engaged in such industries as manufacturing, retailing, oil and gas exploration and production, insurance, real estate development, health care, and food services.

ENVIRONMENTAL LAW AND NATURAL RESOURCES

McKinney, Stringer & Webster has the largest environmental and natural resources law practice in Oklahoma. It earned a leading role in environmental law, in Oklahoma and nationally, as lead counsel for the Hardage Steering Committee in one of the most notable Superfund matters ever tried.

The environmental practice includes local and national-based clients in virtually every type of business confronted by an ever-expanding body of environmental law and regulations. The firm's environmental attorneys include lawyers with degrees in geology, chemistry, engineering, and other technical disciplines.

McKinney, Stringer & Webster attorneys and other employees are active in a wide variety of civic and business organizations, as well as involved in charitable and service activities.

EMPLOYMENT AND LABOR LAW

No area of law has grown faster or become more challenging and complex for today's business managers than the field of labor and employment law. McKinney, Stringer & Webster's labor and employment practice area has not only kept pace with this rapid expansion of regulation, but is a recognized leader in employer- employee relations.

Comprising one of the firm's largest practice groups, experienced attorneys in the employment services section represent clients in federal and state discrimination cases, wrongful discharge lawsuits, and employment matters before federal and state administrative agencies. They also provide representation for many other employment and labor law issues.

PATENT, COPYRIGHT, AND INTELLECTUAL PROPERTY

McKinney, Stringer & Webster is among the few firms in Oklahoma able to provide experienced representation, counsel, and trial support in intellectual property matters including patents, trademarks, copyrights, and trade secrets. The firm's licensed patent attorneys combine valuable technical experience with legal expertise in assisting clients obtaining patents, copyrights, and trademarks under federal and state law, and in licensing and commercially developing these properties.

Clients in this area are individuals, as well as both small and large businesses. Their property interests range from copyrights in musical com-

positions to patents and business trademarks for large complex machinery.

BUILDING COMMUNITY

The firm ensures a multi-specialty, high-quality legal counsel by committing itself to developing every member of the firm to their maximum potential. By promoting loyalty, instilling commitment, and requiring excellence in each person's professional career, McKinney, Stringer & Webster seeks to cultivate a team-oriented approach to the practice of law that, at the same time, permits and encourages individual achievement and recognition.

In fact, McKinney, Stringer & Webster proactively works to build community within the organization through many avenues, including an annual coaching and mentoring retreat involving every firm attorney.

McKinney, Stringer & Webster also strongly believes in its responsibility to contribute to the community and commits its leadership, talent, and resources to the progress and enrichment of Oklahoma. Its attorneys and other employees are active in a wide variety of civic and business organizations, as well as involved in charitable and service activities, Bar functions, and committees that contribute to the improvement of the legal profession and the delivery of legal services to the poor and disadvantaged.

"We feel it's important to put back in the community by sponsoring or contributing to popular civic events and encouraging volunteerism and pro-bono legal activities. These activities further the quality of life in our city and state, and enable our attorneys and support staff to better know, understand, and serve our clients on a daily basis," says Stringer.

McKinney, Stringer & Webster's foundation of law is based on providing cost-efficient, creative solutions to contemporary issues in a complex society. It is dedicated to the aggressive pursuit of this principle. "We take pride that our clients rely on the firm's specialized strengths to help them face today's challenges and accomplish tomorrow's goals," McKinney concludes. ✹

The cornerstone of McKinney, Stringer & Webster's practice is the extensive experience in diverse litigation in every type of forum. Photo by Dwayne Moore.

Coopers & Lybrand L.L.P.

Integrity, teamwork, mutual respect, and personal responsibility are the core values that embody the professional services firm of Coopers & Lybrand.

Coopers & Lybrand is a strong, contributing presence in the Oklahoma business community. The firm has been serving Oklahoma since the 1930s and continues to be a prominent part of the growing business landscape. A staff of approximately 150 employees in Oklahoma City and Tulsa offers companies efficiency, responsiveness, and close personal relationships, combined with the abundant resources of a Big 6 firm.

On a global basis, Coopers & Lybrand brings its clients an outstanding international reputation for quality, service, and leadership. The firm's presence spans more than 100 offices in the United States and 733 offices worldwide, employing more than 71,000 people in 140 foreign countries.

SERVICE AND CLIENTS FIRST

Coopers & Lybrand's five lines of businesses, Business Assurance, Tax, Consulting, Financial Advisory Services and Human Resources, provide relevant, timely, and value-added services. Within this structure, there are approximately 300 subcategories of expertise and services.

Coopers & Lybrand brings this extensive menu of services to clients with a philosophy that puts the client first. Numerous clients are a part of the statewide Coopers & Lybrand community. The strength of the company's credentials and reputation is extended to large, midsized, and small businesses, which vary from private individuals and entrepreneurs to publicly held companies.

The Oklahoma client roster is comprised of a variety of industries, including energy, manufacturing, education, financial service, and health care. Two of the three largest banks owned by Oklahomans—BancFirst and F&M Bank Trust—are Coopers & Lybrand clients. This firm also serves one of the fastest-growing energy companies, Chesapeake Energy Corporation, in the United States and the largest refrigerated trucking company, ROCOR International, in the country.

"We see Oklahoma as a market with solid, steady growth potential over the next several years in entrepreneurial activities as well as existing companies. Our firm is committed to that

Watson Moyers, managing partner.

growth by providing outstanding professional services to our clients," says Watson Moyers, managing partner, Oklahoma City.

"We take pride in the fact that the clients we work with in Oklahoma have strong moral and ethical values," says Jack Short, market managing partner. "With these values in place, it makes it easier for us to assist clients in solving business and technical issues."

Coopers & Lybrand has a set of shared values that stands as a part of the firm's mission. This statement reinforces that these values "will enrich the power of our ideas and enhance our commitment to excellence." The statement outlines the firm's commitment to the organization, its clients, and people.

The Management Group (back row l to r) Paul Jackson, Rick McCune, Chuck Rahill; (seated) Sam McClure, Tina May; (standing r) Brian Hilmes.

Coopers & Lybrand L.L.P. leads in technology.

TRAINING, TEAMWORK, AND TECHNOLOGY

With the caliber of professionals at Coopers & Lybrand, excellence and technical proficiency work hand in hand. "Coopers & Lybrand is basically only as good as its people," says Short.

"Our focus is on engaging the best and brightest from diverse backgrounds with the best technical training. The majority of young professionals who join us come from the University of Oklahoma and Oklahoma State University," says Moyers. "Our professionals are strongly encouraged to balance their professional lives with their personal lives."

Extensive training of new personnel is standard operating procedure for this company. Yet training is not limited to incoming professionals.

Ongoing technical training makes the firm a leader in its use of technology. Coopers & Lybrand knows that technology increases efficiency and its ability to serve clients. Enormous funds are dedicated to hardware, software, and training for its professionals. As an example, the Oklahoma offices are virtually doing away with a heavy library of paper products and utilizing technology in all aspects of the business, from internal communications, to the audit process and technical research.

The paperless audit process not only stresses technology, but teamwork. All team members have access to a document and can work simultaneously in different areas. This very efficient process can give a full, updated picture of the project on a daily basis.

Training also extends to areas that strengthen the work force as a whole. "Coopers & Lybrand is committing more than $3 million on nationwide diversity training for managers and partners with a goal to set the tone at the top," says Short. "We take care of our people—clients and employees—and this training is a commitment to understanding, comprehending, and retaining all types of people in our business."

CONTRIBUTIONS TO THE COMMUNITY

Along with a commitment to help its clients succeed, Coopers & Lybrand is dedicated to the success of the Oklahoma City community.

"We are committed to and are heavily involved in community life through personal involvement by our professionals as well as financial contributions by the staff and the firm," says Moyers. "It is part of a professional's development to give something back to the community."

Some programs benefiting from this company's involvement include adopting families, fund-raising golf tournaments, such as the Jim Thorpe Celebrity Golf Tournament that benefits the Ronald McDonald House, and the Christmas Connection. Several walk-a-thons sponsored by Coopers & Lybrand aid the March of Dimes, Arthritis Foundation, and American Heart Association. A bag-a-thon for Oklahoma City Beautiful assists with the city's beautification efforts, and participation in a corporate challenge athletic competition fosters health for employees and a care-giving organization.

Additionally, Coopers & Lybrand partners and managers serve on the board of directors and in leadership positions for many civic and charitable organizations such as Omniplex, Allied Arts, Oklahoma City Art Museum, Ballet Oklahoma, and Positive Tomorrows.

Coopers & Lybrand is a strength in the Oklahoma City community.

The core values set out by Coopers & Lybrand on a global scale reflect the integrity of its people and services provided to clients in Oklahoma and across the world. Through the use of training, teamwork, and technology, the company never loses sight of its clients' needs and the challenges they face.

Coopers & Lybrand stands firm in its philosophy: "By meeting our clients' needs, our success will follow." ❁

Miles Associates

Miles Associates is a specialized architectural, planning, and consulting firm. It is acclaimed for handling highly complex projects involving technically demanding environments in advanced technology, health-related facilities, and other unique building types. Along with this expertise, the firm's fundamental product is providing its clients with excellence, on schedule and within budget.

With its base of operations in Oklahoma City, the firm's activities have encompassed projects in the public and private sectors throughout various parts of the country, including Hawaii and the Caribbean.

The growth of Miles Associates' technical capabilities, since its formation in 1980, is demonstrated by the specialty experience of the firm. These projects appropriately represent how the firm has expanded beyond traditional design projects to encompass services from conceptual studies, to development planning, to basic design and construction administration, then finally to include facilities start-up and management.

Quality work and unwavering business ethics combine as the cornerstone of the foundation of the Miles Associates Team.

THE TEAM APPROACH

Miles Associates differentiates itself by establishing client-team relationships that consistently involve listening and understanding the situation, then solving the problems. The team resources include the firm, the client, the consulting engineers, and project-specific specialty consultants. A longstanding principle of the Miles Associates team has been to combine complementary talents to achieve each client's objective with quality architecture.

With direct and extensive client participation, each team is directed by the same project manager from conceptual planning through construction completion.

"The firm includes our clients as part the team," says Garrett F. (Bud) Miles, president. "Listening to our clients has been the hallmark of Miles Associates' process in creating architectural solutions. We don't impose our solutions on our clients or the community at large."

QUALITY WORK

Eighteen full-time employees make up Miles

Columbia Presbyterian Center for Healthy Living—Oklahoma City

Acree-Woodworth Building—Oklahoma City

Associates' interdisciplinary staff of architects, designers, managers, and support technical personnel. Fifty percent of the staff are licensed architects—a high ratio for an architectural firm. Because the projects are so technical, "we hire the best specialty consultants and consulting engineers to join the project team, project by project," says Miles. "Our basis of professional services is to staff the projects with the most proficient talent possible."

CLIENT-SPECIFIC EXPERTISE

Special expertise has been developed by Miles Associates in the design and development of technically demanding facilities, including laboratories and health and medical-related projects. This experience has provided the Miles Associates team with a problem solving background that it applies on both corporate and institutional projects such as diverse health and medical facilities, research

laboratory facilities, and other corporate and mixed-use developments.

Major Miles Associates clients have included several non- profit foundations from across the state, such as the Samuel Roberts Noble Foundation, the Oklahoma Medical Research Foundation, the Presbyterian Health Foundation, and the Warren Foundation. Projects range from conference centers to a very technical AIDS research laboratory.

"We try to first listen and think in a specialized mode, thereby producing a specialized service," says Miles.

ADVANCED TECHNOLOGY

Exacting controls and standards essential to research and advanced technology environments are typical parameters for both R&D and corporate facilities. Solutions for these technologically complex and demanding projects have been the mainstay of the firm's experience.

(left) Samuel Roberts Noble Research Foundation Guest House Facility— Ardmore, Oklahoma

(below left) Oklahoma Medical Research Foundation Research Lab— Oklahoma City

(below) Oklahoma Health Center Research Park Building #1— Oklahoma City

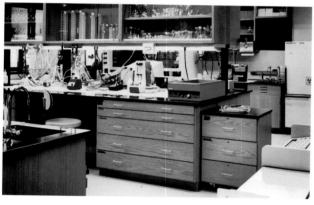

Miles Associates has developed an expertise in this type of building project represented by medical research laboratories, biotechnology laboratories, materials testing laboratories, and numerous technical support facilities. The complexity of the tasks in creating functional and process-driven, user-friendly solutions has developed the firm's range and depth of experience in advanced technology facilities.

"Service in the biotechnology industry is very rewarding . . . . having the opportunity to be even a peripheral part of the discovery process can provide a great deal of professional satisfaction," Miles says.

Examples of advanced technology projects developed by Miles Associates include the Research Park Building One and Building Two, the Oklahoma Medical Research Foundation's Research Laboratories, and the Samuel Roberts Noble Foundation Plant Biology Laboratories and Containment Facilities.

HEALTH-RELATED FACILITIES

Special facilities for health-related activities have become a building type where experience and attention to specific-user demands are essential. The Miles Associates team has established a level of experience valuable to the development of these types of projects, which include medical offices/clinics, dental clinics, ambulatory surgery centers, diagnostic and treatment facilities, as well as rehabilitation and wellness facilities.

The Donald W. Reynolds Children's Center Hospital and the Columbia Presbyterian Hospital Center for Healthy Living, the Medical Office Building, and the Hospital Emergency Room renovation projects are examples of Miles Associates health-related facilities in the Oklahoma City metropolitan area.

"Medical or health-related projects require a level of experience that fits well with our high ratio of professional members in the firm," Miles adds. "It is consistent with our service philoso-

phies to provide an experienced, qualified, and attentive team for each specific project."

ARCHITECTURAL DIVERSITY

Other challenging individual projects for Miles Associates include St. Paul's Cathedral, Xerox Corporation, UroCor, Inc., Seagate Technology, along with the University of Oklahoma, Oklahoma State University, the Veteran's Administration Hospital, and several municipalities across central Oklahoma.

Listening to the client, determining the actual parameters of the projects, and then providing experienced technical assistance are the foundation of excellence for the Miles Associates team in its advanced technology, health-related, corporate, and other technically demanding facilities. The enduring architecture, planning, design, and facilities consulting work of these experts secure expectations and results—making the clients' vision today's reality. ✪

KPMG Peat Marwick LLP

KPMG Peat Marwick is proudly part of one of the largest international accounting organizations creating "firsts" across the globe. The ingenuity and dynamics of the firm's programs set the standard for offices in Oklahoma City and around the world.

Founding partners James Marwick and Roger Mitchell recognized the value of a public accounting firm's ability to serve the worldwide needs of its clients in 1897. To solidify this emphasis, a 1987 merger of Peat, Marwick, Mitchell & Copartners and Klynveld Main Goerdeler created a global network known internationally as KPMG.

KPMG provides professional services—including assurance, tax, and management consulting services—to nearly 27 percent of the world's leading 1000 commercial and industrial companies. The firm's worldwide revenues surpass $7.5 billion, and more than 76,000 KPMG people serve clients from offices in 837 cities in 134 countries.

In all of its United States offices, KPMG is supported by the resources of this extensive international network. Yet each office has a full complement of resources to provide the highest level of local expertise and personal service. This local presence gives KPMG clients invaluable experience on which to draw.

The Oklahoma City office was established in 1950 and presently exceeds fifty client service and client service support personnel. Rocky Duckworth has served as the Oklahoma City office's managing partner since 1987. Other Oklahoma City partners include Jack Finley, Carlos Johnson, and Richard Coshow.

The firm's philosophy is entrenched in integrated Oklahoma City teams, which are matched with the successful industry specialization approach. These "targeted knowledge" teams combine industry expertise with functional knowledge to find inventive, client-focused solutions.

Whether on the local or international level, KMPG is the only firm to deliver market-oriented business solutions through specialized professionals in specific lines of business. The five areas of market concentration for the Oklahoma City office are manufacturing,

KPMG Oklahoma City Partners and Management Group Members

retailing, and distribution; financial services; public services; health care and life sciences; and information, communication, and entertainment.

Areas of expertise provided to these industries are built around the traditional base of audit and tax services, but also includes services in the areas of state and local tax; international tax; trade and tariffs; corporate finance; personal financial planning; business performance improvement; and strategic information technology consulting.

By listening to their clients' needs, by integrating and continually improving on technology, and by focusing on clients' business issues, KPMG's professionals are constantly developing new products and services. And their clients count this innovation as yet another strong advantage to being served by KPMG.

KPMG looks forward to a bright and exciting future as the firm celebrates its Centennial in 1997. Its historical commitment to the success of American business is articulated in its theme for this Centennial year: *A World of Spirit*, which honors the spirit of growth, change, innovation, and leadership that has made KPMG in Oklahoma City and 129 other U.S. cities a long-term success. As the firm's leaders have stated, "We've spent 100 years preparing for tomorrow, and our goal is to create a future worthy of our past. Our greatest moments are still to come." ✷

Jordan Associates

The entrepreneurial spirit is alive and well at Jordan Associates, where every associate has a vested interest in making client service the best it can be—because every associate is an owner of the agency. Employees have the opportunity to contribute to the growth, define the product, and benefit from the success generated by the agency. This driving spirit makes the agency more creative, more dedicated, and uniquely in tune with the challenges facing its clients.

Since 1961, Jordan has provided a full range of communications services, including the creation and placement of advertising, public relations, sales promotion, research, strategic planning, direct marketing, and Interactive Internet Communications. In both staff size and annual billings, Jordan has grown to be recognized as one of the top 25 agencies in the Southwest.

The agency counts among its clients industry leaders in health care, food service, economic development, financial services, retail, entertainment, telecommunications, and education. These clients have benefited with campaigns from complete corporate identity packages to new product introductions, from comprehensive media training seminars to image management, from communications audits to qualitative and quantitative research, from turnkey annual sales meetings to traffic-generating promotional strategies.

Jordan serves clients from its headquarters in Oklahoma City, its service office in St. Louis, as well as through its affiliate offices in the North American Advertising Agency Network throughout the United States, Mexico, Canada, South America, Europe, and the Pacific Rim. This allows the firm to contribute directly to the local economy—bringing dollars into Oklahoma that would otherwise have been spent outside the state.

Agency President and CEO Jeanette Gamba sums up the group's philosophy. "We're an idea resource," she says. "Our approach is to think of everything when it comes to helping our clients accomplish their marketing goals. We are responsible, accountable, and have the drive to help every client succeed. We've structured our company and its services not only to provide excellence in quality, but also maximum return on investment. And our long-term client relationships prove our approach works."

Jordan associates are active community leaders, serving as directors of numerous civic, cultural, and educational boards and volunteering their time, talent, and resources to more than 40 city and state organizations supporting the arts, education, health, business, and the advertising industry.

With a solid core of long-term, established clients and new business growth on an unprecedented upward spiral, Jordan Associates looks forward to continued expansion. ✪

Smart, straightforward selling at Jordan Associates is led by the agency's management team: (seated L-R) Rhonda Hooper, Senior Vice President/Account Service; Jeanette Gamba, President and Chief Executive Officer; Ed Westbury, Senior Vice President/Strategic Planning and Marketing Services; (standing L-R) Sue Dimond, Vice President/Controller; John Santore, Vice President/Executive Creative Director; Rob Andrew, Vice President/Account Service; and Peggy Sealy, Vice President/Public Relations.

Fagin, Brown, Bush, Tinney & Kiser

The Fagin firm has been continually ranked the Number One bond firm in Oklahoma. They have been engaged solely in the practice of municipal bond law since 1935. The Fagin firm is the oldest and most experienced municipal bond firm in the state of Oklahoma and has more full-time bond attorneys than any other firm in the state. Their practice consists exclusively of working with Oklahoma cities, towns, counties, public trusts, universities, and state agencies in structuring their bond issues and counseling with them on other matters of local government law and public finance. In addition, the attorneys are assisted by an excellent, experienced staff of seven employees accustomed to getting the job done.

Over the past 20 years, approximately $17 billion worth of municipal bonds have been approved and delivered through the legal work of Fagin, Brown, Bush, Tinney & Kiser. This translates to nearly 2,200 projects.

To say the least, the corporation is extremely busy at all times.

Four active attorneys and related support staff make up this firm that has been based in Oklahoma City since 1935. The group is generally characterized by youth and experience with the active lawyers accumulating over 75 years experience in the practice of municipal bond law.

Members of the firm include J. Scott Brown, Gary M. Bush, William Don Kiser, and Allan A. Brooks. Two nameplates follow these four at the firm entrance. George J. Fagin, 1908-1994, and Michael D. Tinney, 1944-1989. Fagin was the founder of the firm and Tinney was a longtime partner who died of cancer. As reflected by the firm name and nameplates, both men had a lasting impact on the organization.

The firm is located in the First National Center and has moved only twice within the building over the past 60 years. The Federal Bureau of Investigation (FBI) took the firm's space during World War II. Later, the firm moved to accommodate expansion of its company.

Throughout the years, this established firm has remained loyal to its location and work in the area of economic development for Oklahoma City and the state.

The Fagin firm is the oldest and most experienced municipal bond firm in the state of Oklahoma.

The types of bond projects facilitated by the firm are vast and cover a range of development activities. Monies raised through these financings are used for many public projects, including general municipal improvements; electric, water, and sewer facilities; hospital, nursing homes, airports, colleges, and universities; industrial, manufacturing, and commercial projects; pollution control and solid waste disposal facilities; convention centers and sports facilities; single and multifamily housing projects and historic preservation sites.

The largest number of projects centers around utility revenue bonds, airport, hospital, and various governmental agency indebtedness such as HUD and RECD, formerly known as FHA. In fact, the firm was a part of the establishment of almost every hospital in Oklahoma City.

The second-highest number of transactions is in the areas of industrial development and pollution control bonds.

Additionally, special authority bonds have been transacted for groups such as the Capitol Improvement Authority, Tourism and Recreation Authority, Student Loan Authority, Oklahoma Environmental Protection Authority, and many others. As the attorneys for the Oklahoma Industries Authority, the firm has assisted this organization in the creation of more than 100,000 jobs.

From tax exempt financing to entice existing company expansion or new company location, to the financing of facilities, Oklahoma City has stood to gain from the legal work of this law firm.

Fagin, Brown, Bush, Tinney and Kiser—a group dedicated to the economic development projects of Oklahoma City. Even after facilitating $17 billion in financing, this firm is still going strong, continuing to provide the legal expertise that helps make it happen. ✪

Conner & Winters

Formed in 1933, Conner & Winters offers a full range of legal services. The firm has built its practice by emphasizing quality client service, professional character, reliability, responsiveness, and mastery of the law. The firm's objective is to provide its clients with the quality of service and lawyer expertise and experience equivalent to prestigious national firms at rates which are competitive locally. As one of the largest law firms in the region, the firm has more than 60 attorneys in its offices in Oklahoma City, Tulsa, Northwest Arkansas, and Washington, D.C.

Size, depth, and diversity allow Conner & Winters to provide a broad array of legal services to numerous and varied clients engaged in a wide variety of activities. The firm's services extend from business, general and personal counseling to highly specialized assistance and advice and highlight modern communications facilities and a teamwork approach to solve client problems.

Conner & Winters teaches and practices the principle that achievement of the client's goals usually requires not only a knowledge of the client's situation, but also the business, market, products, personnel, strengths, and weaknesses.

The firm's interest is not casual; it is intense. This is because the firm believes that responsiveness is more than returning telephone calls promptly and delivering work when promised.

It is also giving advice that works. The lawyers at Conner & Winters understand both the challenges and opportunities each client is facing.

The firm's clients include large and small, domestic and international businesses. It counsels sole proprietorships, partnerships, limited liability companies, and corporations, including start-up businesses, emerging companies, and well-established firms. Additionally, the firm represents estates and trusts, municipalities and other public-sector organizations, and charitable foundations and institutions.

Conner & Winters is recognized for its expertise in many different areas of the law. The main areas of practice for Conner & Winters include banking; bankruptcy; complex litigation/antitrust; corporate and securities; dispute resolution; employee benefits; employment and labor relations; energy; environmental; estate planning, trusts, and probate; federal and state taxation; foundations and charities; governmental affairs; health care; immigration; international and real estate.

Part of Conner & Winters' success in achieving this diversity and in attaining its reputation as one of the region's leading firms may be attributed to the importance the firm places on the interest of its clients, and its ability to attract top graduates from well-known, prestigious schools.

The professional character of Conner & Winters is reflected through its attorneys and staff. Its lawyers are dedicated professionals who strive for independent thought and creative, yet practical and cost-efficient, solutions to clients' problems. Above all, they are committed to achieving the client's goals within the bounds of the law, not to throwing up roadblocks or forcing problems into prepackaged solutions.

The firm's support services and systems enable its attorneys to respond quickly and precisely to clients' needs. Every case and every project is administered cost effectively without sacrificing quality.

What results is the exceptional service and sound legal counsel that clients have come to rely on from Conner & Winters. ✦

Reception area of Conner & Winters

(seated l to r) Raymond Tompkins, Peter Bradford and Irwin Steinhorn; (standing l to r) Mark Bennett, Mitch Blackburn, John Funk, Timothy Bomhoff, Kiran Phansalkar, David Cordell and P. Blake Allen.

Day Edwards Federman Propester & Christensen, P.C.

Day Edwards Federman Propester & Christensen, P.C. is a problem-solving law firm. The firm's professional legal services encompass all areas of the business environment. Progressive, creative, and economical business solutions are representative of the core value of the firm's existence. As a professional corporation, the firm not only takes pride in the development of contemporary business solutions that solve clients' current problems, but also strives to provide problem prevention strategies to ensure the future stability and success of its clients. In addition, other core value elements that have established the firm's reputation for excellence in service include responsiveness to a broad range of client needs and a knowledgeable and service-minded work force.

Progressive, creative, and economical business solutions are representative of the core value of the firm's existence.

Day Edwards Federman Propester & Christensen, P.C. was founded in 1994 as a merger of three well-established Oklahoma City law firms. The firm's eight partners/shareholders elected to unite their expertise to improve their client service capabilities. Firm partners/shareholders include Joe E. Edwards, Bruce W. Day, William B. Federman, Rodney J. Heggy, D. Wade Christensen, J. Clay Christensen, Richard P. Propester, and Kent A. Gilliland. The professional services offered by the firm are derived from its broad spectrum of business-related practice areas that include financial and securities transactions, securities litigation, banking regulations and transactions, insurance defense and tort litigation, agribusiness, energy law, franchising regulations and transactions, merger and acquisitions, real estate law, insolvency, employment law, workers' compensation defense, labor relations, health law, arbitration, environmental law, and natural resources law.

Representative clients of Day Edwards Federman Propester & Christensen include some of the most prestigious banking, brokerage, financial, insurance, and industrial businesses participating in local, national, and international economic environments. As well, the firm represents major, publicly owned domestic and international corporations, closely held domestic and international corporations, partnership and individuals.

The firm's clients include some of the most prestigious banking, brokerage, financial, insurance, and industrial businesses.

Although active in the national and global business arena, the firm's investment in Oklahoma City and Oklahoma is of primary importance. Day Edwards Federman Propester & Christensen is involved in the development of a number of private capital funds that are to be invested in Oklahoma business ventures. The formation of the first state specific venture capital fund along with the development and construction of Robinson Plaza are but a few examples of the firm's many positive economic influences on the community. Positions held by the firm's partners, including recent past president of Leadership Oklahoma and former administrator of the Oklahoma Securities Commission, are a reflection of the firm's commitment to its community and civic leadership. The responsibility of this community commitment and civic leadership is the viable force behind the efforts of Day Edwards Federman Propester & Christensen, P.C. to meet the needs of the evolving demands of business and to create economical business solutions for the benefit of firm clients and the advancement of the Oklahoma City economic environment. ✪

Andrews Davis Legg Bixler Milsten & Price

Andrews Davis is a well established, full-service business law firm. Founded over 50 years ago, it is one of Oklahoma City's oldest and most experienced firms. The firm's mid-sized character offers several advantages, including efficiency in approach and personal interaction with clients. Andrews Davis has a diversified national and international business practice.

The firm's strengths are a commitment to excellence, responsiveness, and innovation in serving clients. These qualities have fostered long-term client relationships.

As a founding member of Commercial Law Affiliates (CLA), Andrews Davis sets itself apart in the world of law and demonstrates the firm's vital commitment to maintaining a personal and progressive approach to serving clients.

CLA is the largest affiliation of independent business and commercial law firms within the United States and in selected world markets. Its basic mission is to provide business clients ready access to comprehensive, cost-effective legal services and multi-jurisdictional capabilities. Through this affiliation, Andrews Davis can quickly provide services in all major U.S. cities and around the globe. The firm also can confidently facilitate referrals to a quality law firm in another jurisdiction.

A strict admission criterion for CLA permits only one firm per city and candidate firms must undergo careful screening and evaluation. CLA member firms represent a broad range of business and financial entities in major world markets. As a member firm, Andrews Davis has access to the diverse knowledge and resources of more than 4,000 lawyers worldwide.

In Oklahoma City, the Andrews Davis attorneys are business thinkers and creative problem solvers. Staying on the cutting-edge and keeping up with new developments to meet evolving client needs are the charges of this hard-working team.

The firm's attorneys are known for frequently lecturing and contributing to legal periodicals, in addition to serving as adjunct professors in law school and writing legal treatises. Two members of the firm, Alan C. Durbin and C. Temple Bixler, authored the three-volume work *Oklahoma Real Estate Forms— Practice*, which was originally published in 1987 and is supplemented annually. John D. Hastie is a nationally recognized expert on real estate law and has lectured at American Bar Association national seminars every year since 1969.

The diverse areas of practice for Andrews Davis attorneys include real estate, banking and finance, taxation, corporate and securities, employment law, employee benefits, litigation, and environmental law. Other areas are estate planning and probate, general business, health care, aircraft registration and financing, Indian law, natural resources, and bankruptcy.

The mix of Oklahoma City-based clients with local, national, and international business transactions substantiates Andrews Davis' scope of success.

Andrews Davis not only gives quality service to its clients, it serves the community. The breadth of the firm's community involvement shows its leadership in Oklahoma City. Andrews Davis attorneys lead and support civic, service, arts, educational, and professional organizations. Mona S. Lambird, elected in 1996 as the president of the Oklahoma Bar Association, was the first woman president of such association. J. Edward Barth is president and trustee of the Oklahoma City Community Foundation and serves as chairman of the city's Metropolitan Area Projects Citizens Oversight Board. These are just two among many examples of the commitment of Andrews Davis attorneys to their profession and community.

Andrews Davis Legg Bixler Milsten & Price combines its decades of experience with a progressive reputation. Serving clients from coast to coast, Andrews Davis is leveraging the strengths of its size and service with resources from around the world. ✪

(from left) Don G. Holladay, Elaine B. Thompson, Mona S. Lambird, and Gary S. Chilton, members of the Andrews Davis litigation practice group.

(from left, top row) Mark H. Price, D. Joe Rockett, Charles C. Callaway, Jr., Carolyn C. Cummins, and John F. Fischer, shareholders of the firm.

Cole & Reed, P.C.

All accounting firms, whatever size, buy the same books and follow the same rules. Cole & Reed believes what makes the difference in an accounting firm is its people.

The people of Cole & Reed make their living serving Oklahoma City area businesses and individuals.

Shareholders Sam Cole, Jerry Reed, John Newton, Bill Schlittler, Jim Denny, Paul Nicholson, and Mike Gibson understand the service-oriented, personal nature of the public accounting profession. These professionals possess the qualifications and attitudes to provide responsive and quality service at a reasonable fee. Cole & Reed executives and staff combine a blend of professional experience with an interactive approach to client service.

As an Oklahoma City-based accounting firm, Cole & Reed is located near downtown across from the Civic Center Music Hall. This location makes the firm very accessible. Clients can easily select a parking space directly in front of the building and come right in for a quick delivery or extended meeting.

Cole & Reed's professional practice consists of privately owned and publicly held businesses of various types and sizes. Clients represent most sectors of the business world, including profit and nonprofit organizations and governmental entities.

Audit, tax planning, and compliance services comprise a significant portion of the firm's practice. Tax services include preparation of a comprehensive range of income and estate tax returns, as well as related planning in these areas. Additionally, the firm provides planning and compliance assistance with qualified plans, compensation issues, state and local taxation, and other areas. Unlike many local firms, Cole & Reed has a significant audit practice, serving a wide range of local clients in the area. The firm also provides bookkeeping, computer consulting, litigation support, and management advisory services.

Cole & Reed places significant emphasis on its work with closely held family groups. The firm has consulted with and assisted numerous families and their related entities in maximizing the conservation, transfer, and enjoyment of family wealth and business capital through reduction of their individual, business, and estate taxes. In

Cole & Reed believes what makes the difference in an accounting firm is its people. Photo by Roger Bondy.

fact, the firm's demonstrated effectiveness, trustworthiness, and commitment family clients have made them the accounting firm of generations. Advising family businesses requires extensive knowledge in partnership, estate, and nonprofit taxation, areas in which Cole & Reed has established a respected reputation for its experience and effective service.

The firm also has expertise in providing tax planning services to corporate entities ranging in size from very large to very small firms. Cole & Reed accountants have consulted instrumentally with clients in developing and planning numerous corporate acquisitions, redemptions, liquidations, consolidations, tax-free regoranizations, capital equipment acquisitions, and many more business-related transactions.

Oklahoma City is more than a business environment to Cole & Reed. The firm, its personnel, and their families live and work as part of the local community and are committed to Oklahoma City and its people. A sense of permanence and client continuity are part of the company's history. As a successor to the Billups-Arnn firm that had its beginnings in the late 1940s, Cole & Reed has continued to serve many of the same clients since the 1950s.

As a part of the community, Cole & Reed believes in good corporate citizenship. The shareholders and staff give their time to various local organizations. A unique Thanksgiving meal program put on by the firm has raised thousands of dollars for charity. Annual contributions to the

Oklahoma City Community Foundation support charitable programs in the community, including the Travelers' Aid fund and local homeless shelters.

Cole & Reed's accounting practices step beyond buying the books and following the rules. In the service-oriented and personal business of accounting, the professionals of Cole & Reed are committed to the continuing development of new and existing relationships that are conducive to effective client service. This commitment and the character and integrity of Cole & Reed professionals are reflected in their client relationships and the quality of their service. ✪

Phillips McFall McCaffrey McVay & Murrah, P.C.

Professional excellence and superior client service were the foundations upon which Phillips McFall McCaffrey McVay & Murrah built its full-service business law practice in 1986. From the original group of four young attorneys, Phillips McFall has grown to 27 lawyers and is recognized as one of the leading business law firms in the State of Oklahoma.

"Our Firm was founded during very difficult economic times in the State of Oklahoma. Our goal was not just to survive, but to thrive by meeting the rapidly changing and complex legal environments of our clients, while being one of the most cost-effective providers of business legal services in Oklahoma," says Keith McFall, president. "We have invested heavily in information and communications technology, allowing all of our attorneys to tailor their individual practice areas and services to their clients' specific requirements. This approach allows us to provide our clients with timely, creative, and effective legal representation that addresses their particular business needs."

As a full-service business law firm, Phillips McFall is organized in the following major areas of practice: Banking; Mergers and Acquisitions; Commercial and Consumer Financial Services; Corporate, including limited liability companies and partnerships; Health Care; Estate Planning and Probate; Federal and State Taxation; Administrative;

Employment; Insurance Defense; Complex Litigation; Municipal Finance; Corporate Reorganization and Bankruptcy; Oil and Gas; Real Estate; and Securities. The Firm's clients include entrepreneurs, small businesses, and midsized and large corporations throughout Oklahoma, as well as national and international clients with substantial business interests in the State.

"While maintaining an extensive practice in established areas, we continue to expand our fields of expertise to meet the needs of our clients as their businesses grow or their business environments change. We have seen significant growth in the areas of Technology, Municipal Finance, Health Care, Securities, and Sports Law," McFall says. "By fully and efficiently utilizing our resources, we provide clients with a multi-disciplined team of lawyers able to handle any aspect of their legal requirements."

The focal point of Phillip McFall's practice has been its relationships with clients. The key to surviving in Oklahoma's tough economic environment in the 1980s was the establishment of strong and lasting client relationships with outstanding businesses and individuals. These

relationships are based not only upon the Firm's professional excellence and cost-effectiveness, but also on its commitment to its clients' long-term welfare and business success. The Firm strives to understand its clients' goals and assist them in reaching these goals with a minimum of legal cost and complications. Phillips McFall knows the Firm's success and future depend on the continued success and growth of its clients.

"Another reason for the Firm's success is that we are always open to new ideas and their implementation in our practice. Many of those ideas come from our clients, Firm attorneys, and staff. We are also very involved in the community in which we practice, and encourage our attorneys and staff to participate in and support numerous community development projects and activities."

Phillips McFall is a law firm committed to its clients and to maintaining the quality and professional excellence they deserve. "We are very excited about the future of the State of Oklahoma and look forward to continuing to provide quality legal services to the Oklahoma business community," says McFall. ✪

Attorneys and Staff of Phillips McFall McCaffrey McVay & Murrah, P.C.

Directors of Phillips McFall.

Speck, Philbin, Fleig, Trudgeon & Lutz

John Philbin is President of the Oklahoma City All Sports Association that sponsors Aerospace America each year. This air show brings over 100,000 visitors to the city each year. All Sports also sponsors the All College Basketball Tournament, NCAA Sub-Regional Basketball Tournaments, the College Softball World Series, Big XII Conference Baseball and Softball Tournaments, and many other events.

The lawyers of Speck, Philbin, Fleig, Trudgeon & Lutz practice what they preach. A boutique firm, the practice is restricted to areas that touch upon taxation. Since the firm professionally encourages clients to be involved in the Oklahoma City community through time, support, and leadership, members of the firm try to apply and carry out these same objectives themselves.

The goal of Speck, Philbin, Fleig, Trudgeon & Lutz is to do the limited things it does with a high degree of professionalism. The firm's view is to assist clients in their financial dealings and planning by helping them get the most out of every dollar—a focus created when the firm was established.

The firm was founded by John K. Speck in the late 1940s to practice primarily in the area of tax law. It still concentrates its practice in the tax-related areas of personal, business, and tax planning for professionals, owners of closely held businesses, and highly compensated individuals, in addition to the family law areas of estate planning, probate, trust and estate administration, and litigation.

The Speck, Philbin, Fleig, Trudgeon & Lutz practice also involves real estate law and tax and organizational planning for tax-exempt organizations, private foundations, and governmental units. The firm sponsors and implements, for its clients, prototype and individually designed defined contribution retirement plans as well as other plans for employee benefits and related services for tax reporting.

Its six lawyers, including one certified public accountant, believe strongly in citizen involvement as a part of the community infrastructure. They support with their own efforts, services provided by nonprofit and charitable entities within the Oklahoma City area.

Current named principals of the firm are strongly involved in activities that further the goals of the community.

John Philbin is president of the All Sports Association, a post he has held for many years. This association sponsors numerous sporting events throughout the year, including the annual Aerospace America air show and All College

Speck Philbin attorney, Jon Trudgeon, working with Nancy Anthony, Executive Director of the Oklahoma City Community Foundation. The Foundation, with assets exceeding $152,000,000., offers firm clients opportunities to accomplish their charitable and financial planning goals as well as supporting community nonprofit organizations.

Basketball Tournament, which is the oldest basketball tournament in the country.

As a past president for the local board and past national board member, Jon Trudgeon is heavily involved in the Oklahoma Center for Non-Profit Management, an organization affiliated with the Support Centers of America. This organization provides management assistance and training for the staffs, boards, and volunteers of nonprofit and service organizations. It provides this training to assure that the charitable component of the community is able to get the best value for the dollars available to fulfill its objectives.

Charles Lutz is congregational coordinator for Stephen's Ministry. This program ministers to the personal needs of individuals experiencing difficulties as a result of illness, substance abuse, financial misfortune, and other adversities. Additionally, he serves the community as the municipal judge of Nichols Hills, Oklahoma.

Over the years, the legal and tax framework has grown increasingly complex. Lives, businesses, and relationships are affected by a number of different legal concepts. Speck, Philbin, Fleig, Trudgeon & Lutz analyzes needs, explains options, and develops plans and solutions to legal and tax planning problems. Through this process, the firm and its clients contribute to the betterment of the community. ✸

Grant Thornton LLP

Grant Thornton's professional and industry-specific office in Oklahoma City is distinctive in its service and dedicated to its work. The pioneer spirit that founded this accounting and management consulting firm is taking Oklahoma's middle market companies into the next century through consultation, strategic planning, and technological needs.

Partners Tom R. Gray, III, managing partner; Richard D. Winzeler, tax department head; and Ted R. North, audit department head, lead a staff of 45 people in a service- oriented environment respectful of clients and staff. It offers a full range of services with an emphasis on assurance, tax, and consulting.

"Typically we become business advisors to our clients and really assist them, not only in their normal accounting and tax issues, but also with strategic planning."

On an international scale, Grant Thornton LLP, headquartered in Chicago, provides assurance, tax, and management consulting services to businesses throughout 48 U.S. offices. With $266 million in revenues, Grant Thornton is the seventh-largest accounting and management consulting firm in the country. Worldwide, clients are served through Grant Thornton International, with more than 500 offices in over 85 countries. The companywide mission is to be the preeminent firm serving the middle market.

"Oklahoma City is very much a business community of middle market companies. While we provide assurance, tax, and consulting services to numerous publicly held companies, the crux of our practice is dealing with entrepreneurial, owner-managed businesses," said Gray. "We have strategically placed ourselves as a viable alternative to the giant six accounting firms."

Grant Thornton's emphasis is to understand the client's business, be responsive and accessible. In return, the client receives a high level of partner involvement, national resources associated with the firm's industry specializations, and resources associated with the largest tax department in Oklahoma City. Gray indicated the taxation hot areas are succession planning and state and local taxation (SALT). "Managing state and local tax issues is becoming more and more critical, and basically,

(l to r) Richard D. Winzeler, Eddy R. Ditzler, Ted R. North, Tom R. Gray,lll.

Gary J. Voth, Linda K. Hildebrant

we have put together a group of state and local tax professionals to handle these issues for our clients," Gray continues.

The Grant Thornton practice is reflective of the business community. Its areas of specialization are manufacturing and distribution, financial institutions, real estate, exempt organizations, oil and gas, and retail. "During the 1980s, we saw

our practice evolve as a result in the downturn in the oil and gas economy," says Gray. "Our practice now reflects the transition of the overall Oklahoma City economy to one that is more diversified."

Clients range from family-owned grocery stores to multinational manufacturing and retail concerns.

"Our clients typically have the same philosophy we do—strategic growth and profitability with that pioneering spirit," said Gray.

Along with the practice reflecting the community, Grant Thornton is heavily involved in the community. "We have always been extremely involved in the community, and we actively participate in such activities as Leadership Oklahoma City, Oklahoma City Chamber of Commerce, Children's Medical Research, and various public school foundations, among many others. That's just our way of making sure we give back to the community."

International in scope, but local in scale and personalized attention, Grant Thornton LLP is the only major international accounting firm organized and dedicated to serving midsized, growing companies. Clients recognize this as the "Grant Thornton Difference." ✸

James Farris Associates

President James Farris

James Farris Associates (JFA) is a full-service human resources consulting firm specializing in the three primary areas of executive search, outplacement, and human resource management consultation. Through a highly knowledgeable, personalized, cost-effective, and superior service, JFA develops an understanding of clients and their organizations to become an extension of the staff.

"We are in the relationship building business," said James W. Farris, President. "Our goal is to satisfy our clients with an ongoing human resources consulting relationship rather than a one-time project."

The outcomes of this attitude are demonstrated by the fact that more than 70 percent of JFA's engagements represent repeat business, and a very substantial portion of the new business comes through referrals from former or current clients.

As one of the major firms conducting human resources work in the region and a major provider nationally, JFA works with small independent companies to large international corporations. The company's search work for key positions ranges from management level through top executives.

With the goal of identifying and evaluating executive managers for placement in positions for client companies, JFA differentiates itself through five founding elements of success. These elements incorporate experience from both sides of the desk, superior quality, personalized approach, comprehensive service, and involvement.

Client satisfaction is one of the company's principal goals when it comes to the executive search process. JFA's search completion rate ranks among the highest in the industry.

James Farris Associates also is committed to performing personalized outplacement services for every client in a professional and confidential manner. From resume preparation and interviewing skills to building up self confidence, JFA gives clients the tools to successfully compete in the employment marketplace. Farris indicates that 85 percent of the people JFA has worked with in outplacement have "come out ahead."

Whether it is employee handbooks, teaching sexual harassment seminars, showing managers how to interview, or helping with federal

Staff of James Farris Associates (standing l to r) Susan Litchfield, office manager; James Farris, President; (seated l to r) Jean Ashworth, Associate; Jerry Hoke, Vice President.

regulation compliance, JFA serves client organizations through human resources consulting.

JFA helps client organizations in assessment, development, and implementation of human resources programs, policies, and actions necessary to accomplish strategic and operational objectives. The company is committed to developing practical solutions to personnel and communication issues that confront client organizations in an objective, independent, and experienced manner.

Although the business of James Farris Associates has a national scope, owner James Farris is strongly dedicated to the company's location in Oklahoma City because of the quality of life, ability to be involved in family activities, and location in the center of the United States. He is also impressed by the city's willingness to accept new ideas and people.

"Friendliness, openness, and acceptance are traits found in Oklahoma City that you do not find other places," said Farris. "I enjoy going to Rotary and being involved with such a heterogenous mix. As a small businessperson, I might be sharing a table with a former governor, president of an international company, doctor, lawyer, and another small businessperson."

James Farris Associates assists businesses in the total continuum of human resources needs—from hiring managers to outplacement and consulting on all aspects of personnel administration. By building relationships and offering the highest-quality work, JFA ensures that nothing less than a client's full satisfaction is acceptable. ✪

Coon Engineering, Inc.

Coon Engineering's signature is on projects across central Oklahoma. Participating in the design of numerous landmark ventures through engineering, planning, and land surveying, this multigenerational firm has established standards of practice that serve as benchmarks for the industry.

From engineering design, surveying, and construction management of residential, commercial, municipal infrastructure, and industrial projects, this engineering company strives to offer the best service available.

Coon Engineering, Inc. specializes in all aspects of planning and engineering design for development of residential, commercial, and industrial projects. The firm brings a vast knowledge of local and state codes and practices to the design of earthwork, streets, parking, utilities, lakes, and other features. The company's goal is to provide economical and complete design services. Its size ensures all projects are given personal attention.

The firm offers a complete range of land-planning services, including preliminary investigation, feasibility reports, land use planning, site planning, and cost estimates. Coon Engineering

Coon Engineering, Inc. has been involved in some of the finest golf course communities in central Oklahoma. Photo by David G. Fitzgerald.

Coon Engineering strives to offer the best service available.

has a consistent history of providing prompt, accurate, and reliable service to its growing list of land development, architectural, engineering, and governmental clients.

This industrious engineering firm maintains its own survey crews outfitted with the latest available equipment. The firm's experienced crews can handle all phases of development—from digital mapping and boundary surveys to construction staking.

Coon Engineering has multiple crew capability when it comes to subcentimeter accurate global positioning system (GPS) surveys. These

capabilities complement the engineering services and provide clients with turnkey services from the start of a project to the completion of construction.

This company is a pioneer of new technology and new ideas in land planning—a trend established with Edward M. Coon. He began a noteworthy career in civil engineering in 1948, gaining years of experience that molded him into one of the state's most industrious engineers. Coon and his son Bryan, professional engineer and registered land surveyor, combined their experience and knowledge to establish Coon Engineering, Inc. in 1984.

Respected in the community as a company of integrity and responsiveness, the firm's name is attached to most of the exclusive neighborhoods, golf courses, apartments, and shopping centers in the city, along with community projects such as the Oklahoma City Zoo Cat Forest.

"We are proud of our role in helping with the development and creation of Oklahoma City as one of the finest cities in the United States in which to live and work," said Bryan E. Coon, President. "It is our professional endeavor to help create the finest community and our corporate responsibility to support the community."

Bryan Coon is a member of the Planning Commission's Urban Development Committee. Coon, like his father before him, is heavily involved in reviewing and creating new ordinances that affect the future development of the city. On a regular basis, he also represents clients before the City Council as an expert regarding planning and zoning matters.

Striving consistently to give the best service is a trademark of Coon Engineering, Inc. Its standards of practice used in engineering, planning, and land surveying for the development of projects serve as benchmarks for the industry and assure the growth of a progressive Oklahoma City. ✪

VanStavern Design Group, Inc.

From television broadcast facilities across the nation, to a range of state and local projects, the guiding focus behind every VanStavern Design Group, Inc. project is to create nurturing, responsive commercial interior spaces that meet the everyday needs of their clients.

"As interior designers, we create commercial environments to empower people to be the best they can be," says Vicki VanStavern, owner of the firm. "I believe the services we provide have as much impact on our client's well-being, competence, and self-esteem as any other factor in their lives. As we bring beauty and function into the workplace, we impart the belief that every person is valuable and deserving."

The VanStavern Design Group's primary clients are architectural firms. In fact, several Oklahoma City architectural firms have given this group the ultimate compliment and hired them to do the interior design work for their own offices.

The interior designers approach each project in concert with the architect's approach to the project phases. This team effort with the architect and other consultants results in careful attention to desired concepts, functional requirements, and the image the client wishes to portray.

Since the firm's inception in 1984, the professionals of VanStavern Design Group have taken their work very seriously. "We take on quite large projects," VanStavern adds, "which means we are responsible for substantial budgets and time commitments— that's not something we take lightly."

The firm provides comprehensive interior design services that include space planning, lighting design, custom millwork and furniture, specifications for art, accessories, and furniture, as well as special architectural features incorporated into the floors, walls, and ceilings.

The firm's project list clearly describes its success: repeat and referral work make up a majority of the business. This extensive and varied roster includes corporate offices, healthcare, education, and hospitality installations throughout the country. The Francis Tuttle Vo-Tech Center in Oklahoma City and the Veteran's Administration Long Term Care Facility in Norman are two recognizable projects in the metropolitan Oklahoma City area. Other works-in-progress include the Oklahoma City Zoo's restaurant and Oklahoma Christian University's Student Activity Center.

"We listen carefully to client ideas and

Among the firm's recent projects is Black Entertainment Television's 90,000 square foot corporate headquarters in Washington, D.C. The designers were responsible for all interior details of the project including custom designed china, crystal and flatware.

requirements and then translate these into spaces that clearly reflect their needs," VanStavern says. "The proper functioning of interior spaces, ease of long-term facility management, as well as aesthetic concerns are our directing forces. We have won ten American Society of Interior Designers' Excellence in Design awards in the past six years, indicating our commitment to performing quality work."

VanStavern Design Group has found success through partnership, quality, and meeting the functional needs of its clients. By matching these comprehensive traits with the bottom line of completing projects on time and in budget, the VanStavern Design Group has built a reputation and impressive portfolio of projects that reflect one of Oklahoma's most accomplished interior design firms. ✪

VanStavern Design Group collaborated with Rees Associates in Oklahoma City to design the interior of the architects own office.

Rees Associates, Inc.

Rees Associates, Inc. is a name in architecture that reflects an exclusive approach, international success, national recognition, and client satisfaction in architectural, planning and interior design services. Since 1975, the firm has completed more than 1,000 assignments in 37 states and 10 foreign countries.

Rees is a different type of firm because of a unique process. Founder and President, Frank W. Rees, Jr., AIA, enhanced the firm's architectural strategic master plan process while completing graduate studies at the Harvard Business School. The outcome: an original approach for programming and planning facilities.

This exclusive FACILITY BUSINESS PLAN® is so unique that it holds a registered trademark and copyright granted by the U.S. Patent Office. Utilization of this approach frequently results in completed facilities that operate at the utmost level of operational efficiency.

"We have developed an approach to working with people. It enhances clients' abilities in such a way that the result is they become better than they thought they could be at a chosen activity," Rees says. "The result is people hire us regularly or tell their friends about us."

Rees works for leading companies, government, and service organizations throughout the world. Areas of specialty include broadcast, healthcare, retirement, corrections, office buildings, and educational facilities.

The Rees "difference" is noted through its accumulation of impressive, high profile projects. It has provided architectural and engineering design services for more television facilities than any other architectural firm in the world. Rees is ranked among the top firms in the country in healthcare facilities and has provided healthcare planning and consulting services for the majority of the healthcare facilities in Oklahoma. It has also been listed as one of the nation's top four firms in the design of senior living and long-term care facilities.

The company has won national recognition on the design of its projects, including Gannett

Black Entertainment Television Corporate Headquarters, Washington, D.C.—First Place Corporate Category, 1996, American Society of Interior Designers. Photo by Andrew Lautman.

Company, Inc., Washington, D.C.; National Broadcasting Company, Chicago; Black Entertainment Television, Washington, D.C.; and the IRS Office Building, Federal Aviation Administration VOR/TACAN/DME facility, and Presbyterian Hospital, Oklahoma City.

Comprising more than 70 personnel in Oklahoma City and Dallas offices, the company is proud of its Oklahoma City roots. C. Leroy James, AIA, is the executive vice president and head of the Oklahoma City office.

From the Women and Children's Hospital in Chimkent, Kazakhastan, to neighborhood projects in Oklahoma City, Rees Associates, Inc. clearly offers a unique approach to architectural, planning, and interior design. Rees listens, understands, plans, designs, and solves to create an environment for clients to produce at their best. ✪

(above) Internal Revenue Service Building, Oklahoma City, OK—Special Recognition for Creative Design Utilizing Precast & Prestressed Contrete, 1993, Precast & Prestressed Concrete Institute. Photo by Michelle Wurth.

(left) Oklahoma Veterans Center—Norman Division, Norman, OK— 301-bed replacement hospital for veterans. Photo by Michelle Wurth.

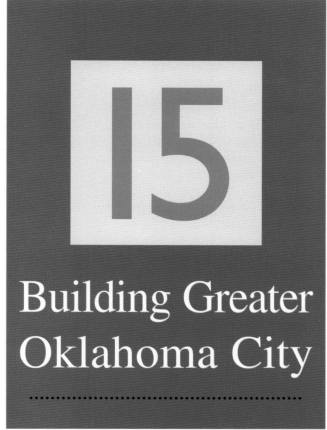

15

Building Greater Oklahoma City

*Photo by Fred Marvel, Oklahoma Tourism
and Recreation Department.*

Lippert Bros., Inc.

Strong structural foundations and buildings parallel a strong family foundation and growth at Lippert Bros, Inc. Photographs of completed projects and family members line the walls of the Lippert Bros, Inc. office, showcasing the quality, diversity, and longevity of this general contracting and utilities firm.

Since 1920, Lippert Bros. has been a driving force in the construction business. With customer satisfaction and investment in people as points of significant focus, the company has been able to successfully complete more than 77 years of a variety of projects.

Lippert Bros., Inc. is dedicated to providing professional construction services responsive to the needs of commercial and industrial markets, utilizing quality materials and craftsmanship. A current emphasis of building commercial, industrial, and government projects helps the firm stay competitive and shows the evolution of a company that started with just skill and desire.

BROTHERS REPRESENT QUALITY

Two brothers, Erick W. and Walter H. Lippert, founded Lippert Bros., Inc. in Boone, Iowa. As brick masons, their first jobs encompassed building fireplaces, pouring sidewalks, driveways, and residential foundations.

The range of construction work gradually increased, so three other Lippert brothers joined the company—Lewis T., Reuben C., and Leo J. Four brothers were trained masons and one brother was a carpenter.

The remodeling of storefronts and buildings, along with the addition to Boone High School, took the company into more large-scale construction and diversification into sewage- and water-treatment plants.

At the outbreak of World War II, the Lipperts built the Camp McCoy, Wisconsin, sewage-treatment plant. This expertise and recognized excellence led to a successful bid on the sewage- and water-treatment system at the new Tinker Air Force Base near Oklahoma City, in addition to construction of numerous military training command projects throughout the Southwest.

These projects guided the brothers to open an Oklahoma City office. When Donald E. and Robert L., sons of Erick, returned from serving

Myriad Gardens Crystal Bridge

Guaranty Bank-Memorial Branch

in the Navy during the war and completed their college education, they joined their father in a permanent move to Oklahoma.

Lippert Bros., Inc. was incorporated in 1947 and became a distinct entity, separate from the Iowa partnership. The firm has since made its mark on the Oklahoma City landscape and the region of Oklahoma, Texas, New Mexico, Arkansas, and Colorado.

DIVERSIFICATION INTO COMMERCIAL AND PUBLIC WORKS

With two main areas of emphasis—a general contracting business and specialized public works projects—the company keeps its broad base while diversifying into its specialty area of wastewater, water, and power plants.

Water Treatment projects include a 30-million-gallon-per day addition to the Lake Hefner Water Treatment Plant in Oklahoma City, $18 million Arcadia Lake Water Treatment Plant, and the $5 million Midwest City plant. Waste Water Treatment plants encompass the $5 million Mustang Waste Water Treatment Plant, $5.8 million Del City plant, and $11 million Cow Creek plant in Oklahoma City, among others across the state.

In regard to commercial and industrial projects, unique and high-profile projects for the firm are the original National Cowboy Hall of Fame and Western Heritage Center, Myriad Gardens Crystal Bridge, Citizens Bank Tower, First Presbyterian Church, and the award-winning renovation of Old Central High School into One Bell Central for Southwestern Bell.

The Cathedral Of Our Lady Of Perpetual Help

Noble Research Center/Oklahoma State University, Stillwater

renovation of existing facilities—perfecting the largest aspect to the finest detail. Renovation projects include the Cathedral of Our Lady of Perpetual Help, First Baptist Church, Federal Reserve Bank of Kansas City—Oklahoma City branch, and grocery stores.

"We are always sensitive to the client's needs," says Rick Lippert, president. "We make sure we understand each client and their operations prior to starting on a project. Whether it is a grocery store open 24 hours a day, or a church conducting weddings and funerals, our work environment has to be flexible within these conditions."

WORK FORCE AND AWARDS

Don E. Lippert, chief executive officer, is semiretired and his son, D.E. "Rick" Lippert, is currently president of Lippert Bros., Inc. Other key officers and third-generation Lipperts are Bruce Barta, vice president; Thomas M. Lippert, vice president, and Joel J. Lippert, estimator/project manager. John Lippert is president of Globe Construction, a Lippert subsidiary company.

"We find and try to keep the best people working for us," says Rick Lippert. "Our company has quality expectations, and these expectations cannot be met without the proper work force. We have had some people working for us for 35 years, as well as a handful of second-generation employees."

Completing between $25 million and $35 million worth of construction projects each year, the Lippert Bros., Inc. employee nucleus is 60 to 70 people and expands as needed. Weekly safety talks on the job site and Oklahoma's strong work force contribute to the company's ability to excel in many areas—including safety.

Lippert Bros., Inc. received the Oklahoma Municipal Contractors Association's First Place Safety Award and is a three-time winner of the Associated General Contractors of America's Safety Award. The American Institute of Architects, Central Oklahoma Chapter, presented Lippert Bros., Inc. the Outstanding Contractor Award.

A history of the company's logos spanning 75 years hangs sequentially on the Lippert Bros., Inc. office walls. The current red, black, and white Lippert Bros., Inc. logo seen boldly on signs at a construction site symbolizes a quality contracting and utilities firm that takes pride in its work and heritage, and strongly believes in continuing to build a three-generational company and Oklahoma City. ✱

With a commercial specialty of cast-in-place concrete work, Lippert Bros., Inc. used nearly 19,000 cubic yards of concrete in the construction of Remington Park's Grandstand Superstructure.

A variety of other projects includes more than a million square feet for the Hobby Lobby corporate offices and distribution center, the aquatic facility at Oklahoma City Community College featuring a 50-meter Olympic lap pool with separate diving tank, *Wall Street Journal* Printing facility, and the Oklahoma County Juvenile Center.

Lippert Bros., Inc.'s proven track record with health care facilities includes the State of Oklahoma Department of Health building, Children's Hospital, Bone and Joint Hospital, Canterbury Retirement Community, Reynolds Children's Center, and Mercy Neuroscience Center. The firm has also made its mark on Oklahoma City's broadcasting community through the specialized construction of KAUT-TV, KOCO-TV, KWTV, and KTVY.

Educational clients in Oklahoma benefiting from the expertise of Lippert Bros., Inc. include the Oklahoma City Public Schools, Oklahoma City University, Oklahoma State University, University of Oklahoma, Oklahoma Christian College, Rose State College, University of Central Oklahoma, and the Oklahoma Education Association.

The company is involved in construction for the Metropolitan Area Projects (MAPS) with improvements to the Norick Arena at State Fair Park.

Company experience also consists of the

Precor Realty Advisors, Inc.

Garage Loft Apartments, Oklahoma City

Precor Realty Advisors, Inc. specializes in *The Business of Real Estate*. Strength of capital, professionalism, and expertise combine to offer a comprehensive approach and a strategy that makes it easy for clients to maneuver within the world of commercial real estate. Precor's service divisions include real estate development, brokerage, leasing, and asset management. Brought together with innovative ideas and hard work, these capabilities allow Precor Realty Advisors to move quickly and be responsive.

"We are small enough to be agile and large enough to deliver," says Nicholas J. Preftakes, president.

Preftakes founded the company in 1989, which now employs 125 people, and takes pride in the company's ability to offer help and make decisions. "The quality of our people, our knowledge of the market, and our track record are vital to making sure a project becomes a reality," says Preftakes.

PROJECT DEVELOPMENT

Joining expertise with capital, Precor is active in the development of new projects, as well as in the acquisition and redevelopment of existing assets. For the business owner who would rather lease than own, but cannot find a suitable facility, or the landowner seeking a developer/partner, Precor has the financial resources necessary to successfully complete build-to-suits for lease, sale/leasebacks, or joint venture projects. Real estate development and investment clients are briefed on what is

available for ownership and building options and can be as involved as they choose to be in site selection and project development.

REAL ESTATE BROKERAGE SERVICES

Precor's real estate professionals closely monitor market trends and movements in Oklahoma City and are thoroughly prepared to assist in leasing, purchasing, or selling commercial sites and buildings. The client list ranges from local business owners who have outgrown their existing facilities, to regional investors looking for commercial properties with promise, to national retailers with plans for several new stores.

Regardless of the challenge, Precor's staff thoroughly analyzes each situation and recommends those properties that match the needs of the investor or user. Whether it is the acquisition of an existing property, or the purchase of a site and the construction of a new building, Precor helps determine the feasibility of a project and follows through until the transaction is completed.

ASSET MANAGEMENT

Asset management is the important task of obtaining the highest return possible on a client's real estate investments. Precor is an aggressive asset manager, determined to increase the cash flow and property value of the projects it oversees.

Whether advising an entrepreneur contemplating the purchase of a new building, or an institutional investor with a portfolio of commercial

Loft Offices, 825 North Broadway, Oklahoma City

property, Precor Realty Advisors applies the same basic principles of success: buy smart, make improvements consistent with market values, manage the property efficiently and economically, and keep both short and long-term objectives in mind.

PRECOR-OWNED PROPERTIES

Precor Realty Advisors makes direct real estate investments in Oklahoma and surrounding states. Its diverse portfolio of properties has grown to include office buildings, apartment communities, retail shopping centers, land developments, motels, service centers, and warehouse facilities.

"We invest in opportunities and the community," says Preftakes. "As a well-capitalized company, we offer flexibility, stability, and the ability to take advantage of those opportunities."

"All of our properties are investments," Preftakes adds. "Every piece of our portfolio is an investment of capital, resources, time, and expertise for the company and our clients."

The company has recently become involved in historical renovation for commercial and residential use. The most visible example of this historical renovation is The Garage Loft Apartments.

In 1930, Norton-Johnson Buick Company built its new Buick and Marquette auto dealership in downtown Oklahoma City. This art deco, registered historical landmark building is currently the home of Oklahoma City's first loft-style apartments. A building that sat empty for many years is now filled to capacity with Oklahoma City residents.

Preftakes indicated the company will continue its efforts for multifamily and commercial development in the downtown and near downtown area. "We see this as a continuing effort and see a great deal of opportunity," Preftakes says.

THE REWARDS

"Opportunities are available with the strength of capital, and there are risks and rewards attached to these investments," says Preftakes. "The knowledge we have as residents of the community is invaluable to investors. The rewards can be both financial and personal."

Preftakes explains that community involvement

Edmond Plaza Shopping Center

Comfort Inn Historic Route 66, Oklahoma City

Pioneer Chemical, Inc., Oklahoma City/Distribution Center (under construction)

is an attitude that is contagious in the Precor office. Through financial contributions and time, Precor Realty Advisors' employees are encouraged to be involved in civic, artistic, and charitable organizations.

The support and encouragement to be involved starts from the top. "I have to be willing to give people time for civic commitments," says Preftakes. "They can't do it all after 5 p.m."

THE OBJECTIVE

Through an understanding of a variety of product types, the ability to deal with complex transactions, the financial capacity to provide the capital necessary for development, and staff expertise, Precor Realty Advisors, Inc. is a driving force in Oklahoma City's commercial real estate market. Recognized for its innovative ideas and comprehensive approach to real estate, Precor Realty Advisors offers the resources, experience, and determination necessary to assure that clients succeed in *The Business of Real Estate.* ✪

Warwick West Apartments, Oklahoma City

Gerald L. Gamble Co.

Gerald L. Gamble Co., one of Oklahoma's leading commercial and industrial real estate companies, has built a solid reputation of competency, expertise, and integrity. This reputation, as well as community involvement, has made the Gerald L. Gamble Co. name one of "the most trusted in real estate."

The Gerald L. Gamble Co. specializes in brokerage of warehouses and manufacturing facilities, shopping centers, apartments, offices, and land for industry and commercial activities. As commercial, industrial, and investment realtors for over 35 years, Gamble and his associates are on the leading edge of economic trends. Their work also allows involvement in professional and civic responsibilities in the economic development activities of Oklahoma City.

The team at Gamble Co. works with many major businesses to establish new and expanding facilities for industry and bring additional jobs to Oklahoma City. Gamble and his associates served as brokers for many of the major commercial properties in Oklahoma City, such as the Waterford Office Complex, Colcord Building, Bryant Square Shopping Center, Prudential Warehouse, Meridian Business Park, Northpoint Commerce Center, plus numerous other investment properties.

Industrial clients for the company include Baker Hughes, Inc., McKesson Corp., Santa Fe/Burlington Northern Railroad Co., United States Government (Tinker Air Force Base), PepsiCo, The Oklahoma Publishing Co., Little Giant Pump, and many others.

The impressive list of accomplishments in the area of industrial and commercial brokerage proves Gamble Co. is a specialized organization committed to excellence. Every detail involved in the transfer of commercial and industrial real estate can be professionally handled by the Gamble Co. team.

By design, the company is a small, yet strong, team with a total of four brokers and a three-person administrative staff. This close-knit, highly

Gerald L. Gamble Co. has built a solid reputation of competency, expertise, and integrity. (standing left to right) Joe Maxey, Mark Patton, Gerald Gamble, and Jim Buchanan represent the leadership for Gerald L. Gamble Co.

competent, and experienced group shows its expertise through professional and volunteer achievements. There is hardly an honor in the real estate community that has not been accorded Gamble or one of his associates. Among the many professional memberships and honors are: President of the Oklahoma City Association of Realtors and the Oklahoma Association of Realtors, Oklahoma Realtor of the Year, and membership in the Society of Industrial and Office Realtors, and American Society of Real Estate Counselors.

Gamble's enthusiasm and unwavering commitment are the driving forces behind his 35-year history of excellence in the real estate profession.

The Gamble Co.'s brokerage team includes Gamble, Mark Patton, James Buchanan, and Joe Maxey. Among the four, there are two law degrees, two MBAs from Harvard Business School, a master's degree from Stanford, and undergraduate degrees from the University of Oklahoma, Oklahoma State University, and Harvard.

In addition to business commitments, the Gamble team is involved on a volunteer basis as a board member or president in a number of civic and charitable organizations including: Last Frontier Council Boy Scouts of America, Food Bank, United Way, Oklahoma Wildlife Commission, as well as local universities and churches. It is no secret that the Gamble Co. assists the community with enthusiasm and unwavering commitment. For example, Gamble served as the 1997 Chairman of the Greater Oklahoma City Chamber of Commerce.

The Gamble Co.'s progress over a third of a century is not the result of professional skill and hard work alone, but is primarily from the confidence of its faithful clients and the reputation attained in the community. Whether it is real estate needs or civic duties, Gerald L. Gamble Co. is a name on which the Oklahoma City community can rely. ✪

Trammell Crow Company

With a vision to be the dominant full-service real estate and investment company in the industry, the Trammell Crow Company is providing service and real estate solutions for Oklahoma City. On the national level, Trammell Crow Company is the country's largest full-service real estate firm providing a wide range of services to owners and users of real estate. It has earned the distinction as the nation's largest property manager for more than a decade, managing assets covering all commercial product types.

Celebrating its 50th anniversary in 1998, the company was founded by Trammell Crow. He is still active in the company at more than 80 years of age, and the roots of his basic customer-focused philosophy continue to be a foundation and perpetuating force for more than 70 markets across the country.

Managing property is one of Trammell Crow's core businesses. It currently delivers facility, property, and construction management services to 400 satisfied tenants, along with the leasing/marketing of over 50 commercial buildings.

As a customer-driven company, Trammell Crow is responsive to their clients and the needs of the market. The company takes pride in its high percentage of customer retention ratings within its client portfolio. "From the maintenance technician to the managing director, we are taught to be attentive and responsive to current needs as well as anticipate our customers' future needs," says Alfred C. Branch, managing director.

Brokerage Services is a growing area of the

The OKC office began operations in 1969 and has grown to include more than 100 employees servicing over 5.5 million square feet of space.

facilities, including corporate headquarters for both locally and nationally-based companies.

The development aspect of Trammell Crow moves with the Oklahoma City economy. "Clearly the expectation is that Oklahoma City is moving forward, and our company will be developing office, industrial, and retail facilities to accommodate that growth," says Branch.

Trammell Crow employees give back to the community that has given to them for nearly 30 years. Along with involvement in numerous community programs, the company's efforts for the past several years include the annual Christmas in April project. Through supply of manpower and expertise, the home of an elderly, underprivileged Oklahoma City resident is totally revamped in a single day. This includes painting, electricity and plumbing, ramping, and other necessities.

By being people-focused, visionary and solution-oriented, Trammell Crow has developed a commanding position in the Oklahoma City real estate market and earned the reputation of delivering high-quality real estate services to a diverse list of loyal, service-oriented customers. The trend will continue. "We are positioned and prepared to deliver real estate services to our market and meet the needs of our customers, which will come as a result of a growing Oklahoma City economy," Branch concludes. ✪

The professionally trained staff with long standing experience in this market offers a variety of leasing, development, property management, brokerage, and consulting services.

The Oklahoma City office began operations in 1969, with 250 acres in the growing Southwest/airport submarket. Since 1987, Trammell Crow has developed from 20 employees managing 2 million square feet of space into more than 100 employees with 5.5 million square feet of space.

The professionally trained staff with long standing experience in this market offers a variety of leasing, development, property management, brokerage, and consulting services.

Trammell Crow business. For a company with multiple locations, the company's national web of 35 offices delivers the same high-quality services in all cities. Locally, Trammell Crow brokers stay attuned to their Oklahoma City customers' growing needs, which may include land for expansion, purchase or lease of property, and build to suit.

In addition to the traditional real estate services, Trammell Crow Company sells its expertise in facilities management to operate corporate-owned

Needham Re-Roofing, Inc.

One satisfied corporate customer after another ensured a stellar launching for Needham Re-Roofing Corporation's first year of business in Oklahoma City. Sales topped $1 million during a nine month period as the company quickly developed an impressive client base by consistently solving customers' roofing construction challenges with timely, innovative solutions using the latest industry techniques.

"We have brought to Oklahoma City the guiding principles Joe Needham has followed since he began Needham Re-Roofing Corporation in the Dallas-Ft. Worth area in 1969," says Deborah Marshall, division manager. As the new division manager, Marshall welcomed the opportunity to return to Oklahoma City. Since she started the office with Debra Penrod, marketing manager, during the summer of 1996, the Oklahoma City staff has expanded to more than 20 employees. "As women, we were determined to demonstrate our knowledge and expertise in the construction industry," Penrod states.

Startup clients include the downtown Liberty Bank, Will Rogers World Airport, and Wiley Post Airport. Needham also secured the City of Oklahoma City contract to supply its roofing materials.

The roofing company's professional, expedient approach has produced continual customer satisfaction. Written testimony from Liberty Bank for Needham Re-Roofing Corporation detailed how fortunate the bank felt to find the roofing company. Liberty experienced minimal interruptions in its daily operations as the resurfacing occurred. Needham was applauded for its ability to deliver what it had promised.

"We don't want to leave a stone unturned," Joe Needham says. "We want customers to know exactly what to anticipate from every aspect of the job."

Joe Needham founded a company that has survived astronomical odds to become one of the most sought-after roofing contractors in Texas. Needham Re-Roofing Corporation's history of success can be traced to its customer service philosophy.

A problem-solving approach assures all needs and concerns are continuously addressed. Special attention is given to the details to take the guess-

From Texas to Oklahoma, Needham Re-Roofing Corporation strives to be the very best. Lone Star Park Race Track in Grand Prairie, Texas is one of Needham's most recent accomplishments.

work out of bidding. Every cost is individually itemized, so the client knows exactly what to expect. "We provide customers with a choice of roofing options, as well as a written schedule of when services will be delivered. We conduct business with a sense of urgency," Marshall explains. "Clients can also expect progress and accountability meetings throughout construction."

Attributes of the Needham Re-Roofing Corporation's customer service approach are now expanded into the Oklahoma City market. "We saw a great opportunity in Oklahoma City with exceptional growth potential throughout the state," Needham comments.

As an industry leader, Needham is frequently ferreting the latest cutting-edge techniques. An outstanding example of this farsighted approach is the exquisite detail on the Lone Star Race Track in Grand Prairie, Texas, and the Dallas County Community College District which now boasts the largest rhine zinc standing seam roof in North America.

Professional integrity and dependable follow-through are the focus for every commercial and residential job Needham Re-Roofing Corporation undertakes. Needham is so confident of its work, it offers the only available Roofing Labor Warranty Package at no additional charge. "We have been known to be out in the rain on the roof doing whatever possible to protect our customer's interior investments. We are committed to delivering the best roofing surface in the most timely, economical manner possible," Marshall concludes. ✪

The thirty-eight story Liberty Tower is our most recent re-roof completion in Oklahoma City.

Photo by Joe Ownbey.

16

Health, Education & Quality of Life

Photo by Fred Marvel, Oklahoma Tourism and Recreation Department.

St. Anthony Hospital

Today, St. Anthony Hospital stands not alone, but rather as a pace-setting partner in a regional network of health care providers. Locally, St. Anthony Hospital collaborates as a partner in the SSM Health Care System (SSMHCS) to deliver care not through one hospital, but several. The Oklahoma network, SSM Healthcare of Oklahoma, includes St. Anthony Hospital, Bone & Joint Hospital, Hillcrest Health Center, Mission Hill Hospital in Shawnee, and Healthcare Systems of Oklahoma. They are affiliated through and governed by one board of trustees.

St. Anthony is also a member of the SSM Health Care System headquartered in St. Louis. The system is sponsored by the Franciscan Sisters of Mary and is one of the largest Catholic health care systems in the United States with more than 4,000 beds.

THE BEGINNING

As the first hospital in Oklahoma City, St. Anthony continues the commitment of its founders to serve those most in need, individually and respectfully, with quality, care, and utmost compassion. On the eve of its centennial celebration, St. Anthony is a regional health care provider with a 664-bed facility offering a continuum of care.

St. Anthony Hospital was founded by a small order of Franciscan sisters who originally journeyed to Oklahoma City in the spring of 1898 to ask for contributions for St. Joseph's Hospital in Maryville, Missouri. When seeking permission to ask for funds from Father D.I. Lanslots, he sent Sisters Beata Vinson and Clara Schaff back to Maryville to ask the order to establish a hospital in Oklahoma City. At that time there was no hospital in Oklahoma City or Oklahoma Territory.

In July, 1898, four sisters returned to Oklahoma City and rented two houses at 219 N.W. 4th Street, the sight of the 1995 bombing of the Alfred E. Murrah Federal Building. One structure served as the Sisters' residence, the other for the hospital. On August 1, 1898, St. Anthony opened with Mother Augustine's mandated, "turn no one away." The hospital was a two-story, frame building with a capacity for 12 patients. Up until November 14, 1898, the small

As the first hospital in Oklahoma Territory in 1898, St. Anthony Hospital remains a pioneer in health care today. Serving Oklahoma City and surrounding communities, St. Anthony offers a continuum of care encompassing primary, secondary and tertiary care; prevention, diagnosis, treatment, rehabilitation, and home health services; and inpatient, intensive outpatient, residential and ambulatory services.

hospital treated 69 patients and had to be closed for the winter because there was no gas or electricity to heat and light the rooms.

Meanwhile, the city continued to grow, and the need for a permanent hospital became paramount. The physicians joined Father Lanslots and the Sisters in a successful appeal to the city council for a hospital. They secured funds for property on the northwest edge of the city and built a 25 bed facility where St. Anthony stands now. Though a vast improvement over the frame house, it was still primitive. Water had to be hauled from four blocks away and there was no sewage system. The first telephone came in 1900, electric lights and power in 1902, and natural gas in 1904. In 1905, the first addition to the hospital was completed bringing the bed capacity to 100. Also in 1905, the state's first school of nursing was started at St. Anthony. From 1900 to 1919, St. Anthony served as the exclusive training hospital for the Oklahoma School of Medicine.

St. Anthony through its physicians, the Sisters and the lay staff continued a commitment to excellence, earning the first Class A designation for any hospital in Oklahoma from the American College of Surgeons in 1921. An example of this commitment was the installation of the first x-ray equipment in 1908, only 13 years after the discovery of x-ray by Wilhelm Roentgen. In 1922, Sister Beatrice Merrigan of St. Anthony, became one of

the nation's first registered x-ray technicians.

St. Anthony continued this tradition of leadership in Oklahoma City with many firsts: a hospital-based pharmacy in 1925, an intensive care unit in 1963, a radioactive isotope laboratory in 1965, a mobile coronary care unit in 1969, an argon optical laser beam in 1971, a kidney transplantation in 1971, a neurological surgery institute and dialysis unit in 1972, a computerized axial tomography (CAT) head scanning unit in 1979, an alcohol treatment unit in 1975, a hospital "wellness" program in 1979, a computerized tomography reconstruction in 1984, a drug screening service for businesses in 1986, a dye laser for eye surgery in 1986, a craniofacial surgical team in 1989, and a non- teaching hospital to offer an accredited family practice residency in 1991. In 1995 the hospital's MRI was judged as the world's best.

St. Anthony maintains 33 centers of excellence including primary care programs, diagnostic services, cardiology, rehabilitation, occupational medicine, diabetes treatment, mental health, and others. In addition, St. Anthony is a partner in CommunityCare, the state's largest provider-owned delivery network offering a full range of managed care products including a Preferred Provider Organization (PPO), a health maintenance organization (HMO), certified workplace medical plans (CWMP) and multiple

option products. St. Anthony is also an equity partner in a rural hospital network offering managed care products, a purchasing cooperative, and educational services.

St. Anthony's health care staff includes 308 physicians on active medical staff, 315 on courtesy staff, 7 certified nurse anesthetists, and 3 physician assistants. Of the physicians on active medical staff, 214 are specialists and 94 are primary care providers. The hospital primarily serves patients from central Oklahoma (87 percent) with others coming from rural areas of the state. The patients represent a broad socio-economic mix. St. Anthony ranks as one of the larger employers in Oklahoma City, with 1,765 people on the payroll. Approximately 46 percent have been with St. Anthony for more than five years.

Located across the street from St. Anthony's emergency room, Bone & Joint Hospital is one of the nation's leading orthopedic specialty hospitals. Bone & Joint is a member of SSM Healthcare of Oklahoma.

COMMUNITY RESPONSIBILITY AND PUBLIC LEADERSHIP

Beyond its primary role as a health care provider, St. Anthony is a leader in renewing the heart of Oklahoma City which has suffered from the effects of the oil "bust" of the 1980's, stalled urban renewal efforts, and most recently, the bombing of the Murrah Federal Building. The hospital helped establish the MidTown Redevelopment Corporation, an economic development resource based on the Main Street model to address these needs. The hospital also takes an active leadership role in numerous community organizations.

In the community, St. Anthony offers a variety of free health education and health screenings to the community each year including diabetes screenings, employer on-site health screenings, osteograms, cholesterol and blood pressure screenings, and dental health screenings for children, as well as seminars on nutrition, stroke awareness, parenting, and substance abuse prevention. The hospital offers health services for the community including vaccinations for children in the public schools, free pharmaceuticals to community clinics, and taxi fares for patients needing transportation to medical care.

St. Anthony Hospital is also the second largest provider of care to the indigent in the community.

As a partner in SSM Healthcare of Oklahoma, Hillcrest Health Center serves patients in south Oklahoma City and is an osteopathic institution whose focus is primary care.

The St. Anthony North campus is located six miles north of the hospital and is focused on ambulatory services which include primary care, occupational medicine, sports medicine, physician clinics, physical therapy, hand therapy, a strength and conditioning program and an ophthalmology center. An ambulatory surgery center is presently under construction.

The Sisters' fund provides medicines to the poor and elderly. In collaboration with the community the hospital provides support services and volunteers to the Redbud Classic, Christmas in April (repairs to homes), the Food Bank, Adopt-A-School, Juvenile Diabetes Walk for the Cure, the Arthritis Foundation, Families First, MidTown Redevelopment Corp., and Neighborhood Housing Services as well as civic leadership organizations such as Leadership OKC, Rotary, Lions, Allied Arts, and Classen Beautiful.

During the events of April 19, 1995 and the subsequent year, St. Anthony was called upon by the community in many ways. The hospital provided sleeping and showering quarters for the rescue workers, $200,000 in care at the on-site triage station, ongoing support and assistance to victims for counseling, medical expenses, and assistance with applications for other sources of support.

St. Anthony and the New Managed Health Care Environment

The dramatic changes in health care reform have created an environment in which cost, access, and assurance of quality are the key issues. These issues will continue to grow in complexity as the baby boomer population ages. Even though the cost per patient may be down, there are many more people consuming health care now and in the immediate future. The question becomes how do we provide access for this increased population and who pays, especially for indigent care.

Secondly, the payer mix is changing with dramatic increases in managed care and a significant increase in capitated plans which restrict access for specialists through primary care physicians.

St. Anthony has positioned itself in this market with both price and service, and has aggressively responded to payer needs for the geographic distribution of primary care services. St. Anthony has developed the community's largest primary care network with 94 physicians located throughout the metropolitan area. The hospital also has its own Family Practice residency program which now in its fifth year of operation trains new physicians to practice primary care in the community.

The Strength and Conditioning program is just one of the sportsciences programs offered at St. Anthony North which focuses on individual needs and goals in a variety of athletic endeavors with an emphasis on education and prevention.

In 1993, the 14 hospitals in the Oklahoma City market were, in most respects, independent, stand-alone facilities, In 1996, there were only two that remained unaffiliated. St. Anthony successfully pursued affiliations with other providers, including one of the nation's premier orthopedic specialty hospitals, Bone & Joint Hospital, physically located across the street from St. Anthony's emergency room and Hillcrest Health Center, an osteopathic institution located in south Oklahoma City with a primary care focus. In the east market, St. Anthony has affiliated with Mission Hill Memorial Hospital in Shawnee.

In addition, St. Anthony is developing outlying campuses focused on ambulatory services. St. Anthony North located six miles north of the hospital offers primary care, occupational medicine, sports medicine, physician clinics,

physical therapy, hand therapy, a strength and conditioning program, and an ophthalmology center. This year St. Anthony is breaking ground for a six suite, two story ambulatory surgery center.

Oklahoma City is also witnessing the reentry of national HMOs to the market. In response to the move towards an environment of managed care, St. Anthony, together with Mercy Health Center and St. Johns and St. Francis Hospitals in Tulsa became equity partners in CommunityCare, the state's largest provider-owned delivery network offering a full range of managed care products, including a Preferred Provider Organization (PPO), a Health Maintenance Organization (HMO), certified workplace medical plans (CWMP), and multiple option products. In just two years CommunityCare has become the fourth largest managed care product in the

Saints Family Health Center South offers primary care services to the community in south Oklahoma City. St. Anthony supplies the largest number of primary care physicians in the metropolitan area.

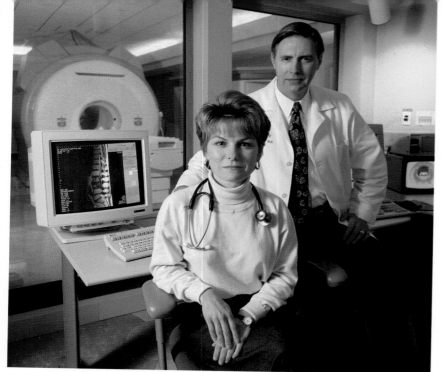

The new outpatient Diagnostic and Treatment Center at St. Anthony houses state-of-the-art diagnostic and treatment equipment. The images produced by the MRI have received international awards and the MRI serves as a world wide training center for physicians and technicians.

market, following the national giants of Blue Cross Blue Shield, Pacific Care, and Cigna.

St. Anthony's physician-hospital organization, Healthcare Services of Oklahoma, founded in 1985, is a for-profit corporation which employs and/or manages the community's largest primary care physician network. Over 300 physicians are currently enrolled in HSO's independent practice association, which in turn contracts with five capitated risk contracts. HSO is the contracting organization for both physician members and the hospital, when selected managed care contracts are negotiated, and it provides clinical and administrative management of these contracts for both entities. In the medical practice management arena, HSO maintains a management services organization which provides comprehensive operational and financial management for over 70 medical practices in 25 locations.

The physicians of HSO and SSM of Oklahoma are about to break new ground in converting the organization into an equity model in which the physicians will become equal business partners. This alignment of interests is designed to facilitate the complex care which will enable the physicians and hospitals to excel in today's market.

A major trend in healthcare locally and nationally is the shift from inpatient to outpatient surgery. In 1995, St. Anthony completed a three-story outpatient center which offers a wide range of outpatient services. Diagnostic and oncological are housed in this area with state-of-the-art technology in magnetic resonance imaging (MRI) and computerized tomography (CT) scanning and radiation therapy. A new outpatient dialysis center whose partner is a major national provider of

dialysis was completed in 1996 and occupies the third floor.

Just as other businesses are providing traditional hospital services, such as ambulatory surgery, hospitals are providing services usually found outside the hospital setting. In response to the need for prevention, health promotion, diagnosis, treatment, therapy, and recovery, St. Anthony provides post-acute services, such as a skilled nursing facility to accommodate the rehabilitative and therapeutic needs of patients in transition from the acute care setting to outpatient settings; a home health service to facilitate the recovery or maintenance outside an institutional setting; fast-tract and occupational injury care modules in the emergency room to provide more timely and appropriate levels of response and cardiac rehabilitation to promote recovery and prevent recurrence in an outpatient setting.

St. Anthony is also addressing health care in the workplace where health care costs have soared. Employers are faced with the decision of providing care, and if so, how they want to pay. In the evaluation of health care plans, cost of care

is a major consideration, but in the long term employers will want to look at health care plans, such as CommunityCare, which provide incentives for employees to stay well. In addition, CommunityCare includes the resources, such as wellness programs, courses, and screenings for education on good health practices.

Cost of work-related injury is also a major consideration. St. Anthony's occupational program is the only multi-site, hospital-based program in the community. In addition, the program lends support and planning in getting the injured employee back on the job as soon as it is medically appropriate to do so.

St. Anthony's strategic vision includes continued development of existing directions in the physician and managed care network. St. Anthony's facility focus will shift to development of ambulatory centers and creating more opportunities to affiliate with hospitals in the secondary market. St. Anthony believes this approach will enhance its market accessibility and delivery of quality health care, and ultimately its competitive standing. Leadership from St. Anthony and SSMHCS is continuing to create an administrative structure principled on collaboration and efficiency which will be crucial in the long-term success of managed growth and the acquisition of new partnerships in the community. St. Anthony's mission is to continue the commitment of its founders for the next 100 years. ✦

St. Anthony's obstetrics department, Joyful Beginnings, not only offers the latest in technology but also the comforts of home in labor and delivery. Construction completed in late 1996 enables labor patients to have drive-up, quick access to the labor and delivery area.

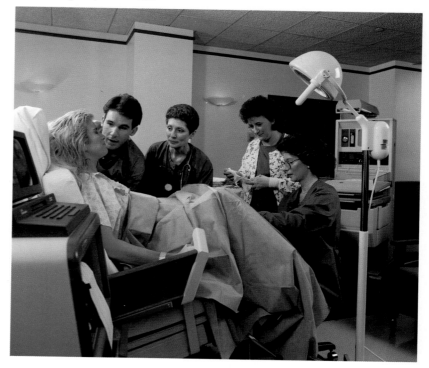

The University of Oklahoma

Former Oklahoma Governor and Senator David L. Boren returned home to become 13th President of the University of Oklahoma.

"Never underestimate the potential of a great university" could be the motto of The University of Oklahoma (OU). The University, which ranks first in the nation per capita among all public universities in the number of national merit scholars enrolled and has developed an Honors College to challenge its best students in small classes, is rapidly becoming a pacesetter in higher education.

On December 19, 1890, 17 years before Oklahoma's statehood, a territorial legislature recognized the importance of a university to any territory with aspirations of becoming a state and

which make up the educational core of Oklahoma Health Center complex.

Together, these campuses continue to build on the tradition of excellence that was established for the university before statehood.

The impact of the University of Oklahoma on Oklahoma City's growth and development ripples from employment to public-private partnerships.

Employment: The Norman and Oklahoma City campuses employ approximately 8,000 faculty and staff who service more than 23,000 students.

Work force: Broadly speaking, OU is producing the next generation of scholars, educators, engineers, and business and civic leaders. Many of

Owen Field, home of the Six-time National Champion Oklahoma Sooners, is an integral part of the proud Sooner athletic tradition.

founded The University of Oklahoma. Since that time, the university has stood as a testament to its mission: OU is a doctoral degree-granting research university serving the educational, cultural, and economic needs of the state, region, and nation.

The main campus is situated on a 3,000-acre campus in Norman, just 20 miles south of Oklahoma City. This campus supports 12 colleges offering a spectrum of undergraduate and graduate degrees. The University of Oklahoma's Health Sciences Center is just a few miles east of downtown Oklahoma City. It consists of a group of 12 public and private health care institutions

these individuals will remain in the area to work.

Economics: Studies have shown that the university has a tremendous impact on the local and state economies. Research grants have produced an eight to one return at OU. Faculty, staff, and students live and shop in the community and pay taxes.

Value: Attending OU is a bargain. In-state tuition at Oklahoma's four-year universities is the fourth lowest in the nation. Oklahoma ranks 11th lowest in tuition for percentage of median household income. OU now has the lowest administrative costs of any state higher education institution.

Research: By continuing to break records in research activity after crossing the $100 million

mark in 1995, OU is helping develop the state and national economy. While federal research funding has been flat, OU continues to increase its research activity through innovative partnerships with private companies that will help them stay ahead in the marketplace.

OU researchers, often in collaboration with private businesses and working side by side with engineers and scientists at public facilities like the Air Logistics Center at Tinker AFB, are developing technologies that will affect virtually all types of industry. Benefits from these technological advancements include developing numerical formulas to predict the weather, drilling for oil, cleaning up the environment, and developing new medical procedures and products to save and improve lives.

Entertainment: Athletics also plays an important role in the state, providing Sooner sports entertainment and attracting crowds that add to the business community's coffers. OU also offers significant cultural entertainment and opportunities.

CULTURAL AND QUALITY INFLUENCE

Opportunities for cultural, artistic, and scholarly enrichment abound through the University of Oklahoma's programs, collections, and events. Cutting-edge activities and collections within the University of Oklahoma environment rank with the best.

Oklahomans do not have to go to New York City for quality entertainment. The OU College of Fine Arts offers a variety of dance, drama, and music performances throughout the year and during its annual SummerWind festival of the arts.

OU's musical theater program is one of the

The beauty of OU's 370-acre landscape provides an impressive backdrop for Oklahoma's premier educational institution.

Oklahoma Press is the oldest in the Great Plains and ranks among the 20 largest university presses in the United States. In keeping with this heritage, more American Indian languages are taught for college credit at OU than any other university in the world.

While the University of Oklahoma has always been "in the business of" education, some of its goals have changed as the state, nation, and world have changed. The university initially taught a student body composed of pioneers with little or no previous formal schooling. Now the university expects its freshmen to come prepared with the basics needed to pursue a quality education that will ready them to excel in the 21st century.

In turn, there is a new focus by the university to widen the horizons for students and prepare them to live and compete in a global marketplace.

very few university programs in the nation to allow students to be in the same cast with professional Broadway actors in brand-new productions. In fact, the theatrical world focused national attention on OU when *Jack*, a musical about the life of John F. Kennedy, had its world premiere on the OU campus in the spring of 1995.

Construction has begun on the new concert hall, which will be one of the finest in the Southwest. OU is also home to the Fred Jones Jr. Art Center, considered one of the premier university art museums in the country. When completed in 1999, the Sam Noble Oklahoma Museum of Natural History will be the largest university-based museum in the world, housing more than five million natural history artifacts.

Scholars and researchers can strike it rich at the university libraries, the largest library system in the state. The libraries house several renowned collections, including the Western History Collection recognized as one of the nation's most important resources on the American West.

Where would one find original work by Galileo in his own handwriting? One of the best of its kind in the country, OU's History of Science Collection holds more than 80,000 volumes, including rare and historical works by the world's most famous scientists like Galileo.

The history of politics is chronicled through several venues at OU. The Political Commercial Archive in the School of Communication is home to the world's largest collection of television political advertisements. The Carl Albert Congressional Research and Studies Center houses the papers of more than 50 current and former members of Congress, making it the nation's leading research center for congressional studies.

Quality classroom instruction is the primary focus of the University's 1,900 faculty members.

LANGUAGE AND LITERATURE

At OU, outstanding language and literature programs are a source of great pride, especially since they extend to an international level. The Neustadt International Prize for Literature is second in prestige only to the Nobel Prize. Nineteen of its laureates, jurors, and candidates have subsequently received the Nobel Prize. *World Literature Today*, OU's international literary journal, is the oldest continuously published international literary quarterly in the United States.

In terms of publishing, OU is the leading publisher of books about the American Indian and Western U.S. history. The University of

An International Programs Center was established and a new curriculum developed to internationalize studies.

OU is the top school in the Big 12 regarding number of international reciprocal agreements, or exchange programs. Seventy-three international exchange programs with 36 countries enable students to study abroad and students from other countries to study at OU. The enrollment of international student population at OU is approximately 1,800, and 103 nations are represented by students and faculty on the campus. A new program called "OU Cousins" pairs American students with new international students.

From international studies to the arts, an OU student has the opportunity to become well-rounded in an academically intensive environment.

In the fall of 1996, the University of Oklahoma enrolled the highest academically ranked student body in Oklahoma history. In fact, the university ranks first per capita in the nation among public universities in the number of National Merit Scholars enrolled. Additionally, OU is the only state-supported university in Oklahoma to be listed in the top 10 percent of all U.S. colleges by the 1997 Fiske Guide to Colleges.

Outstanding students are matched with noteworthy curriculum. In the *U.S. News and World Report*, OU's undergraduate business program was ranked in the top 20 percent in the country, and OU's petroleum engineering program ranked among the top three in the nation. The university

With strong traditions in such programs as Meteorology and Geosciences, OU is poised to become a national pacesetter in research and development.

The University of Oklahoma places a strong emphasis on the Arts, and has received national recognition in such areas as drama, music, and the visual arts.

Oklahoma. As Oklahoma's former governor and U.S. senator, Boren is very active in educating the state legislature, business, and other publics about the university and higher education in general. His work has paid off.

The Reach for Excellence Campaign, a private fund-raising effort that will culminate in the year 2000, has already generated $115 million of the $250 million goal. The campaign is the largest fund-raising drive in the history of the state. Its success is due to the support of more than 35,000 of OU's alumni and friends.

The president and top university officials also give back to the community. They are highly involved at the civic level, holding many key positions in the chambers of commerce, local United Way efforts, and a myriad of other activities. This dedication shows their commitment to the continued progress of the university, the Oklahoma City metropolitan area, and the state.

The University of Oklahoma is truly a great university, ready to become a pacesetter for public higher education in the country. "We must no longer keep it a secret!" said David L. Boren, president, University of Oklahoma. "It's time for us to let others all across our country know about the strengths of our university and our determination to make it even better." ✸

is also noted for having one of the top three meteorological research programs in the nation and is the top university research center in the nation for the study of severe storms.

To ensure students a good start in an academic environment, OU is bringing back retired full professors to teach introductory courses and provide mentoring for first and second-year students. David L. Boren is the 13th president of the University of Oklahoma. He has identified four key areas in which OU can set the pace for American higher education.

1. Preparing the next generation to live and work in an international environment.
2. Challenging students to do their best by attracting and maintaining outstanding professors in settings that allow for the highest standards.
3. Continuing to break records in research activity.
4. Creating a sense of family and community on campus, thereby giving students the experience they need to then go out into the community and create a similar environment.

President Boren's actions prove his commitment to the important role of higher education in

University of Oklahoma Health Sciences Center

Teaching, healing, and discovering are the three words the nationally ranked University of Oklahoma Health Sciences Center uses to describe the facets of its educational operation. From training of tomorrow's health care force to pioneering faculty involved in cutting-edge research, the Health Sciences Center masterfully bridges the best of today with the brightest of tomorrow.

The health care community in Oklahoma City and throughout the state is dramatically impacted by this campus that is the state's major training ground for physicians, dentists, nurses, pharmacists, public health specialists, and a range of allied health personnel.

Fifty-six professional, graduate, and undergraduate degree programs are offered through the seven colleges that make up the Health Sciences Center. Approximately 950 faculty members and 1,900 staff serve the colleges and the more than 3,000 students. These full-time and part-time employees make the Health Sciences Center's programs some of the finest in the country.

The OU College of Medicine has been ranked consistently in the nation's Top 20 Medical Schools by the *U.S. News & World Report*. In a recent national study, OU's counseling psychology program was ranked seventh in the United States.

The Health Sciences Center is the core of a larger medical complex called the Oklahoma Health Center. The Oklahoma Health Center campus houses 18 health care related entities that work in tandem to reach their goals. Within this complex, students and faculty of the Health Sciences Center are able to utilize the clinical, laboratory, and teaching facilities of more than eight other health care institutions.

To broaden the logistical scope of the Health Sciences Center, programs are also offered at two principal sites in Tulsa. Work on the Oklahoma City and Tulsa campuses expands from direct patient care to renowned research activities.

Research programs with both scientific and clinical applications are under way by the hundreds on a consistent basis. Areas of study include Alzheimer's disease, numerous types of cancer, cystic fibrosis, mental health issues, and AIDS.

The OU Health Sciences Center was recently awarded the largest grant in its history by the National Institutes of Health. The five-year, $6,685,000 project focuses on identifying the risk factors and treatment of thrombosis, or abnormal blood clot formation.

Due to the more than 350 percent increase in funding research at the Health Sciences Center over the past decade, new facilities were needed

The Health Sciences Center is the core of a larger medical complex called the Oklahoma Health Center.

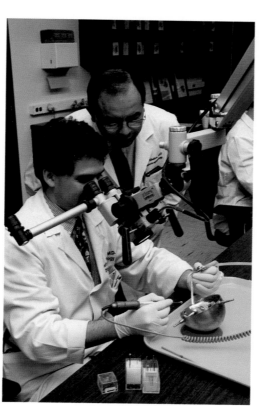

Fifty-six professional, graduate, and undergraduate degree programs are offered through the seven colleges that make up the Health Sciences Center.

to house the projects. A $45.1 million Biomedical Research Center, projected for completion in 1997, will help house the growth in research on campus.

Citizens of Oklahoma and the nation are the direct beneficiaries of the Health Sciences Center training and discovering processes. In addition to patients who come to the Health Sciences Center for care, the $4.3 million Oklahoma Telemedicine Network links nearly 40 rural hospitals and 15 regional hospitals with the center.

This statewide electronic network provides Oklahoma's rural area with access to consulting and diagnostic services in areas including radiology, pathology, and cardiology. It is believed to be the largest medical communications system of its type in the world.

In addition, many programs of the Health Sciences Center provide quality health care for little or no cost to the underprivileged.

Through the education and research in many fields of medicine, the University of Oklahoma Health Sciences Center stands as an icon of excellence . . . striving for the best in health care and hope. ✪

The University Hospitals

Ensuring the healthy heartbeat, grasping the child's tiny finger, roaring helicopter blades, programming for a toddler's developmental disability, giving instruction for health care providers, and conducting research for leading-edge treatment—these components describe the everyday work at The University Hospitals (TUH).

TUH are comprised of University Hospital (UH) and The Children's Hospital of Oklahoma (CHO). A five-member board governs the hospitals through The University Hospitals Authority, which operates as a separate state agency. Each hospital serves as a beacon of health and hope for Oklahomans and the nation. Collectively, they have earned national acclaim for their distinctive work and are honored with many of the nation's finest health care institutions.

RECOGNITION OF EXCELLENCE

The University Hospitals ranked in the top 40 nationwide for five medical specialties and outscored all other Oklahoma hospitals in 12 categories in a special *U.S. News and World Report* guide. The *1995 America's Best Hospitals* guide ranked TUH in the top 40 nationally in orthopedics (No. 25), endocrinology (No. 29), cancer (No. 30), neurology (No. 30), and gynecology (No. 37).

"No other publication or organization objectively ranks the quality of hospitals on a nationwide basis," says Lane Simpkins of *U.S. News.* "Any institution listed among the top 40 medical centers in any specialty should be considered a leading center. . . ."

The University Hospitals are the primary teaching hospitals for the University of Oklahoma Health Sciences Center (OUHSC) and trains the health care force of tomorrow. To enhance this teaching mission, all of The University Hospitals' physicians are faculty members at the OUHSC.

The strength of service provided by TUH is also exhibited through the number of clients they treat on an annual basis. For 1996, University Hospital handled 10,522 admissions, 135,474 clinic visits, and 26,545 emergency department visits. Children's Hospital admitted 4,686 children, treated 86,750 patients through office visits, and managed 20,498 emergency department visits.

University Hospital was created in 1917 to meet the needs of a growing new state and to be a paragon of teaching, research and patient care.

More than 26,000 patients from across Oklahoma are treated each year at University Hospital's Emergency Department.

Nearly 600 physicians at the University Hospital and more than 250 at Children's Hospital cover nearly every specialty and subspecialty. They work with a team of nurses, LPNs, and other employees nearing a total of 2,250 Oklahomans to make a difference in the life of each patient. In short, The University Hospitals are making the grade—and history.

UNIVERSITY HOSPITAL

In 1917, University Hospital was created with $200,000 in seed monies appropriated from the legislature, with funds originally intended to build the State Capitol's dome. The hospital was established to meet the needs of a growing new state and to be a paragon of teaching, research, patient care, and hope. More than 75 years later, University Hospital still stands dedicated to those ideals.

A national leader in providing quality health care for adults and newborns through both inpatient

and outpatient services to all Oklahomans, UH is staffed by highly trained physicians, nurses, and allied health professionals who benefit daily from the innovative environment that only a teaching hospital can provide. Additionally, many programs within the hospital create a powerful overall standard for the state and nation.

CARDIOLOGY RANKING NO. 1 IN STATE

Cardiologists and cardiovascular surgeons, members of the University Physicians Medical Group, provide the personalized, integrated patient care at University Hospital that has earned the facility a national reputation for innovation and excellence.

These doctors diagnose and treat heart and circulatory disorders ranging from arrhythmias, or abnormal heartbeat, to congestive heart failure and heart attacks. Cardiac surgeons perform coronary bypasses and other heart surgeries. At

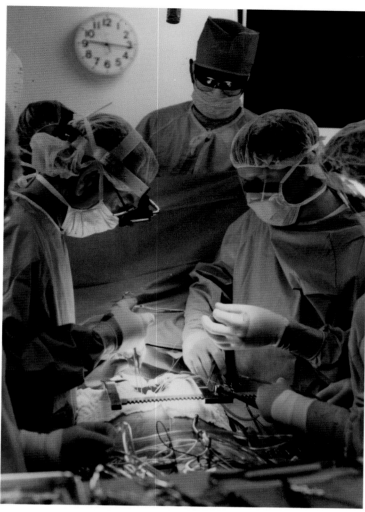

Surgeries performed at The University Hospitals range from repairing hearts of tiny newborns to organ transplantation.

University Hospital, the education and research components of the OUHSC join to offer state-of-the-art patient care.

University of Oklahoma heart specialists are experts in treating unusual or especially complicated cases. Dr. Warren Jackman cures certain types of heart arrhythmias with a procedure known as radiofrequency catheter ablation. Dr. Ronald Elkins has perfected the Ross procedure, where a patient's own heart tissue is used to replace a failing aortic valve. Both physicians have national and international reputations and have performed the procedures on hundreds of patients from around the world.

The major advantage for heart patients at University Hospital is that a "whole treatment" approach is taken regarding their care, simplifying what is often a complex process. A primary cardiologist can consult with other "superspecialists"—OU physicians and surgeons who have a great deal of advanced knowledge and expertise in the many different and complicated aspects of cardiology.

The team of OU cardiologists, thoracic surgeons, and vascular experts brings all this knowledge and skill to bear on one focus: the successful treatment of each patient. This broad-based expertise means heart patients at University

Hospital don't need to be referred elsewhere in the state or nation for the best possible care.

HEAD AND NECK CANCER PROGRAM No. 1

Treating cancers of the head and neck is a complex effort, requiring skill and expertise of several specialties. Patients benefit from an organized, multidisciplinary team of experts dedicated to providing care expected from a nationally ranked cancer program.

This team treats cancers of the head and neck involving tumors in the throat, mouth, and sinuses, along with skin cancers that form on the head

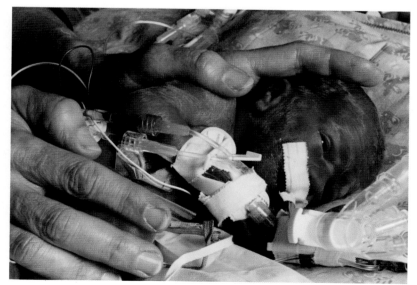

The Neonatal Intensive Care Unit at Children's Hospital provides state-of-the-art care for the tiniest patients.

and neck. They also treat tumors in the salivary glands and thyroid gland.

The Head and Neck Cancer Program excels in part because of its progressive treatment planning, which involves representatives from specialties such as radiation oncology, medical oncology, oral surgery, dental prosthodontics, pathology, plastic and reconstructive surgery, and radiology.

Due to the frequent need to combine these treatments in various ways, it is imperative for the specialists to communicate on the spot as the patient is initially evaluated and as the treatment progresses. Few treatment centers—besides internationally known places like the Mayo Clinic—utilize this team method of evaluation and care. The mainstay of the approach is threefold: surgery, radiation therapy, and chemotherapy.

Additionally, the strength of the clinical research helps expand frontiers in head and neck cancer. Research includes the use of lasers on cancerous tissues and cooperation with other departments for basic science research regarding genetics and the prevention of cancer.

CHILDREN'S HOSPITAL OF OKLAHOMA

Children's Hospital is dedicated to providing the best in health maintenance as well as giving seriously ill and injured children the best care. As the state's only full service pediatric health care provider, CHO has offered the latest in medical technology and research for more than 60 years. Specially-designed clinics and units meet the needs of children—from tiny newborns to adult-size 21-year-olds—and their families.

The hospital's main goal is to deliver the best pediatric medical care in an emotionally supportive environment. To further enhance the child's sense of well-being, parents are encouraged to spend as much time as possible with their child.

Outstanding programs of CHO

include the Child Study Center, the Children's Psychiatric Center, Children's Orthopedic Center, Neonatal Intensive Care Unit, Pediatric Intensive Care Unit, the Children's Hospital Emergency Department, and Child Life Program.

THE JIMMY EVEREST CENTER FOR CANCER AND BLOOD DISORDERS IN CHILDREN

From the playful addition to Garrison Tower that resembles children's building blocks in shape and color, to the toy train motif inside, the Jimmy Everest Center for Cancer and Blood Disorders in Children is designed to improve the treatment experience for a child and the child's family when faced with cancer or a blood disorder.

James "Jimmy" Christopher Everest died July 21, 1992, at age 17, after a brave bout with bone cancer. Memorial donations were made to this project in honor of Jimmy's life.

The center is clearly staffed to be a child's best resource for cancer and blood disorder treatment, along with the emotional and mental stresses that illness may cause. Many great innovations at the facility include the ability to provide quality care in an outpatient setting. Televisions and video-cassette recorders are also installed to entertain patients and families—as well as educate them about their illnesses and treatments—during their time at the center.

This child-friendly center is located in the Hematology-Oncology Service of Children's Hospital. It is a referral center for children

Medi Flight, Oklahoma's statewide air ambulance service, averages more than 100 flights per month.

throughout the state. The three main programs of the center include the Cancer Treatment Program, Oklahoma Hemophilia Treatment Center, and Sickle Cell Anemia Program. Other rarer blood disorders are also given treatment at this center.

The Jimmy Everest Center for Cancer and Blood Disorders in Children is a tribute to the care and concern for the children of Oklahoma and the progressive, quality patient care offered through Children's Hospital.

CHILD STUDY CENTER

When the Child Study Center was established in 1958, services to children with developmental disabilities were nonexistent or fragmented. With the foresight and concern for these exceptional children, the Oklahoma Legislature, Department of Human Services, OUHSC, and a private donor collaborated to build the present Child Study Center within the medical complex in 1973.

The Child Study Center is administered as an outpatient facility of CHO, affiliated with the Department of Pediatrics, University of Oklahoma College of Medicine. Professionals staffing the center are from the fields of child psychology, speech pathology, occupational and physical therapy, pediatrics, social work, and education who specialize in the multidisciplinary evaluation of the infant or child with developmental and/or behavioral problems.

There are five major programs in which this team of professionals assesses a child's cognitive, learning, motor, language, social, behavioral, and emotional function within the context

The colorful playroom in the Jimmy Everest Center for Cancer and Blood Disorders offers a cheerful environment to patients awaiting treatment. More than 70 percent of Oklahoma's pediatric cancer patients are treated at the center.

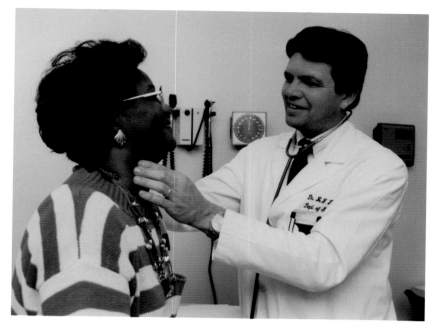

More than 135,000 patients are seen each year at the outpatient clinics at University Hospital.

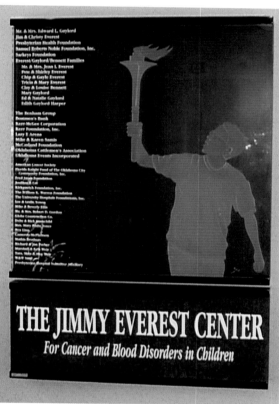

of the family. These programs include Early Childhood Programs, Neuropsychology Program, Psychosocial Program, Communication Disorders Program and Attention Deficit Hyperactivity Disorder (ADHD) Clinic, and the Behavior Clinic.

Individual specialists evaluate the child and then meet to coordinate their findings and recommendations. This comprehensive evaluation is then shared with the family. Team members assist the family in networking with schools and other community resources for possible treatment programs. The center also provides treatment programs in certain cases.

Through knowledge and understanding, interdisciplinary clinical evaluation, short-term treatment and programming, support to state and local agencies caring for children with a handicapping condition, plus as an advocate for children, the Child Study Center continues to play an important role in the care of Oklahoma's children.

MEDI FLIGHT OKLAHOMA

The hum of the Medi Flight helicopter is the whir of hope for residents in distress across Oklahoma. From the neonatal team taking care of an hour-old baby boy, to the pediatric/adult team members flying in to get a burn patient, Medi Flight Oklahoma makes this lifesaving service available.

The experienced Medi Flight team members, coupled with the latest in medical equipment placed on board the sophisticated helicopters, make up the emergency medical air transport system of the Oklahoma Medical Center. It supplies regional helicopter service to the citizens of Oklahoma and is the longest-serving civilian hospital-based emergency medical air system in the state.

Serving the entire state, Medi Flight operates 24 hours a day, 7 days a week, taking the emergency room to the patient. Before joining Medi Flight, the team members must have worked for at least two years in an emergency room or intensive care unit. While speed is important, the primary goal of the service is to quickly stabilize the patient by skillfully administering life-sustaining procedures at the local hospital or scene and continuing the treatment en route.

Medi Flight helps physicians, community hospitals, ambulance services, and law enforcement agencies in the stabilization, treatment, and rapid movement of critically ill or injured infants, children, and adults. Although based at the Oklahoma Medical Center, the service is prepared to transport patients from a referring hospital, from a disaster site, or directly from the scene of an emergency to the closest, appropriate Oklahoma City area receiving hospital that is best able to treat the patient's illness or injury.

Medi Flight Oklahoma is a friend in high places.

A BEACON OF HEALTH AND HOPE

By weaving treatment, teaching, and research into the healing process, the gamut of physicians, medical care providers, and staff delivers nationally ranked medical specialities to Oklahomans through the state's hospitals. The everyday and extraordinary work at The University Hospitals is a beacon of health and hope for Oklahoma. ❂

Deaconess Hospital

Among Oklahoma City hospitals, Deaconess was the first to open an outpatient surgery center.

Deaconess Hospital is committed to its mission, its patients, and its community. As an independent hospital, not part of a larger health system or hospital network, Deaconess offers progressive medical programs, advanced technology, and compassionate Christian care. This is why patient satisfaction surveys continually rank Deaconess in the highest percentiles.

"Through its history, Deaconess has been swift to recognize and respond to the needs of the patients we serve," says Paul Dougherty, administrator.

Deaconess offers a commitment to patient care that Oklahomans believe in. It is a comprehensive hospital offering 98 percent of all hospital services used by the general population. The two percent of services not offered at Deaconess consist of burns, transplants, and some neurological procedures.

An independent patient satisfaction survey for the third quarter of 1996 ranks Deaconess in the 99th percentile among 353 hospitals in the United States in the following areas:
* likelihood of recommending the hospital to others.
* overall cheerfulness of the hospital.
* nurses' friendliness.

"At Deaconess, we place a premium on patient care," Dougherty adds. "That's why Deaconess' nurse-to-patient ratio is among the highest of all Oklahoma City hospitals."

Along with patient satisfaction, Deaconess has grown to become one of the region's most respected hospitals and a pioneer in a lengthening list of important medical advances. Among Oklahoma City hospitals, Deaconess was the first to:
* open a critical care unit.
* open an outpatient surgery center.
* offer a lithotripsy center serving central and western Oklahoma.
* open a center dedicated to the geropsychiatric care of adults 65 and older.
* offer single-room maternity care that features labor, delivery, recovery, and postpartum in one room.

From humble origins predating Oklahoma statehood, Deaconess continues its story of firsts. Following are just a few of the many areas of excellence in which its 600 physicians practice medicine.

DEACONESS CANCER CENTER

The Deaconess Cancer Center is among an elite 25 percent of hospital programs fully accredited by the American College of Surgeons' Commission on Cancer. Its team of physicians, nurses, and other health care professionals use the latest treatments, including radiation therapy, in the war against this once undefeatable disease. Patients benefit from the center's participation in ongoing cancer research studies designed to determine the most effective cancer medications and treatment strategies.

CARDIAC HEALTH CENTER

Doctors at Deaconess have performed more than 1,625 open heart surgeries in the past decade and average nearly 200 each year. The first Minimally Invasive Direct Coronary Artery Bypass (MIDCAB) in Oklahoma was performed at Deaconess Hospital. With only 18 percent of the hospitals in the United States offering angiography, Deaconess was the first in the nation to offer equipment with a high-speed angiography C-arm. Rotating 15 degrees per second, it is twice as fast as traditional systems. Deaconess is equipped with two state-of-the-art angiography suites for vascular imaging and cardiac intervention. Procedures including angioplasty, atherectomies, and coronary stent placement are performed.

The Cardiac Health Center provides Deaconess physicians with the equipment and facilities needed to perform the total scope of cardiac evaluation, treatment and rehabilitation for their patients.

DEACONESS CENTER FOR WOMEN'S HEALTH

The Deaconess Center for Women's Health is dedicated to promoting the importance of health and wellness for women of all ages and their

The Birth Center provides Deaconess patients with all the comforts of home and physicians with the latest technology available.

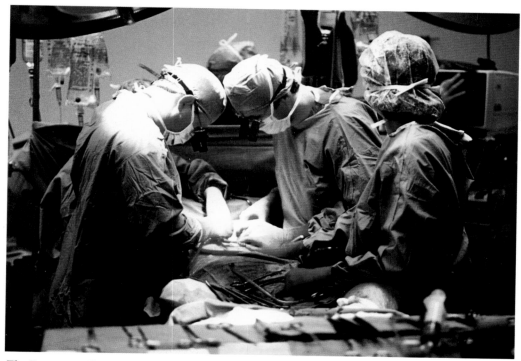

The Deaconess Cancer Center is among an elite 25 percent of hospital programs fully accredited by the American College of Surgeons' Commission on Cancer.

families through superior educational services and programs throughout the community. It includes the Birth Center of New Generations, the Mammography Suite, wellness programs, the Osteoporosis Center of Oklahoma, the Women's Cardiac Health Center, and a resource library.

BIRTH CENTER OF NEW GENERATIONS

The Birth Center provides Deaconess patients with all the comforts of home and physicians with the latest technology available. The center is further complimented by a Level II Neonatal Intensive Care Unit, which is staffed by a board certified neonatologist. Its "birthing rooms" are elegantly appointed, offering an embracing environment for the birth experience.

OKLAHOMA UROLOGY CENTER

Located on the campus of Deaconess Hospital, the Oklahoma Urology Center is the first and only center of its kind serving central and western Oklahoma. The center offers lithotripsy, the nonsurgical removal of kidney stones, in Oklahoma City and several locations around the state through a mobile unit. More than 40 urologists from across Oklahoma use the center for diagnosis and treatment of a wide range of urological problems.

Deaconess offers progressive medical programs, advanced technology and compassionate Christian care.

SENIOR DIAGNOSTIC CENTER OF OKLAHOMA AND ADULT MENTAL HEALTH UNIT

The Senior Diagnostic Center was the state's first facility devoted exclusively to geropsychiatric evaluation and treatment. It is the only facility with an area specifically dedicated to care of dementia patients. The goal of the center is to return senior adults to the most suitable, independent living situation possible.

The Deaconess Adult Mental Health Unit offers persons age 18 or older diagnosis and treatment of a variety of problems, including depression, schizophrenia, paranoia, and sexual trauma.

EMERGENCY DEPARTMENT

More than 20,000 times yearly, when an emergency situation occurs, people in central Oklahoma turn to Deaconess. The Deaconess emergency department staff stands ready—poised to respond to any medical crisis within seconds, 24 hours a day, 365 days per year.

Nine physicians comprise the Deaconess emergency medical team. The four full-time physicians have more than 50 years of combined experience in emergency medicine.

DEACONESS HOME CARE

Patients ready for discharge from the hospital, but not yet fully able to care for themselves, turn to Deaconess Home Care for continued medical treatment and health services. In 1996, just two years after beginning, Deaconess Home Care provided more than 42,000 home visits and continues to provide the care that Deaconess is known for, in the home.

CONTINUING TO SERVE WITH EXCELLENCE

The mission of Deaconess Hospital is to serve with excellence the health care needs of all persons as a nonprofit, family practice hospital. Deaconess standards make it a pacesetter in the medical community and an institution poised for the future. "As an independent hospital, Deaconess continues to grow to meet the needs of patients and physicians," says Dougherty. "We achieve this through a strong base of family practice physicians, alliances with other area hospitals for the purposes of managed care contracting and ownership of one of Oklahoma's premier managed care networks—WellCor America."

As America's health care system faces dramatic change, Deaconess will meet it with a strong financial foundation and a staff dedicated to the difficult job of providing exceptional health care by incorporating the touch beyond the technology. ✸

Oklahoma Urology Center

The Oklahoma Urology Center is a continuing success story that exemplifies the positive outcomes of coalition building. Urologists in the Oklahoma City area started with a dream to develop a urological center of excellence that would house a state-of-the-art lithotripter designed to eliminate kidney stones without surgery. Collaboration, hard work, and alignment with Deaconess Hospital brought this plan to fruition and even greater dreams for a comprehensive urology center.

A state-of-the-art center now provides central and western Oklahoma nonsurgical relief from kidney stones . . . and much more. This is how it began.

INVENTION, TREATMENT, AND OKLAHOMA

German engineers with Dornier, an aerospace technology group, were discussing damaged tiles on their spacecrafts. It seems when a craft reentered the Earth's atmosphere, the water droplets and electrical energy shock waves caused pitting on the spacecraft's tiles. A doctor present during this discussion noted the value of garnering this energy and focusing it on kidney stones.

Subsequently, the lithotripsy procedure became reality. Lithotripsy, a word meaning stone crusher, is a revolutionary kidney stone treatment. It is the safest method of kidney stone removal, causing fewer complications than any other procedure. Oklahoma City urologists knew there was a great need for this cutting-edge equipment in the United States. Nine southeastern states, including Oklahoma, are in a region known as the "Stone Belt." This designation is held by the states with the most kidney stones. Due to the heat, climate, and outdoor work of its residents, more kidney stones are found in individuals in this region than any other part of the country.

The lithotripsy procedure became available in Oklahoma when a network of 28 urologists in the Oklahoma City metropolitan area and seven communities in western Oklahoma unified to form the Oklahoma Lithotripter Associates. They pooled resources to buy the $2 million lithotripter. Alignment with Deaconess Hospital finalized the partnership and the location of a facility.

Urological Surgery Center Associates Chairman William J. Miller, M.D., and Oklahoma Lithotripter Associates Chairman Jon C. Axton, M.D., cut the ribbon during the grand opening of the Oklahoma Urology Center. These two physician-owned corporations, representing more than 45 urologists, have combined to provide a state-of-the-art treatment facility specializing in urology disorders.

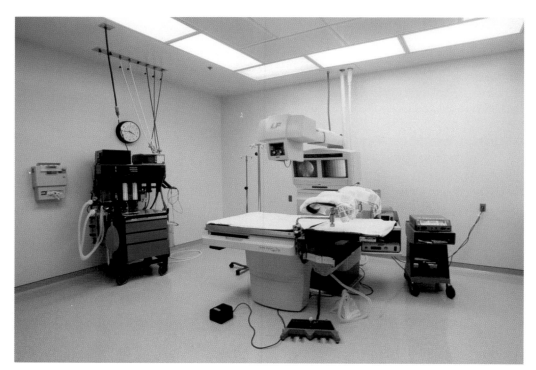

The cystoscopy suite enables urologists to visualize the urinary system under flouroscopic examination or with endoscopic cameras.

Through the pro-physician attitude of Deaconess, the 6,000- square-foot addition for the Oklahoma Lithotripsy Center was completed on the campus of Deaconess Hospital. It started operations in November, 1986, and was the first facility in Oklahoma to use the revolutionary lithotripsy treatment.

The Oklahoma Lithotripsy Center, recognized as a premier center in Oklahoma for the treatment of kidney stones, is the first such center in the country to receive certification by the Accreditation Association for Ambulatory Health Care (AAAHC). This physician-owned center's success is attributed to the specialized staff, state-of-the-art equipment, and dedication to quality care.

Soon, following the opening of the center, Deaconess Hospital and the Oklahoma Lithotripsy

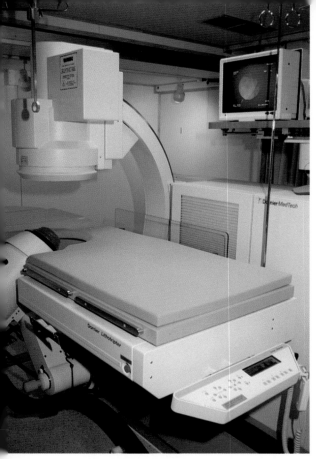

The Doli 15 lithotripter is the latest technology in extracorporeal shock wave lithotripsy. Housed in a 45' medical coach, this unit travels throughout Oklahoma and Texas providing lithotripsy service to rural and metropolitan hospitals.

Center physicians joined to form the Oklahoma Kidney Stone Research Foundation. This foundation provides ongoing research to identify causes and treatments for kidney stone disease.

REVOLUTIONARY PROCEDURE

The lithotripsy procedure offers patients an unprecedented alternative to surgery and recovery. An overnight stay in the hospital may be required, however most patients have the procedure performed on an outpatient basis.

An anesthetic is given in preparation for the lithotripsy treatment. The patient is placed in a specially designed tub filled with water. Two x-ray systems show the precise location of the kidney stone on monitors viewed by the physician.

Shock waves are then ignited at a frequency in time with the heart beat. Approximately 1,500 ignitions are usually required to pulverize a kidney stone. The entire procedure takes an hour. After the procedure, the patient is taken to recovery. Pain medication is given as needed, although many patients do not require it. The small particles, which were once a kidney stone, are passed out of the body spontaneously over several days or weeks.

In addition to the shock wave lithotripter, a laser lithotripter is also available. This lithotripter can fragment kidney stones lodged in the lower and mid-ureter.

LITHOTRIPTER ON THE ROAD

As the technology for the lithotriptsy improved over the next three to four years, a new piece of equipment became available without the tub, with the benefit of being portable. This technology allowed the Oklahoma Lithotripter Center's mobile unit to move into action across the state.

The mobile unit consists of a 45-foot semi-tractor trailer, which houses the lithotripter equipment. A truck driver and technician are a permanent part of this mobile unit team.

With the trailer set up just like a clinic, it can pull up in back of a hospital and hook up to an electrical outlet or utilize its onboard electrical generator. The patient is then wheeled into the trailer for treatment by doctors from the local hospital. The mobile unit provides services to Enid, Stillwater, Ardmore, Altus, Ada, Norman, and Tyler, Texas, and the list is growing.

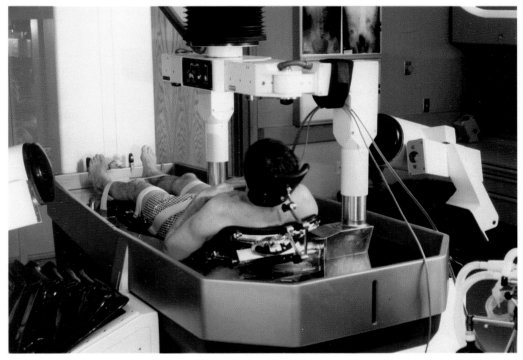

In the Dornier HM3 lithotripter, shock waves travel through a water bath to crush the kidney stones. Although recent lithotripsy technology has eliminated the water bath, the HM3 is still considered the gold standard of lithotripters.

OKLAHOMA UROLOGY CENTER BEGINS

Because of the success of the Oklahoma Lithotripsy Center in recognizing and conquering kidney stones, a second company was formed by 20 urologists. Urological Surgery Center Associates created a partnership with the Oklahoma Lithotripter Associates to form the new Oklahoma Urology Center in 1997.

The Oklahoma Urology Center's mission is to expand urology services and be recognized as a premier "center for excellence" for urology. It provides surgical services, as well as lithotripsy and related urodynamic procedures. The center also renders ambulatory urological services to meet the specialized and growing needs of urology patients in an outpatient setting. The Oklahoma Urology Center offers the highest quality of care and introduces the latest and most effective diagnostic and treatment technology. It occupies 18,000-square-feet on the first floor of a three-story medical office building on the Deaconess Hospital campus.

Donald Albers, M.D. continues to act as the center's Medical Director, and Marie Lee, R.N., C.N.A. is the Executive Director. The boards of the Oklahoma Lithotripter Associates, L.C. and Oklahoma Urological Surgery Center Associates, L.C. serve as the Oklahoma Urology Center's governing body.

"Deaconess and the Oklahoma Lithotripsy Center have worked together in a mutually beneficial relationship since 1987," says Lee. "The development of the expanded, state-of-the-art Oklahoma Urology Center significantly enhances the availability of services for Oklahoma's urological patients." ❂

Presbyterian Health Foundation

The story of the Presbyterian Health Foundation is one of deep caring and concern underlying the fundamental mission of improving humankind through medical education and research. Each dollar from the foundation is given with the belief that no Oklahoman should have to leave the state in order to receive the best medical care in the world.

As a private, independent, nonprofit organization, the Presbyterian Health Foundation is a part of the Oklahoma Health Center, dedicated to excellence in patient care, education, and research to ensure today's investment in tomorrow's health. "Oklahoma is no longer a dust bowl. We are now a center of scientific and medical excellence, with a wealth of brain power and ingenuity. We are Oklahomans with the vision and intelligence to make this a state where biotechnology is our next oil field," says Jean G. Gumerson, president.

A FOUNDATION TAKES LIFE

With $67 million in proceeds from the sale of Presbyterian Hospital to Hospital Corporation of America, Presbyterian Health Foundation was born on October 3, 1985. Trustees of the Presbyterian Hospital became the trustees of the new foundation. Work of the foundation began, with a focus on the Oklahoma Health Center and community health related programs in Oklahoma. Since its inception, managed assets of the foundation have increased in value from $67 million to more than $157 million. The foundation is currently giving approximately $6 million in grants each year.

TODAY'S INVESTMENT IN TOMORROW'S HEALTH

The purpose of the Presbyterian Health Foundation is to provide resources to encourage the development of medical education and medical research programs primarily conducted in Oklahoma. The foundation concentrates its support in five areas: medical research, medical education, clinical pastoral education, community health related programs, and resource development through technology transfer.

The purpose of the Presbyterian Health Foundation is to provide resources to encourage the development of medical education and medical research programs primarily conducted in Oklahoma.

The foundation currently focuses a large portion of its resources on the Oklahoma Health Center and the University of Oklahoma Health Sciences Center (OUHSC) with the goal of developing a premier regional health center for research, education and patient care.

Located on 200 acres, the Oklahoma Health Center consists of 19 institutions including the OUHSC. The Oklahoma Health Center is the destination for more than a million patient visits each year. In addition to the direct benefit of

health care, Oklahoma benefits from the research funded by the Presbyterian Health Foundation.

Another area of emphasis for the foundation is the support of endowed chairs at the OUHSC. The Oklahoma Legislature pledges to match each $500,000 in private funds with $500,000 in state funds for endowments.

Contributions to the Presbyterian Health Foundation help secure the future of the foundation so that its grants will continue to support these areas of emphasis.

The work of volunteers and staff, through the Presbyterian Health Foundation, realizes a healing ministry dealing with the physical, spiritual, and educational health needs of people.

FOCUSING ITS RESOURCES

The foundation continues to support the highest standards of excellence in the sciences related to medical arts and human healing and maintain a clear, moral, and ethical concern for enduring human values.

A summary of how the Presbyterian Health Foundation funds have been spent illustrates the conscientious giving, solid vision, and focused mission of its leaders. General categories in education and research include the resident physician medical education program, medical student research scholarships, seed grants to medical researchers, the M.D./Ph.D. program, physician scientist research fellowship, endowed chairs, and clinical pastoral education.

In the realm of technology transfer, projects include the UroCor Building and program investments in medical start-up companies. The remaining 10 percent of the Presbyterian Health Foundation's giving provides numerous smaller grants to nonprofit organizations in the community and state.

The Presbyterian Health Foundation is governed by a board of up to 40 corporate members, including 18 trustees who donate their time to the foundation and its goals. A small, efficient staff of three people oversee the foundation's more than $157 million in assets.

Corporate officers are Stanton L. Young, chairman; Michael D. Anderson, vice chairman and chairman-elect; Jean G. Gumerson, president; Dennis McGrath, vice president; Fred Zahn, secretary; William M. Beard, treasurer, and Melissa Kizer, assistant secretary.

The synergy of the foundation's volunteer

Oklahoma is now a center of scientific and medical excellence, with a wealth of brain power and ingenuity.

structure fuses the various investment aspects of the organization into one vision. Volunteer committees include the executive committee, the investment, technology transfer, nominating, grants, and funds development committees.

The work of these volunteers and staff, through the Presbyterian Health Foundation, realizes a healing ministry dealing with the physical, spiritual, and educational health needs of people. It also brings goals and dreams to reality for generations of Oklahomans.

PRESBYTERIAN INVESTS IN THE FUTURE

The Presbyterian Health Foundation is noted for its responsive trusteeship, compassionate concern, and pioneering vision. "It is inspiring to contemplate what 10 and 20 years of support by our foundation to the future health needs of the people of Oklahoma will mean to our children and grandchildren," says Stanton L. Young, chairman.

Development of the Oklahoma Health Center Research Park is an initiative of the Oklahoma City Urban Renewal Authority and the Presbyterian Health Foundation to provide support areas for research and development, technology transfer, and economic development. This connection is critical to the upswing in business and technological advances in Oklahoma and to position Oklahoma as a world leader.

"Oklahoma high school graduates are receiving scholarships and offers of advanced placements from OU and OSU as well as the Stanfords, CalTechs, Harvards, and MITs of the nation. Many of our best and brightest will earn master's and Ph.D. degrees. Where will their future be?" said Young. "In constructing the UroCor Building in the research park, the trustees placed as a priority the support of developing the biomedical technology industry. This will provide our state's most valuable asset—young, well-educated Oklahomans—with the option of remaining in Oklahoma."

It is through unselfish giving and loyalty by the people of Oklahoma that the Presbyterian Health Foundation's vision for the greater good continues. It instills the love of discovery in the hearts and minds of students throughout Oklahoma, leads to medical breakthroughs that touch the lives of individuals and families, and brings exciting economic opportunities by positioning Oklahoma at the forefront of genetic research, biotechnology, cancer research, and pediatrics.

Through the generosity of the Presbyterian Health Foundation, the future holds great hope. ✪

Oklahoma Health Center Foundation

The Oklahoma Health Center Foundation exists as a conduit to perpetuate the long-term success of the Oklahoma Health Center (OHC), Oklahoma's largest medical and academic health center. OHC, composed of 19 institutions, is a major provider of health, educational, and research services. Located just south of the state capitol, this 200-acre complex offers medical services to more than a million patients, students, and other clients each year. The foundation serves as the OHC link.

"The Oklahoma Health Center Foundation is a forum and conduit for all the CEOs that make up the fabric of the Health Sciences Center," says David Rainbolt, president. "We work on common issues and problems, from concepts as simple as crosswalks and locations of bus stops to the future of health care."

ALLIANCE OF EXCELLENCE

The Oklahoma Health Center includes major hospitals, professional schools, research laboratories, physicians' clinics, and specialized care facilities, in addition to a number of other public and private health care related agencies. Institutions affiliated with the OHC are the American Red Cross, Children's Medical Research, Columbia Presbyterian Hospital, Dean A. McGee Eye Institute, Medical Technology and Research Authority, Oklahoma Allergy & Asthma Clinic, and the Oklahoma City Clinic. Oklahoma Department of Health, Oklahoma Department of Mental Health & Substance Abuse Services, Oklahoma Medical Research Foundation, Oklahoma School of Science and Mathematics, and the Oklahoma State Medical Examiner are also a part of the Oklahoma Health Center complex.

Rounding out the prestigious mix of institutions at the center are the Sylvan N. Goldman Center/Oklahoma Blood Institute, Presbyterian Health Foundation, University of Oklahoma Health Sciences Center (OUHSC), University Hospitals, UroCor, and the Department of Veterans Affairs Medical Center.

COLLECTIVE IMPACT

In addition to its many medical benefits, the OHC is an economic enterprise with a far-reaching impact on the economies of the state—benefiting far more than those it serves directly.

Oklahoma Health Center, composed of 19 institutions, is a major provider of health, educational, and research services.

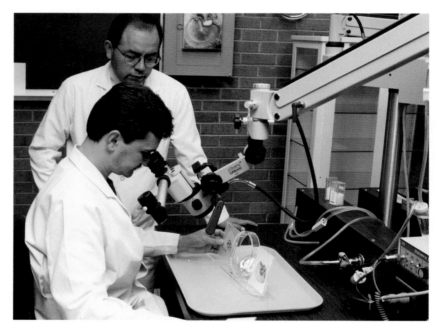

The foundation wants to help bring more research and biomedical applications to the Oklahoma Health Center, resulting in more jobs, more prestige, and more healthy lifestyles for the people of Oklahoma and the nation.

Stimulating growth in Oklahoma and the metropolitan Oklahoma City area, the Oklahoma Health Center employs more than 11,500 people, third in the state only to Tinker Air Force Base and American Airlines. These health care providers, educators, and researchers—along with their supporting staffs—produce services that initiate a chain of spending throughout the state.

Figures from a 1992 economic survey illustrate that the spending chain begins with the OHC's annual expenditures of $796 million, which leads to further spending on supplies, services, equipment, and construction. These expenses create additional spending on employees, suppliers, and contractors, which add up to a total spending

impact of $1.7 billion. In all, total expenditures by the OHC result in a payroll of $690 million and jobs for more than 34,000 Oklahomans.

The OHC payroll also has an impact on the state's fiscal health by generating income and sales taxes. In fact, this payroll produces taxable income resulting in $33 million in state income tax collection. It supports purchases by wage and salary earners that generate an additional $22.5 million in sales taxes.

BUILDING RESOURCES CENTER

One key element affecting the economic impact of the Oklahoma Health Center is capital formation—the accumulation of resources that

Room to grow — OHC Child Development Center.

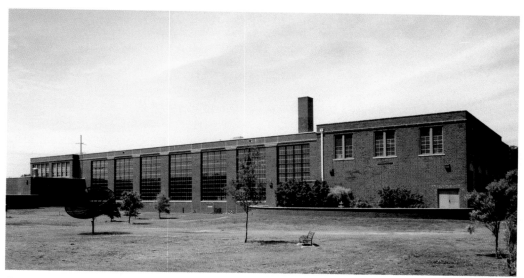

Progress in action — Oklahoma School of Science and Mathematics, part of the Oklahoma Health Center.

LEADERSHIP AND EXCELLENCE

The Oklahoma Health Center's institutions are constantly challenging the status quo—then finding rewards for their work.

* The OUHSC is one of only four Academic Health Centers in the nation with seven professional schools. It has one of 26 schools of public health and one of 54 dental schools in the nation.

* World-class clinical and research areas include the Cardiology and Arrhythmia Center, rheumatology program, infectious disease, and genome research centers.

* Research expenditures are at $54 million per year with a growth of 350 percent in the past 10 years.

* Research enterprise has led to the development of a Biomedical Research Park with future expansion under way. New jobs in this area average $40,000 per year. This research park is a priority area of development by the Greater Oklahoma City Chamber of Commerce and the Oklahoma Department of Commerce.

The Oklahoma Health Center Foundation recognizes these achievements and acts as a mouthpiece and communications link with OHC and the corporate and outside community.

"At the Oklahoma Health Center, we have accumulated a critical mass that is acting as the genesis and catalyst for continued development," Rainbolt says. "As we continue to be successful in transferring research and the academic and health care capabilities into commerce, our success is going to be noted around the world."

The Oklahoma Health Center Foundation is working to enhance the ability of the Oklahoma Health Center to develop and achieve its academic, research, and biomedical commercialization goals. In short, the foundation wants to help bring more research and biomedical applications to the Oklahoma Health Center, resulting in more jobs, more prestige, and more healthy lifestyles for the people of Oklahoma and the nation. ✪

will be used in the future to produce health, educational, and research services.

The largest and most important portion of OHC's assets is derived from human capital—the value of professional health care, educational, and research employees. Estimated on expected future returns, this currently accounts for the largest portion of OHC's assets—exceeding $5 billion. The value of buildings, plants, and fixed equipment is estimated at its replacement cost and is more than $1 billion.

All told, the Oklahoma Health Center has the resources to deliver $6.3 billion in health and educational services.

FORUM FOR GROWTH

Chartered in 1965, the Oklahoma Health Center Foundation was established to coordinate the planning of the Oklahoma Health Center complex. The foundation's primary objectives are to provide an interface with the OHC and the business community, as well as to inform Oklahomans about the progress being made in medical technology and discoveries.

The foundation's original board was comprised of civic leaders headed by Dean A. McGee, Stanton L. Young, and James L. Dennis, M.D. Other driving forces included E.K. Gaylord, E.L. Gaylord, George Cross, E.T. Dunlap, Nate Ross, Donald S. Kennedy, John Kirkpatrick, W.W. Rucks, Jr., M.D., and Harvey P. Everest. The efforts of these men resulted in a 1960s growth boom. More than half the structures at the Oklahoma Health Center have been built since this time.

"Everything we do is looking forward," Rainbolt added. "The coalition comes together to plan for the future of the entire complex by coordinating, facilitating, and having an intellectual exchange of ideas."

Columbia

Columbia Presbyterian Hospital, established in 1910, is Columbia's flagship hospital in Oklahoma City.

All-encompassing health care is a must for the future. Columbia is ahead of the curve in delivering this full-circle care with quality, patient satisfaction, fair price, and compassion.

The vision statement of Columbia is to work with its employees and physicians to build a company that is focused on the well-being of people, that is patient oriented, that offers the most advanced technology and information systems, that is financially sound, and that is synonymous with quality, cost-effective health care.

This vision began in 1987 with two hospitals. Columbia since has grown to become the largest provider of medical services in the United States. With hundreds of facilities across the country, England, and Switzerland, Columbia's employees and associated physicians care for more than 100,000 patients each day.

Columbia has a presence across the entire state with nearly 50 facilities and growing.

Columbia facilities have been delivering excellence in Oklahoma for years. An example is the oldest institution—Columbia Presbyterian Hospital—founded in 1910, as Wesley Hospital. Renowned for research, patient care, and education, Columbia Presbyterian Hospital and many of its doctors are recognized around the world for their contributions to the health care industry.

Columbia combines these longstanding traditions with the everchanging world of health care for a product that "has never worked like this before."

DEVELOPING INTEGRATION

In Oklahoma, Columbia is building a comprehensive network by providing a diverse base of all-encompassing health care facilities.

"The concept—integrated healthcare delivery system—means bringing together all methods of health care into one system of hospitals and services," said David Dunlap, FACHE, president of Columbia Oklahoma Division. "This not only provides health care for a person at all stages of his or her life, it also carries the concept of efficiency, accessibility, and total health care within one entity."

Seven main areas run the gamut of health care options at Columbia. Through the hospitals, a system of community is established. These acute care centers provide all the traditional and innovative hospital services. Outpatient surgery centers

offer the options of convenient, cost-effective, short-stay facilities. Nurses and technicians give care to patients through home care.

Outreach provides quality, convenient medical services to residents in nursing homes and assisted living centers. An intermediate level of care, between hospital and standard nursing home care, is found through the skilled nursing units. Family medicine and specialty clinics provide medical benefits for everyone from expectant mothers to senior citizens. Rehabilitation programs help patients return to normal activity after sickness or injury.

"In addition to hospitals and physician offices, an integrated health care delivery system such as Columbia offers home health care to people who require long-term care following hospitalization or rehabilitation," said Dunlap. "Certain primary care clinics stay open late in the evening to provide routine medical care for all members of a family. Outreach services bring primary and specialized health care to people in common settings such as nursing homes."

In what might seem like a large tertiary care environment, Columbia strives to create a small community of care.

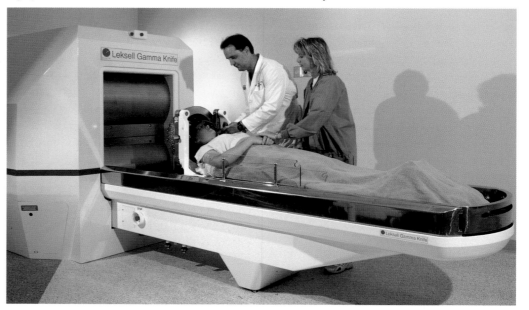

The Columbia Gamma Knife Center houses the only Gamma Knife between Dallas and Kansas City and is an example of Columbia's emphasis on providing state-of-the-art technologies. The Gamma Knife is used to treat brain tumors and other brain abnormalities without an incision.

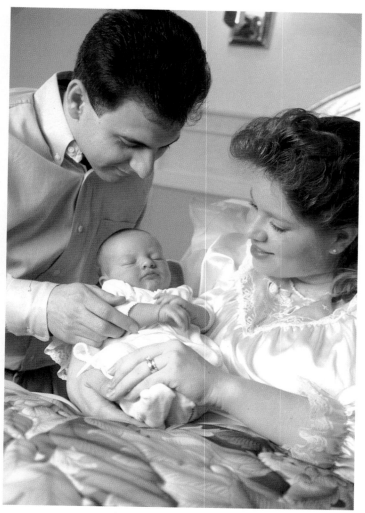

Providing quality, cost-effective, compassionate care is what health care is all about.

COLUMBIA'S COMMUNITY

Columbia serves the communities in which it lives by providing needed medical care. But not everyone has the resources or insurance to deal with a medical crisis. In 1996, Columbia facilities in Oklahoma provided more than $44 million in uncompensated care to individuals and families in need. Nationally, Columbia provided $1.4 billion in uncompensated care for 1996.

Columbia takes an active part in the business community by lending its time and money to support the activities and projects of the chambers of commerce in many Oklahoma communities. This company believes in contributing to the growth and well-being of the business community to provide jobs and resources for its citizens.

Columbia cares about people in their daily lives. Employees, volunteers, and physicians are compassionate people. They share their community spirit and teamwork through meaningful services in the Oklahoma City area by donating time and money to a number of worthwhile projects.

Columbia is a long-time pacesetter company with the United Way of Oklahoma. Employees donate time, money, and resources to assist and promote the United Way efforts, not only within the company but throughout the community.

Employees' care and compassion are further demonstrated by active participation in "Columbia Community Day." This day of community activism is designed to create partnerships with community leaders and organizations, unite employees through teamwork and community spirit, plus improve the quality of life in the communities Columbia serves.

Columbia's community is all-encompassing and represents the future of health care. Disease prevention, transitional care, long-term care, and wellness through education and behavior changes are among the more modern methods Columbia follows in achieving a healthy, quality of life for patients.

"With the shift toward managed care in many parts of the country, health providers are looking for ways to make health care available that is all-encompassing—from preconception to the final stages of life—as well as efficient and cost-effective," said Dunlap. "The goal is to lower the cost of health care in this country and help Americans attain their dream of a greater quality of life."

Columbia plans to be the most comprehensive provider of health care in Oklahoma by continuing to expand all of its services in the Oklahoma City area and throughout the state well into the 21st century. This growth will be based on the four driving factors of quality and compassionate care, fair price, and patient satisfaction. ✪

FOUR DRIVING FACTORS

It is not enough just to provide medical services. Columbia is committed to four driving care factors: quality, patient satisfaction, fair price, and compassion.

At Columbia, quality is measurable. About 36 percent of Columbia hospitals have received Accreditation with Commendation, the highest evaluation given by the Joint Commission for Accreditation of Healthcare Organizations. This rating is the highest in the health care industry. In keeping with this excellence, Columbia finds the best example of practices and outcomes, then tries to replicate those practices across the nation.

In regard to patient satisfaction, Columbia received a 95 percent "satisfied" or "very satisfied" approval rating in a recent national survey of patients. To maintain Columbia's goal to be the most comprehensive, cost-effective provider of quality health care in the state, the organization is continually looking for ways to lower cost through improvements in the operations of the facilities.

Compassion is the underlying theme at the heart of Columbia activities.

The health care needs of seniors will steadily increase into the 21st century.

Oklahoma State University

Oklahoma State University (OSU) is the university of the future, providing Oklahomans and Oklahoma businesses a competitive edge through shared expertise, cutting-edge research and the technology linking solutions to people. From lasers that let manufacturers trim at the precision of a single atom to processing labs that help food producers meet the strictest standards, OSU is a living example of the collaboration that works for Oklahoma.

With campuses in Oklahoma City, Stillwater, Tulsa and Okmulgee, 66 county offices, 12 research stations and 27 telecommunications sites, OSU is Oklahoma's only comprehensive land-grant university. That distinction means OSU must offer more than teaching and research, but also services that extend the resources of the university to the state and nation.

OSU is the choice of more than 1,000 high school valedictorians from every county in Oklahoma. It is home to the nation's largest asynchronous transfer-mode computing network. Programs in accounting, business administration, animal science, mathematics, chemistry, engineering and hospitality management are nationally ranked.

OSU is the choice of more than 1,000 high school valedictorians from every county in Oklahoma.

"State" is OSU's middle name, and while OSU serves all of Oklahoma, the Oklahoma City community is the beneficiary of several specialized programs. More than 2,000 employees at Lucent Technology have received training through ETOP (Enhanced Training Opportunity Program), developed by OSU-OKC and the IBEW trade union, and the effort has become the national model. General Motors, the United Auto Workers and OSU-OKC are partners in another in-plant training program. The campus has developed a

special fast-track program with the Oklahoma City Police Department that has allowed 145 officers to earn associate degrees in Police Science. A similar effort has just debuted in the Oklahoma City Fire Department.

DISTINGUISHED GRADUATES AND INTERNATIONAL PROGRAMS

OSU students range from coeds seeking the classic college experience to corporate professionals planning for the next promotion. They come from every county in Oklahoma, every state in the nation and 90 nations—enjoying degrees in nearly 200 majors.

Among distinguished OSU graduates, the university counts 89 generals and admirals, following only the service academies and Texas A&M. OSU is the choice of 42 Fulbright scholars. Half of Oklahoma's congressional delegation earned degrees from OSU. Two of the nation's largest oil companies, Phillips Petroleum Company and Kerr-McGee Corporation, prospered under the leadership of OSU graduates.

OSU is the international headquarters for fire training and safety. Its Fire Protection Publications is the largest producer and distributor of fire and emergency response materials in the world. International Fire Service Training Association accreditation from OSU is the world standard.

OSU's expertise in telecommunications is second to none. That means advanced courses in subjects from Physics to Russian are available to students in small high schools from Oklahoma City to Alaska. It offers on-site master's degrees

OSU students range from coeds seeking the classic college experience to corporate professionals planning for the next promotion.

to computer scientists at Seagate plants in Oklahoma City. Researchers at the Oklahoma Medical Research Foundation can link high-tech equipment at OSU through the magic of telecommunications. In fact, OSU's newest degree program was developed in consultation with businesses from Southwestern Bell to American Airlines to blend engineering, computer and business expertise into a master of telecommunications.

OSU AND ECONOMIC DEVELOPMENT

Oklahoma City relies on OSU for everything from product testing to economic forecasting. The MAPS project downtown was modeled by the OSU School of Architecture. Nurses at Integris Health are given specially tailored training through OSU-OKC. In backyards across the metro area, OSU Master Gardeners help the grass grow greener and tomatoes ripen redder. The horses at Remington Park even run faster, thanks to the latest in Equine Science from OSU's College of Veterinary Medicine.

The university delivers more than 200 business services through the XTRA center—eXtending Technology, Research and Assistance. And the services are deliverable by satellite, fiber-optic cable or the Internet.

Nationally recognized faculty at OSU work with Oklahoma City businesses to find solutions to today's problems. The 3M Company saved nearly $1 million when an OSU scientist used $1,000 of spare parts and a leftover photodetector to make a detector of microscopic flaws in layers of the patented Matchprint film. Scanning devices commercially available at $1 million each were no match for the cost-efficient prototype designed by OSU.

OSU is the prime contractor for the military's Computer Assisted Technology Transfer (CATT) project. It is a new idea that uses computers to give small and medium-sized businesses a chance to successfully bid on government contracts. The entrepreneurs team together to form "virtual corporations" that meet the specifications for a single part or project. Tinker Air Force Base is the national model for CATT and OSU's partner in an idea that could benefit up to 4,000 Oklahoma companies.

PRODUCING THE EXCEPTIONAL

Research at OSU supports Oklahoma priorities in telecommunications, biomedicine, natural resource preservation and product processing.

The horses at Remington Park run even faster, thanks to the latest in Equine Science from OSU's College of Veterinary Medicine.

The OSU Center for Laser and Photonics Research ranks in the top two or three in the world.

The OSU Center for Laser and Photonics Research ranks in the top two or three in the world. Laser scientists there are conducting research toward laser solutions in communications, manufacturing, defense and medicine. The potential for blue laser applications in the next generation computer chips and OSU's international reputation in that field are attracting the interest of computer engineers worldwide. Cancer treatments that match colored lasers and dyes

absorbed only by cancerous cells are being tested at OSU's College of Veterinary Medicine.

Oklahoma's scientific community shares the benefits of the state's only 600-MHz, high-field, nuclear magnetic resonance (NMR) spectrometer located at OSU. One of the most powerful research instruments in the world, the NMR is critical in bio-medical and agricultural research. An Internet connection makes OSU's NMR available to researchers at Oklahoma's universities, research hospitals, research foundations and industrial research and development departments.

The Oklahoma Food and Agricultural Products Research and Technology Center at OSU was created to stimulate and support the state's product processing industry. Its labs can help Oklahoma businesses comply with new food safety standards, design equipment to increase profits, or even develop new product lines.

A new Advanced Technology Research Center at OSU offers similar support for the state's technology-based potential. The 165,000-square-foot center contains a series of modular labs where OSU scientists and the best minds from the industry can work side by side to develop new products or better ways of producing old ones. Energy conservation, energy storage and conversion, manufacturing, materials processing, hazardous and industrial waste management, plus laser applications are top priorities for this cutting-edge facility.

OSU is the university of the future—sharing its incredible resources to help build the future of our state and nation. ✪

Oklahoma City Community College

The Oklahoma City Community College philosophy encompasses student success, technology, community, and cooperation. These four elements, inspired by President Robert Todd, provide an education reinforced by quality and innovation to approximately 12,000 students per semester and individuals in the community.

Celebrating its 25th anniversary in 1997, Oklahoma City Community College is the youngest public institution of higher education in the state. Through academic focus, flexibility, on-site state-of-the-art facilities, and collaboration, the college is steadily setting its course for the future.

Oklahoma City Community College is a two-year public institution located in the southwest quadrant of Oklahoma City. Thirty-eight associate of arts and associate of science degree programs are offered in areas ranging from biology to theater arts. Another 35 associate in applied science degree programs vary from accounting to nursing. More than 10 certificate of mastery programs are available.

Faculty members at Oklahoma City Community College annually receive teaching excellence awards from the National Institute for Staff and Organizational Development and other organizations. The current faculty-to-student ratio is 20 to 1.

In addition to day and evening classes, Oklahoma City Community College has moved into an extremely flexible course schedule. Weekend College takes place on Friday night, Saturday, and Sunday afternoon. The 24-hour college provides late night and early morning classes convenient for shift workers and others unable to attend classes during traditional class hours.

The College also helped initiate The Greater Oklahoma City Higher Education Consortium, which offers a variety of classes in downtown Oklahoma City. Oklahoma City Community College, Oklahoma State University-OKC, Redlands Community College, Rose State College, and the University of Central Oklahoma make up the consortium.

Celebrating its 25th anniversary in 1997, Oklahoma City Community College is the youngest public institution of higher education in the state.

Oklahoma City Community College offers an extremely flexible course selection.

The Office of Business and Industrial Development (BID) is another flexible program that addresses the educational needs of area businesses and industries. Microcomputer classes, continuing education for professionals, customized training, and business-related classes are available. The training courses can be arranged on the campus or at the business location.

The College's Office for Distance Education bridges the distance gap and offers courses over the 12-LIVE two-way audio and video interactive system. This cutting-edge fiber-optic network links the community college and five school systems, a university, and a vocational-technical school.

Technology is also a predominant feature in the newly constructed $9 million library, which is a landmark in the Oklahoma City area. The goal of the library is to provide students and members of the community with the ability to complete research by using one computer that gives them access to all available electronic reference material.

This state-of-the-art library has good company on the Oklahoma City Community College campus with the Aquatic Center, one of the finest Olympic aquatic facilities in the nation.

The campus is also the location for Arts Festival Oklahoma. Hosted by the college, this arts festival features art crafts, the Oklahoma City Philharmonic Orchestra and fun for the whole family and attracts more than 60,000 people on the Labor Day weekend.

Whether offering on-site classes at all hours, enhancing the lives of individuals in the community, focusing on technology, or coordinating programs with other educational institutions and businesses, there is no doubt that Oklahoma City Community College helps knowledge cross traditional boundaries and its students achieve success. ✹

The Henry G. Bennett, Jr. Fertility Institute—INTEGRIS Baptist Medical Center

The medical team at the Henry G. Bennett, Jr. Fertility Institute has one clear goal for each patient: to achieve pregnancy and delivery of a live-born baby. After a decade of dedicated service, approximately 140 babies have been welcomed into the world as a result of the institute's commitment.

In 1985, the institute was created in order to provide state- of-the-art reproductive technology to infertility patients. David A. Kallenberger, M.D.; Royice B. Everett, M.D.; Fenton M. Sanger, M.D.; Mike Seikel, M.D.; Tony Puckett, M.D.; J. Clark Bundren, M.D.; and Ed Wortham, Ph.D. created the institute as a nonprofit foundation at the Women's Center located in the INTEGRIS Baptist Medical Center. The birth of their first baby from in-vitro fertilization occured on December 6, 1986.

The facility's services and personnel have continued to expand over the years. When Eli Reshef, M.D. was recruited as Medical Director in October, 1995, he brought advanced experience to the institute, moving it to a new level of sophistication.

Providing superior reproductive medical services to Oklahoma City at INTEGRIS Baptist Medical Center's Henry G. Bennett Fertility Institute has brought new options to numerous couples hoping to achieve pregnancy.

The extensively trained and talented staff of 11 physicians, nurses, and support personnel offer an expedient and thorough evaluation for the infertile couple. An extensive range of treatment options includes the most recent technological advances for achieving a successful pregnancy. Couples are assured of personalized, supportive care as various treatment alternatives are considered.

"We understand the frustration many couples have when they are unable to become parents," says Dr. Kallenberger, the program director "To help them

achieve a successful pregnancy and delivery, we continually add new technologies as they become available. These range from a thorough evaluation up through reconstructive surgery, ovulatory stimulation, treatment of male factor, and assisted reproductive techniques."

The Henry G. Bennett, Jr. Fertility Institute has recently offered intracytoplasmic sperm injection for the treatment of severe male factor infertility. Like every laboratory procedure conducted at the institute, this technique is subject to the strictest quality control measures.

"Our medical team is committed to strict protocols with no shortcuts in medical standards,"

says Dr. Reshef, BFI's medical director. "Every year, our laboratory is completely reconditioned to eliminate potential obstructions in clinical procedures and outcomes. We will not compromise the integrity of the embryo under any condition." Among its future goals, the institute plans to continue expanding pregnancy rates by providing affordable and effective treatment for the infertile couple. Additionally, the team hopes to work with insurance companies to include fertility treatment as a covered expense.

"The physicians and staff of the Henry G. Bennett, Jr. Fertility Institute will continue to provide compassionate, complete, and up-to-date care for infertile couples," Dr. Kallenberger concludes. "We have made a commitment to constantly improve our techniques, so all fertility services available globally can be found in Oklahoma City. We will continue to operate with the highest levels of integrity, keeping our patients' interests as our top priority." ✜

Bennett Fertility Institute TEAM (bottom row, l-r) Eli Reshef, M.D., Medical Director; David A. Kallenberger, M.D., Program Director; J. Edward Wortham, PhD., Laboratory Director. (second row) Kathy Lynn, Administrative Secretary; Marcia Anderson, IVF technician; Corlis McCleod, RNC, Lead Nurse Clinician. (third row) Tommy Waugh, lead IVF technician; Gary Strebel, M.D., staff physician; Jenny Price, RN, staff nurse. (fourth row) Cindy Newcomb, RNC, nurse clinician; Fenton Sanger, M.D. and Michael Seikel, M.D., staff physicians.

Ashton Edwards, BFI's first In Vitro fertilization baby born December 6, 1986 and Rachael Shadid a recent GIFT (Gamete IntraFallopian Transfer) baby born April 22, 1997.

Southern Nazarene University

Southern Nazarene University (SNU) is an institution with deep conviction. Through focus on academic excellence, service, Christian faith, and lifelong learning, SNU's mission to build responsible Christian persons is achieved.

This four-year college embraces the Wesleyan theological perspective and services the South Central region of the Church of the Nazarene. An average student body of more than 1,800 graduate and undergraduate students represents 38 states, 38 countries, 60 of Oklahoma's 77 counties, and 50 different religious denominations. The university's motto is Character, Culture, and Christ.

In 1998, SNU will celebrate 100 years of vital Christian education and service. This centennial recognition honors the heritage and future of Oklahoma City's oldest, independent liberal arts institution.

Although classified as a liberal arts university, a selection of undergraduate majors offers more than 70 major fields of study.

The university's educational programs are designed to achieve two major educational goals. The first is to help students become critical and creative thinkers who can clearly discern and communicate a Christian perspective in every aspect of life. The second goal is to prepare students for successful professional careers. Programs of excellence help students reach these goals.

SNU is home of the state's only undergraduate laser research program and one of only two Oklahoma schools offering an undergraduate level cadaver study program. Throughout its 50-year history, the school's premedicine program boasts a 90-plus percent admittance rate of its graduates into medical schools.

The Society of Physics Students has named SNU's Physics Department among the Top 20 Physics Chapters in the nation for six consecutive years. The School of Business touts five Intercollegiate Management Business Gaming Team national and international titles during the past six years.

Educational diversity and community commitment are exemplified through the School of Adult

In March 1997, Southern Nazarene's Lady Redskin basketball team won their fourth consecutive NAIA National Championship. The Lady Skins have also won an all-time women's record 114-straight games at home. Most consider SNU to be the most dominant basketball program at any level in the decade of the 1990s. Photo by Berry J. Yarbrough, courtesy of Southern Nazarene University.

Since Southern Nazarene University President Dr. Loren Gresham took office in 1989, SNU has shown a 24 percent increase in enrollment. This four-year liberal arts university is considered to be one of the finest in the Midwest, and now boasts of an enrollment of more than 1,800 undergraduate and graduate students. Photo by Toby C. Rowland, courtesy of Southern Nazarene University.

Studies, PEER LEARNING NETWORK with community businesspeople and CEO's from major corporations, and the EXCEL camp for children grades 4-12.

Distinction is also found in SNU's elite athletic program in a Christian atmosphere. Six NAIA National Title have been championed by the Redskins. The Lady Redskins women's basketball program is the only program to win four consecutive national championships.

In addition to development of mind and body, SNU emphasizes spiritual development. Chapel services on Tuesday, Wednesday, and Thursday are coupled with 12 hours of course requirements in religion before graduation.

A strong emphasis is also placed on service to others. Incoming students participate in a major service project as part of their orientation week at SNU. In fact, Southern Nazarene has the only collegiate chapter of Habitat for Humanity in Oklahoma.

SNU President and Fulbright Scholar Loren P. Gresham, Ph.D., established the goal for each SNU student to have international awareness and experience. Global service projects help students reach this goal.

Mission programs in Bulgaria, Sicily, Moscow, and the Ukraine are available for students. An annual Mexico mission with alumni, students, and faculty allows the SNU community to help build medical clinics, low-cost housing, and churches.

Reaching from service to academic excellence, SNU has developed into one of the nation's outstanding Christian colleges. It is an institution with deep conviction—a university that is growing into the next 100 years of building responsible Christian persons. ✿

UroCor, Inc.

Leading edge. Large U.S. market share. Progressive. Anchor tenant for the Oklahoma Health Center's Biomedical Research Park. This is hardly a description for a fledgling company struggling to reach financial and marketing milestones in 1987.

That was phase one. Since then, with a new president and chief executive officer, company name and building, UroCor, Inc. has transformed into a dynamic pacesetter with substantial direction for the future.

UroCor is one of the most highly evolved and successful companies in the emerging field of disease management with a chosen specialty in urology. It provides a broad range of diagnostic and information services to urologists and managed care organizations across the United States.

The company assists urologists in the management of patients with urological cancers and other complex urological diseases. UroCor has also advanced to provide quality and comprehensive capabilities throughout an entire disease cycle. The goal is to improve patient outcomes while reducing the total cost of managing the diseases.

This integrated approach works. UroCor has grown in annual revenues from under a million in 1990 to nearly $30 million in 1996. The number of employees has jumped from 15 to more than 260, with 30 senior managers including 12 M.D.'s and 6 Ph.D.'s in early 1997.

Urocor was included in the *Inc.* 500 ranking of the fastest-growing private companies in the United States for 1993 through 1996. In 1995, the company was also the recipient of the prestigious, national Blue Chip Enterprise Award that recognizes companies for outstanding development and growth. In May, 1996, UroCor transitioned from a private to a public company through an initial public stock offering.

The foundation of UroCor's performance is based on four operating units. The UroDiagnostics Group provides primary diagnostic services to urologists through its nationwide technical sales and managed care specialist organization. The UroSciences Group develops, acquires, and applies advanced diagnostic technologies directed at improving disease management.

The third group is UroTherapeutics, which focuses on licensing or acquiring rights to distribute therapeutic products. The Disease Management Information Systems deliver clinical management tools ranging from decision support through practice management analysis through a secure computer network.

William A. Hagstrom, president, chairman and chief executive officer, credits a number of agencies and programs, and the Oklahoma Health Center for providing a "tremendously supportive environment" to build a business. This region did not have existing infrastructure to support a biotech company.

Hagstrom is now part of a team working on a long-term vision to cultivate the Biomedical Research Park into a flourishing home for biotech industries. The park immediately adjoins the 19-institution Oklahoma Health Center campus, is one mile from the downtown business district, and one-half mile from the state Capitol. Hagstrom noted this park is ideally situated so that key resources and support networks are available for a biotech company to thrive.

"The biotech industry is important to all of our futures," said Hagstrom. "It represents an incredible opportunity for our future as it relates to the ability to better diagnose human diseases. It brings high-quality, highly skilled jobs. This alliance of success cannot help but breed success."

Success is UroCor, Inc.

UroCor is one of the most highly evolved and successful companies in the field of disease management.

UroCor is located adjacent to the Oklahoma Health Center which is a great resource and support network for the biotech company. Photo by J. D. Merryweather.

Metro Tech

Metro Tech is one of Oklahoma's premier vocational schools. Training, high-tech equipment, modern facilities, low tuition, small class sizes, and personalized attention strongly empower Oklahoma City residents with "Education for Success!"

Oklahoma's vo-tech system is among the top two in the nation, and Metro Tech is one of the largest service providers among the 29 area vocational-technical districts in this statewide network. As a multicampus vocational training center, Metro Tech operates seven training sites in the Oklahoma City area.

The main Springlake campus is a 95-acre property, formerly the site of Springlake Amusement Park. Opened in 1922, this park was an ideal spot for entertainment, dancing, and roller coaster rides for some 60 years. Where there were shrieks and laughter from breathtaking rides on the Big Dipper roller coaster, students now learn exciting new career skills for the future.

From "The Thrills of Yesterday to the Skills for Tomorrow," Metro Tech knows its mission. Serving approximately 13,000 students annually, the school district is unique in its ability to provide quality education, training, and related services dedicated to preparing diverse populations for successful employment in a competitive, global economy.

With everything from aviation technology, computer science, and electronics to practical nursing, Metro Tech offers programs for high school students and adults, as well as specialized training programs for business and industry personnel. The course selection involves more than 30 daytime programs, over 200 short-term courses, and specifically designed industry training courses.

High school students in the 11th and 12th grades attend vocational education classes for half a day, at no cost, and receive credit toward graduation, as well as college credit for most programs. Metro Tech provides bus transportation to all Oklahoma City and Crooked Oak high schools at no charge.

Adults may enroll in daytime programs on a full-time or part-time basis. Hundreds of short-term evening or weekend classes focus on skills and interests that can make an immediate difference in someone's life.

Metro Tech's nationally acclaimed Aviation Career Center offers a quality training experience for aviation-related careers. The Aviation Maintenance Technician program is an 18-month course that is certified and approved by the Federal Aviation Administration.

Through the division of Adult and Continuing Education (ACE), Metro Tech offers customized business and industry training programs. Businesses benefiting from services include General Motors and Columbia Presbyterian Hospital, the City of Oklahoma City, and the Oklahoma Supreme Court system.

Large corporations and small businesses can be accommodated through specific training at the business location. Hundreds of new business owners take advantage of courses offered through the Small Business Assistance Program. Special programs are available for single parents, teen parents, displaced homemakers, students interested in nontraditional careers, and those who are vision or hearing impaired. Language interpreters are also available. Ability and interest assessments, computerized skill-building learning centers, and job search assistance are other special services offered by Metro Tech.

A community resource in every respect, Metro Tech's Springlake Business Conference Center and outdoor amphitheater offer the metro area a popular choice for meetings and special events, providing flexible, affordable facilities in a parkland setting.

Metro Tech's vocational-technical training centers, located throughout Oklahoma City, are the front doors to the future of work force development for this dynamic metropolitan community. Metro Tech is Education for Success! ✪

Oklahoma City University
Vision Becoming Reality

The national and statewide attention Oklahoma City University receives should come as no surprise. With a fall 1996 enrollment of 4,697, OCU remains the largest independent university in the state. Truman scholars, All-College *USA Today* Academic First Team members and national NAIA athletic championships are reaffirming that OCU is a university whose time has come.

OCU has been making a positive impact on Oklahoma City since its establishment in 1904. The University's outstanding undergraduate programs and highly rated professional and graduate programs continue to earn recognition: *The National Review College Guide* lists OCU as one of America's top 58 liberal arts colleges and universities, and the University is included in *The Student Guide to America's 100 Best College Buys.*

During the past 18 years, OCU has invested nearly $40 million in the construction of six new academic buildings and renovation of existing facilities while incurring no debt. This continuing and dramatic campus development is required to accommodate expanding academic programs and a student enrollment that has nearly doubled in the past decade.

Not only does OCU play a major role as an educational leader in our community, state and nation, the University has a strong economic and cultural impact in Oklahoma City. According to a 1996 analysis by the B.D. Eddie Business Research and Consulting Center, OCU's economic impact upon Oklahoma County was $233,618,567, a 19 percent increase over 1995. In the 12 years that this study has been conducted, the 1995-96 fiscal year figure reflects the greatest increase ever in OCU's economic impact.

Oklahoma City University, a vital industry without a smokestack, is among the top 15 non-manufacturing employers in metropolitan Oklahoma City. The total payroll for 1996-97 will exceed $20.6 million.

In April 1994, the University began its most ambitious campaign to date, the 10-year, $100 million Era of Emergence campaign. Three years into the campaign, nearly $42 million, or 42 percent of the goal, has been achieved. OCU will celebrate its centennial in 2004 with the conclusion of the Era of Emergence campaign.

To complete the campus master plan over the next eight years, the University will invest an additional $20 million in facilities (including a new Health, Convocation, and Wellness Center and Panhellenic Quadrangle), raise $58 million in endowment funds, greatly increase faculty

salaries and scholarships for students, re-equip classrooms and labs with technology to keep them competitive, and assure ongoing fiscal stability by providing sufficient annual operating funds. Accomplishing these objectives will require an investment in excess of $100 million by the end of the century.

As it approaches its centennial, Oklahoma City University maintains a tradition of excellence by responding to the educational needs of a wide variety of students, and through a strong partnership with the church and the community it serves. ✪

As Oklahoma's largest independent university, OCU had a quarter of a billion dollar impact on the Oklahoma City economy in 1996.

Oklahoma Medical Research Foundation

Progress toward discovery—medical breakthroughs—then treatments and cures. This pattern defines the more than half-century mission of the Oklahoma Medical Research Foundation (OMRF), committed to medical research that more may live longer, healthier lives.

As long as there are diseases, the fight against them must continue. That is why OMRF was founded in 1946, by Oklahomans with a vision for the future. They recognized the value of medical research and gave it a place to thrive in this great state.

"It's a story of people—both the researchers who have dedicated their lives to discovering the causes of disease in man and the citizens of Oklahoma who continue to open their hearts in support of this important work," says William G. Thurman, M.D., president and scientific director.

OMRF scientists represent the best in their fields. As an example, OMRF is home to Oklahoma's only two distinguished Howard Hughes Medical Institute (HHMI) Investigators. Dr. Charles T. Esmon was the first person named from outside a university setting when he was designated an HHMI investigator in 1988. Dr. Joan W. Conaway was named an HHMI investigator in 1997, along with 70 other professionals from across the United States.

Individually and collectively, OMRF scientists have earned worldwide recognition for research in the following areas:
* Arthritis and Immunology—lupus, rheumatoid arthritis, scleroderma
* Cardiovascular Biology—heart attack, blood diseases, stroke, septic shock
* Children's Diseases—cancer, leukemia, genetic disorders, cystic fibrosis
* Clinical Pharmacology—FDA drug trials for hypertension, cholesterol, asthma, diabetes
* Free Radical Biology and Aging—Alzheimer's disease, Parkinson's disease, stroke
* Immunobiology and Cancer—leukemia, Hodgkins' disease, lymphoma, osteoporosis, breast cancer
* Molecular and Cell Biology—ALS, muscular dystrophy, multiple sclerosis, new antibiotics, cancer genes
* Protein Studies—HIV, AIDS, diabetes, ulcers

OMRF is literally its own world of research. The foundation supports the work of approxi-

Knowledge gained today in laboratories such as those at OMRF will impact the health of generations to come. Drs. Linda Thompson and Katherine Kelly work to unlock the secrets of cancer and other diseases of the immune system. Photo by David Fitzgerald.

Dawn Schweitzer, left, Yukon Mid-High School science teacher and former OMRF Foundation Scholar, leads students Jake Caldwell, Macara Roberts and Sarah Smith on a journey through the Internet courtesy of OMRF's Oklahoma TeleScience Project. The TeleScience Network links the Foundation's research library and scientific staff with high school science teachers and students across the state of Oklahoma. This online computer access enriches science education through direct phone lines to state-of-the-art information, resources and contact with scientists otherwise unavailable in the classroom. Photo by David Fitzgerald.

mately 350 full-time employees and provides a working environment for 125 additional research personnel who are employed by other institutions of the Oklahoma Health Center. OMRF has 140,000 square feet of laboratory space, equivalent to nearly three football fields.

The funding necessary for OMRF to fulfill its mission comes solely from competitive grants and public and corporate donations. Every dollar donated to OMRF is spent only on research.

A 1996 publication highlighting the 50th anniversary of the Oklahoma Medical Research Foundation gives specific examples of OMRF's excellence and impact on the medical community. It highlights discovery of a second gastric enzyme and development of a treatment for patients with protein C deficiency and other blood abnormalities. The publication acknowledges discoveries regarding the role of reactive oxygen species (ROS) on damage to the brain that occurs during stroke, Alzheimer's disease, Parkinson's disease, and other neurodegenerative

diseases of the central nervous system.

One scientist recognized the vital importance of the protein portion of the lipoprotein and devised a classification and naming system that was adopted worldwide in 1972. OMRF boasts the first scientist in Oklahoma to receive a Fellowship Award from the Jane Coffin Childs Memorial Fund for Medical Research, which annually awards fellowships to top young scientists conducting research into the cause, origins, and treatment of cancer.

A discovery of several different autoantibodies produced by lupus patients has enabled doctors to diagnose the disease and provide information about its expected course. Another finding is a major step forward in understanding the process that changes a normal cell into a cancer cell.

These examples illustrate the Oklahoma Medical Research Foundation's work. The answers are there, and foundation scientists stand ready with the brightest minds and the highest hopes—"*...that more may live longer, healthier lives.*" ✻

Oklahoma Christian University

Since its beginning in 1950 and its transition to a four-year college eight years later, Oklahoma Christian University (OC) has been a national leader in innovation and liberal arts education in a Christian environment.

The university's development of its prototype Mabee Learning Center with banks of electronically linked study carrels for each student revolutionized the productiveness of student study time in the 1970s.

In the 1980s, Oklahoma Christian took the national stage for free enterprise education with the showcase for American business genius, Enterprise Square, USA.

In 1994, Oklahoma Christian seized a unique opportunity to expand the dream of its founders by establishing a branch campus, Cascade College, in Portland, Oregon.

Today, the university continues its tradition of excellence and has received national acclaim for its performance. For three consecutive years, OC has been recognized by *U.S. News & World Report* as a "best value" in higher education, and the John Templeton Foundation has selected the university to its Honor Rolls for Character Building Colleges and Free Enterprise Teaching.

Oklahoma Christian University is a private, coeducational university where excellence in classroom teaching is an institutional priority. The university's 200-acre campus consists of 33 contemporary buildings coordinated in architectural design.

Committed to academic quality, the university seeks to prepare students for leadership in the church and service in their careers. Oklahoma Christian has a full-time faculty of more than 65, with the majority holding a doctorate or terminal degree.

Oklahoma Christian offers more than 70 degree options, including preprofessional programs such as prelaw and premed. Studies are offered in the colleges of Bible, business, education, liberal arts, and science and engineering.

Oklahoma Christian is accredited by the North Central Association of College and Secondary Schools. Its engineering programs are accredited by the Engineering Accreditation Commission of the Accrediting Board for Engineering and Technology. Specialized accreditation has been received in the areas of music, business, and teacher education. The university encourages international study through semesters abroad in Ibaraki, Japan; Vienna, Austria; and South America.

Oklahoma Christian is a member of the National Association of Intercollegiate Athletics and the Sooner Athletic Conference. Sports for men are soccer, cross-country, baseball, basketball, golf, tennis, and track. Sports for women are cross-country, basketball, soccer, softball, tennis, and track.

An all-time enrollment record of 1,817 was established in the fall of 1996, representing 45 states and 24 foreign countries.

It is the individual growth and success of each of those students that fuels Oklahoma Christian's mission. In his September, 1996, inaugural address as the fourth president of Oklahoma Christian University, Dr. Kevin Jacobs affirmed the university's commitment to the mission and purpose for which it was established.

"Oklahoma Christian places God at the center of its being," Jacobs said. "The study of the Bible is the center of our core curriculum. Daily chapel is central to the campus experience."

"This is a community of believers—from board to student. That is our purpose. That is our heritage. That is our distinctiveness." ✱

Dr. Kevin E. Jacobs serves as the fourth president of the University.

The Allison Biblical Studies Center serves as the campus backdrop for Spring Commencement.

Integrated Medical Delivery

Taking a proactive approach in the delivery of health care services is the forward-thinking determination of a group of physicians who founded Integrated Medical Delivery (IMD). A diversified medical management company, IMD is designed to ensure the success of a physician-owned integrated health care delivery system. IMD was founded on the belief that managed care is an effective method of delivering health care, as long as the physicians are managing the care. An independent, physician-owned, multispecialty provider network, with expert management and information systems, IMD contracts directly with purchasers of health care service.

IMD was formed in 1995 by a group of Oklahoma City physicians. Stanley Pelofsky, M.D., a neurosurgeon, states that "a network of the highest quality physicians and facilities can deliver what the consumer wants, effective affordable health care." He believes that the competitive nature of health care has compelled physicians "to develop a physician-owned health care company to ensure quality patient care, access to specialists, all at a reasonable cost."

The focal point of IMD is its provider network, PhysiciansCHOICE. "The development of PhysiciansCHOICE was a unique opportunity. It was the first time that physicians and payers worked in tandem to develop a provider network," says Dr. Pelofsky. PhysiciansCHOICE is a network of more than 400 physicians and facilities located throughout 72 Oklahoma counties. Servicing Oklahoma's self-insured companies, PhysiciansCHOICE's principle of intensive management of health care cases and ensuring one-on-one interaction with the patient and the physician, results in tremendous patient satisfaction. Satisfied patients create an environment of reduced health care costs.

IMD has recruited a diverse staff of outstanding health care and business professionals. "The business of health care is changing rapidly; new ideas and approaches will be critical to future success," says Burt J. Loessberg, Chief Executive Officer of IMD. "The owners of IMD have outstanding vision. I believe that working for the physicians provides me with the insight to better manage this process."

IMD provides maximum efficiency by

(seated l to r) Bob J. Rutledge, M.D., Burt J. Loessberg, CEO of Integrated Medical Delivery; (standing l to r) Robert L. Remondino, M.D., Stanley Pelofsky, M.D.

reducing costs associated with managing facilities, physicians' practices, and physician networks, while expanding growth and service opportunities. The consolidation of services, patient billing, managed care contracting, purchasing, accounting, marketing, and data processing allows the IMD companies to complete more effectively in the marketplace. The key to maintaining the affordability of health care is the management of utilization. Loessberg believes "by developing a network of providers with historically efficient utilization outcomes, immediate cost savings can be achieved."

Managed care mandates that to deliver effective health care a network needs the highest-quality physicians and sound business principles. "As physicians, we needed a partner who understood the business of health care," says Dr. Pelofsky. To ensure the development of this network, IMD enlisted the support of Dorchester Capital, a local investment company. "The purchasers of health care want the controls afforded by managed care. What makes IMD unique is that the providers of care are developing the controls and taking the economic risks," says John T. Perri, President of Dorchester Capital.

What started out as one physician's dream has grown into an effective partnership of multiple physicians and facilities, all sharing the common belief that physicians can manage health care. The future of IMD is through the development of strategic partnerships with new providers and provider-owned facilities. "The development of Integrated Medical Delivery is in its infancy," says Loessberg. "Through quality providers, strong case management, and strong employer relations, this company is taking control of health care delivery in Oklahoma." ❂

Oklahoma City Clinic

The Oklahoma City Clinic's founding fathers appear in a group photo in 1919.

The Oklahoma City Clinic has been advancing health and healing since its inception in 1919. Six physicians, dedicated to providing quality health care and services, founded the clinic using a concept known as "group practice."

Working together at a hospital during World War I, surgeons Abraham Lincoln Blesh, M.D., and William Ward Rucks, M.D., were impressed with the idea of physicians working side by side, helping each other to better help their patients. When these two doctors returned from military service, four other specialists joined them to establish the Oklahoma City Clinic.

As one of the largest physician groups in Oklahoma, a team of approximately 100 doctors and 450 affiliated physicians represents more than 20 medical specialties that make up the Oklahoma City Clinic. Designed around the successful Mayo Clinic, these dedicated teams of physicians work in tandem toward the common goal of giving health care at the highest possible standards, while keeping patient costs down.

This concept of managing health care has been providing Oklahoma City the ultimate in health care for more than 75 years —through the Oklahoma City Clinic.

As a multispecialty physician group, the Oklahoma City Clinic physicians combine their specialized training and expertise to deliver patients the finest health care available. Specialty care areas include aviation medicine, behavioral medicine, cardiology, cardiovascular/thoracic surgery, dermatology, endocrinology, otolaryngology (ear, nose and throat), family medicine, gastroenterology, and general surgery.

Other areas of specialty care are hematology/oncology, infectious diseases, internal medicine, neuroradiology, obstetrics and gynecology, ophthalmology, optometry, pediatrics, podiatry, pulmonary diseases/critical care, radiology, rheumatology, and urology.

For the patient's convenience, ancillary services include laboratory and X-ray, pharmacy, optical shop, hearing shop, audiology, electrocardiographic computer services, Sexual Health Center, Diabetes Management Center, and nutrition counseling. The Surgery Center allows the clinic to provide a high level of care without the additional costs associated with hospital surgical centers.

The Oklahoma City Clinic is privately owned by the physicians. The clinic is affiliated with MedPartners. "Today, with a team of the area's finest physicians and the strength of our corporate partner MedPartners, Inc., the Oklahoma City Clinic has grown to provide multispecialty services across much of the state," said Denise Semands, executive director.

The Oklahoma City Clinic is committed to developing and maintaining strong ties with physicians who practice outside the metropolitan area. The clinic recognizes the importance of supporting the health care facilities in rural areas and the benefits of enabling patients to receive health care in their hometowns.

With a central facility on the Oklahoma Health Center campus, five satellites, and numerous remote outreach clinics, the Oklahoma City Clinic's physicians work together to furnish convenient, quality care to Oklahomans.

"On any day, I can walk through the halls of the Oklahoma City Clinic and see our team of physicians consulting with each other in private on patient care issues. Obviously, our founders knew what they were talking about," said Semands. "While other physician groups may struggle to find a niche in today's health care market, we can spend our energies providing caring, state-of-the-art medical services for the patients who are our neighbors—and our friends." ✪

Oklahoma City Clinic patient William Long(seated) has experienced first hand the value of the multispecialty "team" offered by the Clinic. He is surrounded by his physicians: (from left) Tony Ringold,M.D., gastroenterology; John Bell,M.D., ophthalmology; Thomas Russell,M.D., cardiology; and Peter Young,M.D., internal medicine.

Oklahoma Allergy & Asthma Clinic

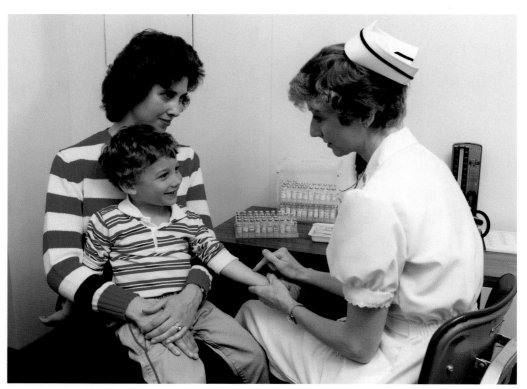

Medical treatment, individualized attention, and community impact perpetuate the clinic's dedication to provide the highest quality of care.

The Oklahoma Allergy & Asthma Clinic is one of the oldest, largest, most respected, and advanced allergy and asthma centers in the United States. This clinic is dedicated solely to the research, diagnosis, and treatment of allergies, asthma, and other allergic disorders. It is one of several privately owned member institutions of the Oklahoma Health Center.

Medical treatment, individualized attention, and community impact perpetuate the clinic's dedication to provide the highest quality of care.

The Oklahoma Allergy & Asthma Clinic, nationally recognized for excellence within the medical profession, has built an outstanding reputation on the ability of its staff to help people of all ages with their allergy and asthma problems. Almost one fourth of the clinic's patients are referred from outside Oklahoma and travel many miles for the sophisticated, high level care provided by clinic physicians.

Each of the clinic's eight active physicians is certified by the American Board of Allergy, Asthma, and Immunology and is a teaching faculty member at the University of Oklahoma College of Medicine.

Techniques and equipment used at the Oklahoma Allergy & Asthma Clinic include the latest advancements in the field. Since the clinic was founded in 1925 by Ray M. Balyeat, M.D., a pioneer in the specialty of allergy medicine, clinic physicians have been participating in or conducting research projects and new drug studies. These endeavors help improve tests and treatments available for asthma and allergies.

Through the recently established Clinical Research Department, the clinic plans to greatly increase its participation in new allergy and asthma drug studies. The clinic also is one of only a few allergy groups in the country with a full-time registered, licensed dietitian specializing in food allergies.

Benefits of having such a wealth of experience, skill, and scientific technology under one roof are recognized daily by patients.

First-time appointments at the Oklahoma Allergy & Asthma Clinic usually last an entire day.

The Oklahoma Allergy & Asthma Clinic stands today as a nationally respected center dedicated to improving the health of those who suffer from allergies and asthma.

"We know spending time with a patient realizes better results. We take a complete allergic history, do a physical examination, order lab work, do allergy testing as indicated, and then spend 30 minutes to an hour with each patient," says G. Keith Montgomery, Executive Director. "By the time a patient is ready to leave, he or she has the doctor's instructions in writing, so there is no question about what is supposed to be done."

A patient leaving the doctor's office with an average four-page written instruction summary is unique. This type of service is indicative of the clinic's level of dedication to the principle that the patient comes first.

The education and outreach extend far beyond the clinic walls. As a Certified Aeroallergen Network Station for the central region of Oklahoma, the clinic gives a pollen and mold information report each morning to local and national media and physician offices.

The clinic physicians also take their role as experts in the fields of allergy and asthma treatment very seriously. They honor numerous requests to speak at schools, parent meetings, organizations, and medically oriented groups on this subject.

Because of the initial dream and determination of Dr. Balyeat more than 70 years ago, the Oklahoma Allergy & Asthma Clinic stands today as an advanced, caring, and nationally respected center dedicated to improving the health of those who suffer from allergies and asthma. ❂

Bone & Joint Hospital & McBride Clinic

From the child's broken arm to the professional with arthritis, specialized orthopedic and arthritis care is found at the McBride Clinic and Bone & Joint Hospital. By exceeding customer expectations, providing nationally recognized care, and committing to keeping costs down, the advancement of medicine at these facilities continues to enhance the quality of life for Oklahomans.

For more than 75 years, the McBride name has been associated with innovation and high standards. Orthopedic surgeon Earl McBride, M.D., became one of the first physicians to do a partial hip replacement in 1947, and he authored a book on rating disability injuries for workers' compensation, a system still used today. The McBride Clinic was the first nationally to combine orthopedic surgeons and physicians specializing in rheumatic diseases.

As an extension of this clinic, a hospital was opened in 1926 to meet the orthopedic needs of Oklahomans, especially crippled children and polio victims. Later named Bone & Joint Hospital, it is only one of 10 orthopedic specialty hospitals in the United States.

Bone & Joint Hospital provides surgical, medical, and comprehensive medical rehabilitative services. Each year, Bone & Joint performs more joint replacement procedures than any other hospital in Oklahoma. The outpatient surgery center gives patients a convenient option when an overnight stay isn't necessary.

This facility was the recipient of the 1996 Oklahoma Quality Award, and the first and only hospital in the state recognized with this prestigious honor. The hospital also received Accreditation with Commendation, the highest level of accreditation awarded by the Joint Commission on Accreditation of Healthcare Organizations.

In 1995, the hospital was named to the *U.S. News and World Report*'s "America's Best Hospitals" list of 40 quality hospitals that perform orthopedic procedures. Ninety-nine percent of Bone & Joint Hospital patients indicate they would recommend the hospital to others.

The Bone & Joint Hospital Rehabilitation Center provides comprehensive medical rehabilitation to promote every patient's return to a functional, independent lifestyle. The emergency department is available 24 hours every day to treat orthopedic, arthritic, and occupational emergencies.

The McBride Clinic Occupational Health Center meets the needs of more than 200 companies in Oklahoma. Respected for conservative care and returning the employee to work as quickly as possible, the center provides a full range of services, including injury care, physical examinations, substance abuse screening, inoculations, and a 24-hour emergency room.

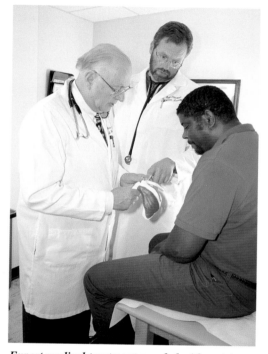

Expert medical treatment coupled with a stringent return to work philosophy are why Oklahoma businesses have relied upon McBride Clinic's Occupational Health Center for 75 years. Richard Hess, M.D., and Bob Havener, P.A.-C., explain an injury to a worker.

Olympic gymnast and gold medalist Shannon Miller chose McBride Clinic and Bone & Joint Hospital for sports injury care. The sports medicine program furnishes preventive, surgical, and rehabilitative services to professional and weekend athletes. The group also provides athletic trainers and physician coverage for a number of high schools and universities across Oklahoma.

Physicians and staff stay abreast of the latest technology through innovations in arthroscopy, laser surgery and micro surgery. The latest technological addition is the DEXA machine that screens women and men for osteoporosis or decreased bone density.

"We are committed to improving the health status of our patients and our community," said James A. Hyde, president, Bone & Joint Hospital. Health care workers provide free community education on arthritis and osteoporosis. Other areas of emphasis include physical exams for school children and neighborhood redevelopment efforts.

Precision, innovation, tradition, and a national reputation for health care excellence portrays McBride Clinic and Bone & Joint Hospital. These facilities are dedicated to providing quality, economical orthopedic, arthritis, and rehabilitative treatment for Oklahomans of all ages. ✪

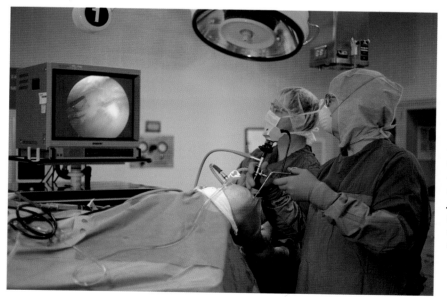

With the goal of returning a golfer to her sport, Joseph F. Messenbaugh, III, M.D., assisted by Don Flinn, P.A.-C., performs an arthroscopy at Bone & Joint Hospital.

The Dean A. McGee Eye Institute

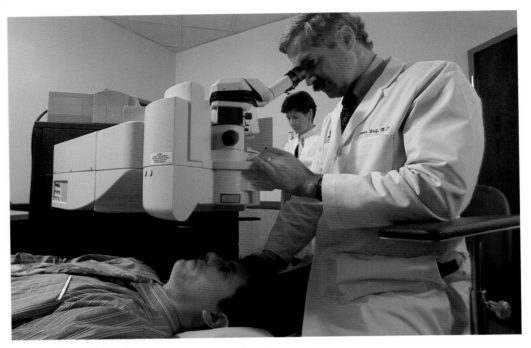

The Dean A. McGee Eye Institute (DMEI) is dedicated to providing the highest quality patient care and maintaining a renowned investigative center for vision research. The Institute is also home to the University of Oklahoma's Department of Ophthalmology and conducts training programs for medical students, residents, and clinical fellows. It is a young facility with a proven record.

"Established in 1975, the McGee Eye Institute has grown to nearly 30 ophthalmologists and 10 basic scientists, making us one of the largest eye institutes in the country," notes David W. Parke II, M.D., president and CEO.

Patient care at the McGee Eye Institute is offered in all the major subspecialty areas of ophthalmology. DMEI provides clinical services for nearly 60,000 patient visits each year. It represents the only facility between Dallas, Phoenix, and St. Louis that can offer this complete spectrum of subspecialty eye care for everything from tumors to macular degeneration.

The international reputation of DMEI ophthalmologists in areas such as glaucoma, retinal disease, cataract surgery, pediatric ophthalmology, and neuro-ophthalmology draws patients from throughout the United States to Oklahoma. (Over one-third of DMEI ophthalmologists are listed in *Best Doctors in America*.) The Institute also operates an ultramodern laser refractive surgery center—the Stephenson Laser Center, which is used by ophthalmologists from throughout Oklahoma.

The Institute has met some specific challenges in the delivery of eye care with great success. Oklahoma has the highest number of Native American citizens of any state, and these citizens have a relatively high prevalence of diabetes and diabetic retinopathy.

To help address this problem, the McGee Eye Institute ophthalmologists developed remote diabetic retinopathy screening programs using telemedical software and systems developed at the institute. The innovative nature of this program has been recognized by the Smithsonian Museum, in its inclusion in *24 Hours in Cyberspace*, and in national news magazines.

DMEI is one of an elite group of facilities in

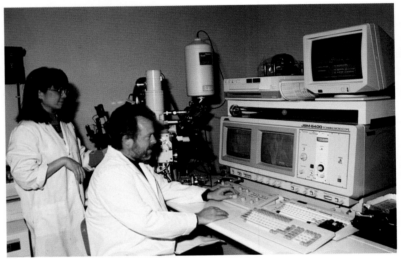

The Dean A. McGee Eye Institute (DMEI) is dedicated to providing the highest quality patient care, combining talented physicians with state-of-the-art medical equipment.

DMEI's vision research laboratories have recruited world class scientists to focus in areas such as molecular biology.

the United States granted "Unrestricted Grant Status" by Research to Prevent Blindness, Inc.—the largest and most prestigious vision research foundation in the world. The Institute's basic vision research is funded by the National Eye Institute of the National Institutes of Health and private foundations. It also currently participates in numerous national collaborative clinical research projects and contributes millions of research dollars annually to the Oklahoma economy.

"We are committed to using our abundant research and clinical talents to ensure that every patient has access to the latest technology and medical knowledge," states Dr. Parke. "It is also our responsibility to communicate this knowledge to our trainees and colleagues."

Nearly 40 percent of Oklahoma ophthalmologists have completed a portion of their instruction at the McGee Eye Institute. The department trains nine ophthalmology residents (three graduates/year) with a faculty-to-resident ratio of more than four to one.

Residents participate in patient care at DMEI, Oklahoma City Veterans Affairs Medical Center,

Childrens' Hospital of Oklahoma, University Hospital, and Columbia Presbyterian Hospital. Their clinical and surgical experience exceeds national standards by as much as 400 percent. Ophthalmologists also come to DMEI to spend one or two years of subspecialty clinical fellowship training and research after residency. Many subsequently pursue academic careers.

Community service remains a major aspect of the Institute's mission. DMEI conducts programs annually for approximately 1,300 students from across Oklahoma. Institute ophthalmologists give about $1 million yearly in care to indigent patients.

The McGee Eye Institute has a strong sense of its roots and responsibility to the Oklahoma community. Although one of the largest eye institutes in the country, it is less than a quarter-century old in a state yet to celebrate its centennial. With the leadership and dedication of its Board, faculty, and staff; support of alumni; and trust of countless thousands of patients, the Dean A. McGee Eye Institute pledges to continue its contributions to the science of ophthalmology and the art of vision care into the new millennium. ✪

YMCA of Greater Oklahoma City

The YMCA is much more than youth sports, adventures in camping, health and fitness facilities, swimming, field trips, and child care. Intertwined in each activity is the common thread of character development—utilizing teachable moments that reinforce positive values.

The goal of today's YMCA is to build strong kids, strong families, and strong communities. Through teaching values such as caring, honesty, respect, and responsibility, plus involving the entire family, the YMCA works to develop skills and character while providing fun.

Nine YMCA branches deliver the organization's programs and services. Child care, health and fitness, camping, youth sports and aquatics, are the YMCA's core programs. Through each program, there is a gentle, yet sharp focus on the YMCA's mission: To put Christian principles into practice through programs that build spirit, mind, and body for all.

The YMCA's branches reach across all segments of the Metro Oklahoma City area. "One of our challenges as an organization is to respond to each community's unique and changing needs," said Mike Grady, president and CEO. "Working to be a part of each community, and addressing the unique issues that exist, means recognizing different types of kids, families, and cultures."

It is a policy of the YMCA that no one is excluded or denied an opportunity to participate in its activities because of financial limitations. This service is made possible by the funds contributed directly to the YMCA each year by local individuals and businesses, as well as the United Way.

Since it was organized in 1889, the YMCA of Greater Oklahoma City has been "blessed" with the foresight and support of some of the community's top leaders such as E.K. Gaylord, Harvey P. Everest, C.R. Anthony, Sen. Robert S. Kerr, John Kirkpatrick, B.D. Eddie, and many more. The current board of directors continues to be comprised of the community's most outstanding leaders.

The YMCA faced one of the most serious challenges imaginable on April 19, 1995, destruction of the downtown branch when a bomb exploded one-half block away at the Alfred P. Murrah Federal Building. Now the goal is to restore the YMCA's presence downtown.

An outpouring of support is helping the YMCA of Greater Oklahoma City rebuild. The

The goal of today's YMCA is to build strong kids, strong families, and strong communities.

historic *Oklahoman* building and adjacent property was donated to the organization by The Oklahoma Publishing Company. The historic building will house the YMCA's corporate office as well as a number of other not-for-profit headquarters. The adjacent site will be home for a new YMCA Wellness Center and Child Development Center.

"The YMCA has effectively served the people of Oklahoma City for more than 100 years, and we're proud of our record. But we're also excited about the future and the opportunities to develop new and innovative ways to make our community a safer, healthier, better place to live and work," said Grady. "Be assured that the YMCA will continue, well into the next millennium, to nurture the healthy development of children, encourage positive behavior in teens, help adults live longer, more productive lives and give families the support they need to succeed." ✪

The YMCA works to develop character and skills while providing fun.

Blue Cross and Blue Shield of Oklahoma

Ronald F. King, president and chief executive officer.

FOUNDED ON EXPERIENCE, BUILDING FOR THE FUTURE

Years ago, a good insurer simply provided peace of mind— protection from unforeseen health care expenses.

Today, families depend on Blue Cross and Blue Shield of Oklahoma for security, unparalleled customer service, and benefit plans to meet their changing needs.

From tiny rented offices in Oklahoma City and Tulsa, Blue Cross has grown from 10,000 members in 1940 to nearly 500,000 members. The company's work force numbers more than 1,100 in Oklahoma City and Tulsa.

Key to Blue Cross and Blue Shield of Oklahoma's growth is the integration of local, personalized customer service with advanced information technologies. Innovative computer systems provide fast, accurate claims processing, with virtually no paperwork. And at their website, Blue Cross and Blue Shield is just a keystroke away.

These tools and many others have allowed Blue Cross to exceed customer expectations, building a reputation for quality, caring, choice, and value.

A REPUTATION FOR QUALITY

Blue Cross and Blue Shield of Oklahoma is fast achieving a national reputation for quality and efficiency. In 1996, Blue Cross and Blue Shield of Oklahoma was one of only 16 Blue Plans to receive the Brand Excellence Award for outstanding customer satisfaction, service, and financial performance.

That same year, BlueLines HMO received three-year, full accreditation from the National Committee for Quality Assurance (NCQA), meeting its rigorous quality standards.

And BluePreferred PPO was first in Oklahoma—second in the nation—to receive National Network Accreditation by the Utilization Review Accreditation Committee (URAC).

A REPUTATION FOR CARING

Community service is a vital part of the Blue Cross corporate culture. Blue Cross associates are active in many local organizations, including the United Way and Junior Achievement.

In 1995, Blue Cross created the Caring Program for Children to help tackle one of the state's most critical problems— uninsured children. Approximately 45,000 Oklahoma children come from working families earning too much to qualify for Medicaid, but not enough to afford health benefits.

In Oklahoma City and throughout the state, Blue Cross and Blue Shield of Oklahoma builds better communities by supporting the United Way and the company's own Caring Program for Children.

The Caring Program provides free primary and preventive outpatient care for hundreds of these children. Blue Cross donates administrative services and matches every contribution, dollar for dollar, up to $100,000 annually. More than 5,000 health care providers also contribute by drastically reducing their fees.

A REPUTATION FOR CHOICE AND VALUE

From a single hospitalization plan in 1940, Blue Cross has expanded its "family of companies" to meet nearly every insurance need.

Employee groups choose from the entire managed health care spectrum, including two health maintenance organization (HMO) plans, two preferred provider organization (PPO)

plans, and a traditional plan.

This year, Blue Cross introduced BlueWorks, a certified workplace medical plan providing case management for workers' compensation. Blue Cross subsidiaries offer third-party administration, life and disability plans, and auto, home, and other personal insurance plans.

The Blue Cross and Blue Shield of Oklahoma mission statement sums it up: "Our mission is to provide our customers with a wide range of managed care health plans, insurance products, and financial services which feature competitive pricing, quality provider networks, and first-class administration."

Blue Cross and Blue Shield of Oklahoma will continue to expand customer choices, providing a wide selection of plans created and administered by Oklahomans, for Oklahomans. ✦

ZymeTx, Inc. The company name is as innovative as its product line. Now moving into international markets, ZymeTx has been able to harness an area of biotechnology long considered impossible to penetrate: the timely, efficient diagnosis of potentially life-threatening viruses.

First conquering the elusive detection of viral influenza, the company has plans to quickly expand to develop and manufacture similar diagnostic devices for numerous diseases, including herpes simplex, cytomegalovirus, respiratory syncytial virus, adenovirus, and parainfluenza. Similar programs are planned for viral treatments.

Statistics demonstrate the dramatic impact viruses can have on human health. Consider influenza. In a typical year in the United States, approximately 20,000 preventable deaths are associated with the flu. Complications from the disease are major contributors to respiratory illness and have been the sixth leading cause of death in this country. Despite these alarming statistics, until the ZymeTx breakthrough, more than 80 percent of physicians did not test for influenza, citing such reasons as a lengthy wait for results, difficulty with existing testing methods, and high costs.

ZymeTx is committed to changing this situation completely. In about an hour, a physician is able to report to a patient whether influenza is the culprit responsible for those dreaded aches and pains with the product ZymeTx's scientists have patiently and diligently developed.

"What we're doing was long considered impossible in the world of science," says Peter Livingston, President and CEO. "We have beaten the odds and proved the possibility of a quick, easy diagnosis and treatment for viruses."

Visionary researchers and business leaders committed to unraveling the mystery surrounding influenza and other viruses were the keys to ZymeTx's development as an upcoming leader in the biotechnology industry. Av Liav, Ph.D., section head in Synthetic/Organic Chemistry; Joyce Hansjergen, section head in Hybridoma Development and Viral Studies; Craig Shimasaki, Ph.D., Vice President of Research; and Peter Livingston joined the Oklahoma Medical Research Foundation (OMRF) in founding the company in 1994. A firm commitment from OMRF and with help from the Presbyterian Hospital Foundation, ZymeTx chose to remain in Oklahoma because of very strong community business and pro-industry support.

Until recently, when it established its own facility, the company operated as a virtual corporation within OMRF, an internationally recognized independent medical research institution. ZymeTx continues to collaborate with OMRF, while maintaining its own distinct group of scientists, laboratories, and business staff.

From this setting, the company has been able to accomplish many tasks with equipment and personnel, usually beyond the means of start-up companies. ZymeTx uses these resources to supplement its own proprietary process. This relationship with OMRF will continue, allowing the company to focus investment capital on scientific research and product development.

ZymeTx, Inc.

At a time when substantial interest has developed in an immediate, accurate diagnosis linked with effective treatment, ZymeTx is at the forefront with products directly responsive to this emerging trend. Keep a close eye on ZymeTx's growth in the coming years. Its contribution to world health, through diagnostic devices for life-threatening diseases, promises to bridle what many thought unconquerable. ✪

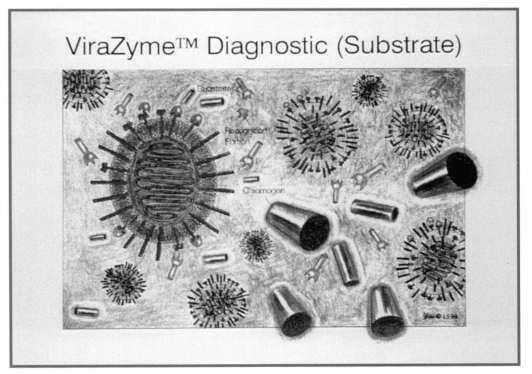

ZymeTx scientists unlock the unique differences of viral enzymes. Their approach links the development of diagnostic measurements to effective treatment methods.

ZymeTx is at the forefront with products directly responsive to emerging trends.

Epworth Villa

Three words capture the guiding philosophy of Epworth Villa, a retirement center: Resident First Care. Every staff decision focuses first on the physical, emotional, social, and spiritual needs of its nearly 350 residents.

"We provide quality care to people for the rest of their lives," says Joe White, President and CEO. "Each person is assured of continuous care at whatever level is needed—from independent living to extensive nursing."

Prayer, hard work, and dedication transformed the dream of this state-of-the-art retirement center into a reality for the Oklahoma Conference of the United Methodist Church. Groundbreaking for the 35-acre facility began in 1989. The first residents moved in a year later. By 1992, all available apartments were sold—18 months ahead of schedule.

Currently operating from an annual budget of 8.6 million, dollars, the 197-member staff claims an annual payroll of 3.9 million. Within its short history, Epworth Villa has set new standards for retirement living in Oklahoma City.

"By keeping Resident First Care as our basis, the business fell into place," White says.

Residing at Epworth Villa has all the advantages of being at home without the multiple responsibilities associated with home ownership. Grounds care, housekeeping, and maintenance repairs are routinely provided as a continuing service.

On-site features offer an extensive list of amenities not often found in private residences, including a lake with gazebo, walking trails, convenience store, woodworking shop, indoor heated pool, exercise and fitness center, and extensive library. Lounges are located on every floor providing residents with opportunities to gather for any of the many planned social activities, or for their own game of cards, book discussion, or television/movie viewing.

Personal services include scheduled transportation, two beauty/barber shops, and meal preparation. Everyone is assured of a 24-hour emergency call system, which is installed in the bedroom and bathroom of each residence. When activated, help is immediately summoned.

Overnight visitors stay with residents in their homes or in one of the guest rooms in the community. During their stay, guests are invited to share all the outstanding services and elegant

Lounges are located on every floor providing residents with opportunities to gather for any of the many planned social activities, or for their own game of cards, book discussion, or TV/movie viewing. Photo by Joseph Mills.

Epworth Villa has set new standards for retirement living in Oklahoma City. Photo by Joseph Mills.

features of Epworth Villa—from white linen-covered dining tables to quiet strolls along the lakeshore.

Within the scope of the resident contract, on-site around-the-clock nursing care is provided at the Susanna Wesley Center at no addition to the monthly fee. The center's primary goal is to provide exceptional services focused on quickly returning residents to their own homes and active life-styles.

Long-term strategic planning for Epworth Villa includes a special care center for an additional 48 residents challenged with Alzheimer's disease. All direct care employees will have up-to-date training to assure residents' needs remain the top priority.

Coupling its loving, sensitive philosophy with plush surroundings, Epworth Villa provides a vital, positive direction for the senior population. People at all stages of life can now anticipate the joys of a fulfilling retirement.

"Our primary emphasis at Epworth Villa is on service," White concludes. "Resident Care First will always remain a part of our mission and our ministry." ✣

Casady School

Fideliter et Fortiter — "Loyalty and Strength." For students, faculty, parents, and alumni, these words emblazoned upon the Casady School crest embody the mission and the important place Casady has held in Oklahoma City for 50 years. One of the southwest's premier independent college preparatory schools, Casady was established in 1947 by The Right Reverend Thomas Casady, then Episcopal

A broad array of opportunities in the visual and performing arts, vocal and instrumental music, foreign language, and computer science complement the traditional focus on composition, literature, history, science, and mathematics.

Bishop of Oklahoma. The focus of the school since its founding has been on the full development of each student in mind, body, and spirit from prekindergarten through grade 12. With this goal comes the challenge of preparing Casady students for success at the nation's most selective and demanding colleges and universities, and for leading compassionate and productive lives well after the Casady experience.

In educating the whole child, Casady School draws from the energy and experience of dedicated teachers who recognize the individual strengths of each student and who work with uncommon devotion to develop the child's full potential. Casady also benefits greatly from an alumni and parent body that provide the support and assistance necessary, both in time and resources, to ensure that the educational endeavor enriches the school community as a whole. The benefit of a Casady education can extend well beyond the student—to family, friends, and to those with whom the young person comes in contact. In this way, and in many other

ways as well, Casady School has left a positive imprint on the Oklahoma City community.

In the life of any school, it is the students and teachers who define what that school is and what it does best. Casady School students and their families reflect a diversity of backgrounds, beliefs, and experiences. Similarly, Casady faculty offer an impressively wide range of educational backgrounds and interests that enhance their effectiveness as educators. These characteristics are particularly valued by families who relocate to Oklahoma City from around the nation, and who seek the best options for meeting the educational needs of their children.

Class sizes at Casady are small, and the interaction between students and teachers is lively and productive. The school places a strong emphasis on academic excellence and seeks to promote a learning environment in which every boy and girl can achieve the highest level of success of which he or she is capable. A broad array of opportunities in the visual and performing arts, vocal and instrumental music, foreign language, and computer science complement the traditional focus on composition, literature, history, science, and mathematics. Most importantly, the joys of learning, discovery, and achievement are reflected in myriad ways by Casady students of all ages.

Casady is unique among independent or public schools in the Oklahoma City area in its focus on the development and maintenance of each child's mental and physical health through a daily, comprehensive program of recreation, exercise, and interscholastic sports. The Casady Cyclones compete in the Southwest Preparatory Conference, comprised of member schools of the Independent School Association of the Southwest located in Tulsa, Oklahoma City, Dallas, Fort Worth, Austin, San Antonio, and Houston. At the end of each athletic season, after regional competition, teams meet in the Dallas-Fort Worth area for conference championships and engage in spirited competition in which doing one's best is as valued as being the best.

One constant in the life of Casady School is that of St. Edward's Chapel and the chapel services that all students in grades one through twelve attend daily, and kindergarten students attend weekly. It is through the chapel program that the school affirms its Episcopal tradition, emphasizes the basic tenets of Judeo-Christian values and the Christian faith, and also welcomes those of other faiths.

Casady School embarks on its second 50 years with the strong assurance that its students, faculty, parents, and alumni will continue to be enriched by its mission, with *Fideliter et Fortiter*. ✺

Class sizes at Casady are small, and the interaction between students and teachers is lively and productive.

Westminster School

Westminster School provides three year olds through eighth graders a safe and caring environment in which children are encouraged to explore and learn.

A new playground and nature center provide more opportunities for active learning.

Westminster School, nestled in an historical neighborhood close to downtown, Bricktown, the state capitol, and the medical center, provides three year olds through eighth graders a safe and caring environment in which children are encouraged to explore and learn. Founded in 1963 as a Montessori preschool, Westminster expanded gradually to serve children throughout the Oklahoma City community. This nonsectarian, independent school has benefited from stable leadership; the founding Head of School served 29 years, and the current Head has been at Westminster over 20 years.

Westminster School recognizes children as valuable and capable people. It provides an educational experience in which questions are encouraged, creativity and curiosity are nurtured, academic skills and self-confidence are developed, and thinking is highly prized. Westminster also believes that children and schools should be filled with joy, so it is an informal, child-centered place where children and teachers know each other well. It is a place of high expectation that encourages children to try out new things and a place that values the freedom to make decisions and the importance of taking responsibility for them.

The school operates physically and administratively in three divisions in separate, adjacent buildings. Each division has its own director and faculty, and teaching teams have the freedom to create the most appropriate and challenging curriculum possible for their students. The Montessori primary division for children three to six years old includes three-day preschool, five-day preschool, and kindergarten. The lower division for six to eleven year olds includes first through fifth grades. The middle division includes sixth through eighth grades and is affiliated with Brown University's Coalition of Essential Schools. After-school programs are also available in each division.

Westminster's facilities support its active learning philosophy. A new playground and nature center provide more opportunities for active learning and enjoyment for students of all ages. The nature center hosts time and measurement experiments, water studies, and horticulture

projects; a greenhouse allows for year-round experiments. There is also a butterfly garden, and an energy area and archaeological dig are still to come. The soccer field, basketball court, and imaginative play equipment appeal to any child.

What makes Westminster a special place, however, are its dedicated people and its progressive philosophy. Youth and experience blend to create a developmentally appropriate curriculum that is still challenging to the most talented of students. On average, Westminster teachers have been at the school for over 11 years and understand and support its philosophy and mission. They celebrate children and believe strongly in a free, tolerant, and diverse educational community,

a community that willingly provides financial assistance to meet demonstrated need. The academic, athletic, artistic, and social success and leadership of Westminster graduates have been well documented in the community and beyond, and while the school is justifiably proud of the success of these fine young people, it holds firmly to its belief in the importance of children and its commitment that "the end depends upon the beginning." ✪

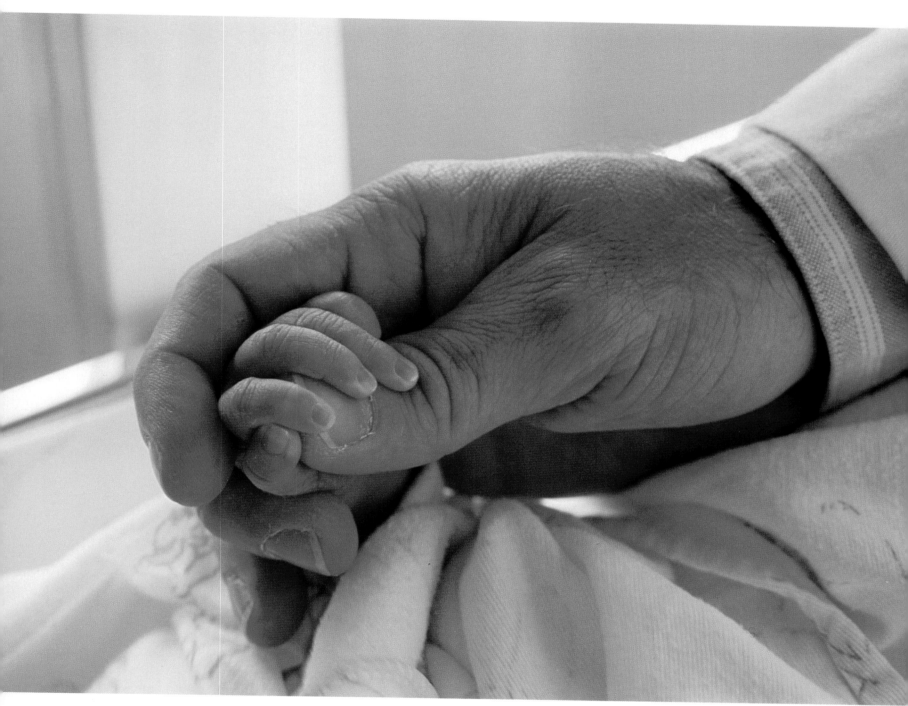

Photo by Joe Ownbey.

17

The Market Place

Photo by Joe Ownbey.

Fifth Season Hotel and Conference Center

If travelers are looking for value in hotel accommodations, the Fifth Season Hotel and Conference Center offers a special blend of comfort, convenience, and amenities at truly affordable prices. It's like having the big city benefits with the small- town price and personal touch.

The Fifth Season Hotel is a contemporary facility built in 1984. It features 202 sleeping rooms, which are oversized in comparison to the standard hotel. Most rooms surround the indoor garden atrium with a fountain, heated pool, and whirlpool spa. This hotel is one of three in Oklahoma City offering this year-round pool-side capability in an outdoor setting that translates into a restful indoor spot for leisure time enjoyment.

EXTRA VALUE

Each morning, guests receive a newspaper and are offered a complimentary full breakfast buffet brimming with eggs, bacon or sausage, biscuits and gravy, hash browns, fruit, bagels, cereal, and juice. In the evening, a complimentary manager's reception is hosted in our garden atrium to help guests unwind after a long day.

Each sleeping room is equipped with a coffee maker, and all local telephone calls are free. For the fitness-minded guest, the Santa Fe Fitness Center is available without a fee. Guests can work out on the aerobic equipment including a cycleplus, stairmaster, and treadmill. Access to state-of-the-art Cybex and free weights are also a part of the exercise package.

In addition, handicapping sessions are available for guests each race day morning prior to the opening of the track at Remington Park, Oklahoma City's state-of-the-art horse racing facility. During these sessions, a horse racing expert describes the various aspects of a race, including horse and jockey history. This overview gives guests additional knowledge and insight when trying to determine race winners.

Free transportation is an added bonus for overnight guests. Shuttle buses to and from Will Rogers World Airport, Remington Park racetrack, and other nearby destinations show the hotel's commitment to value and hospitality for both the business and leisure traveler.

Our indoor garden atrium is the perfect location for year-round leisure or social events, with an indoor whirlpool spa and a swimming pool. Complimentary health club memberships are available.

Join us for an entertaining or relaxing moment in Dr. Sages Lounge which is open seven days a week.

For your convenience, our full service atrium grill is open daily for a relaxing lunch or dinner.

EXCELLENT LOCATION

The Fifth Season Hotel is conveniently located in the heart of the Oklahoma City area off I-44 via the Broadway Extension and N.W. 63rd Street. This setting provides easy access to all major interstates, including I-35, I-44, and I-40, plus many of Oklahoma City's finest attractions.

Fifth Season Hotel is only two miles from the National Cowboy Hall of Fame and Western Heritage Center, Oklahoma City Zoo, the Omniplex, and Remington Park.

Shopping meccas Penn Square Mall and 50 Penn Place are within three miles of the hotel. Also within three miles are Frontier City, the National Softball Hall of Fame, and the Amateur Softball Association Hall of Fame Stadium.

The hotel's professional staff is also available 24 hours a day to assist with information about local destinations.

IN-HOUSE OPPORTUNITIES

Venturing from the hotel is an option of luxury, not a necessity for Fifth Season Hotel guests. The facility offers individual suites, dining and evening entertainment options, along with a broad range of meeting facilities.

Twenty-five junior executive suites expand the traveling horizons and comfort level of Fifth Season Hotel guests. A semi-private living area touts a refrigerator and built-in wet bar. A 25-inch television, two telephones, and separate bedroom with a king-size bed bring the comforts of home to the traveler.

The Fifth Season Hotel is known for its unique bridal suites that feature in-room Jacuzzis. In addition to the ballroom, the hotel's atrium is an ideal setting for a wedding reception.

A full spectrum of events come to life within the more than 6,500 square feet of varied meeting space at the hotel's conference center.

The 4,500-square-foot Remington Ballroom can seat up to 350 for a banquet or 600 for a meeting. It can be configured for a business meeting, small convention, or a trade show. The Remington Ballroom can also be divided into three smaller rooms when needed.

Perfect for more intimate events, the Crossover and Orchid Rooms hold up to 50 people for a sit-down banquet. Two smaller board rooms can comfortably seat up to 12 people for committee or staff meetings.

In-house catering and full audiovisual capabilities assist event planners by providing everything needed for almost any event. From centerpieces to timing, the Fifth Season Hotel works with its clientele on every planning detail. A full-time catering manager is on staff to exclusively provide expert assistance regarding the many aspects of banquet, meeting, convention, and conference planning.

Catering, room service, restaurant dining, and cocktail lounge hors d'oeuvres afford hotel guests flexibility and variety when it comes to food service.

Special event menus range from dinner theme buffets of a Mexican fiesta or Hawaiian luau to luncheons offering a full deli, and lighter side buffet or an array of plated luncheon choices. Breakfast suggestions encompass a continental buffet of juice, fruit, and pastries to a tantalizing four-ounce sirloin steak with scrambled eggs.

When it comes to business, we have the space and service to compliment your meeting needs and create a successful event.

The full-service Atrium Grill offers a casual choice for lunch or dinner. The family or corporate traveler can choose from a wide selection of reasonably priced dishes, from sandwiches and pasta to steaks and chicken. Of course, room service is available during restaurant hours. The cocktail lounge is open daily and features popular hors d'oeuvres Monday through Friday.

So, if the itinerary demands early morning or late-night scheduling, the Fifth Season Hotel ensures the same type of accessibility and convenience a guest would find at home.

HOME IN OKLAHOMA CITY

The Fifth Season Hotel and Conference Center is closely allied to the people and places of the metropolitan Oklahoma City area. A dedicated management team is led by Michael J. Nocula, CHA, general manager.

Nocula came to Oklahoma City from the West Coast and has spent his career in cities like Chicago, Philadelphia, San Francisco, Washington, D.C., and Atlanta.

"It's refreshing to come into a southwestern environment," said Nocula. "I really enjoy this city on both a personal and professional basis."

With anywhere from 75 to 100 full-time employees in the areas of sales, food and beverage, housekeeping, accounting, maintenance engineering, and administration, Nocula pays tribute to the work force in the Oklahoma City area.

"The biggest asset Oklahoma City has is the quality of the people," said Nocula. "The people are outgoing, friendly, and excellent workers. They are motivated with a good work ethic. Our goal is to build an environment that continually elevates levels of guest service and satisfaction."

As the host hotel for the Oklahoma City Cavalry, Oklahoma Thoroughbred Association, and the Sooner state games, the Fifth Season Hotel and Conference Center has an important, yet easy job. The workers convey Oklahoma City's genuine, friendly, and personable attitudes to visitors who make Oklahoma City their location of choice, if only for a short period of time.

In winter, spring, summer, and fall, the Fifth Season Hotel and Conference Center works to make guests feel at home while they are away from home. With the special blend of comfort, value, and service, the Fifth Season Hotel is "truly an affordable luxury." ✪

We have over 6,000 square feet of banquet and meeting rooms including the Remington Ballroom with comfortable banquet style seating for 350.

The Fifth Season has 202 Guest Rooms, including Honeymoon Suites with hot tubs and twenty-five 2-room Executive Suites. All of our accomodations are spacious, comfortable and attractively furnished.

The Fifth Season is your choice northwest location with its close proximity to Remington Park, the National Cowboy Hall of Fame and Western Heritage Center, the Oklahoma City Zoo, the Omniplex and many other fine local attractions. Several fine shopping malls and plazas are within a 10 minute drive. In addition, the Fifth Season Hotel is located in the heart of several corporate office centers and is easily accessible to all major highways in the Oklahoma City metro area.

Join us in our all season garden for a complimentary evening managers reception. Consider this spacious lovely garden setting for your next formal or informal event.

The Waterford Marriott Hotel

Whether it is four stars or four diamonds, the Waterford Marriott Hotel has earned top drawer ratings. This boutique hotel can be enjoyed in many ways—by staying overnight, attending a wedding, or business luncheon, dining in The Veranda Room or The Waterford Dining Room, and more. From the full-time florist on staff to the large swimming pool flanked by four lion fountains, all facets of the Waterford Marriott experience reflect a style and service unique to this individualistic facility.

The Waterford Marriott Hotel features 197 guest rooms, including 32 suites. Located on the ninth floor, The Top Floor suites include upgraded amenities and concierge services.

Political leaders, including George Bush, Dan Quayle, and Al Gore, have stayed at the Waterford Marriott Hotel. Many movie stars and directors graced this hotel during filming in Oklahoma City. Along with Dustin Hoffman and Tom Cruise from *Rain Man*, actors Helen Hunt, Bill Paxton, and director Jan DeBont were guests at the hotel during the filming of *Twister*.

Entertainers from the music industry also selected the Waterford Marriott Hotel for overnight accommodations when bringing their talents to Oklahoma City. From rock to classics, superstars such as the Eagles, Neil Diamond, Grover Washington, Jr., Van Cliburn, John Tesh, Reba McEntire, and Vince Gill were all guests at the hotel.

Companies in Oklahoma City that are recruiting or working with out-of-town business clients and partners also use the Waterford Marriott Hotel for their corporate guests. The Waterford Marriott Hotel management is proud of the fact their property reflects Oklahoma City as a cosmopolitan destination and helps to break down any stereotypical images.

For business or for pleasure, the Waterford Marriott Hotel serves up excellence to all guests. Its distinction for service prevails in all areas of the hotel, including the many rooms available for community meetings and events.

From Governor Keating's personal thank you victory party to community events throughout the year, the meeting and banquet facilities at the Waterford Marriott Hotel offer Oklahoma City 8,000 square feet of function space.

The garden and terrace views complement the

There is no doubt that the Waterford Marriott Hotel is making its mark of excellence on the Oklahoma City community.

The Waterford Marriott Hotel management is proud of the fact their property reflects Oklahoma City as a cosmopolitan destination and helps break down any stereotypical images.

(above) The garden and terrace views complement the wood paneling, book shelves, and marble fireplace of The Board Room.

(right) The Veranda Restaurant is open for breakfast, lunch, and Sunday brunch.

wood paneling, book shelves, and marble fireplace of The Board Room, which can host as many as 50 guests. The Chapter Room accommodates medium-sized conferences and meetings.

Crystal chandeliers reflect within The Waterford Ballroom, which serves as an elegant choice for weddings, receptions, large conferences, and other special events. For a private dinner party of 12, small meetings, and special celebrations, the Crystal Room offers an intimate room choice.

Culinary delights served from The Veranda Restaurant and The Waterford Dining Room reflect a tradition of excellence from renowned chefs. With a goal to exceed guest expectations through presentation, service, creativity, and quality of food, Waterford Marriott dining offers a unique culinary experience for a variety of tastes.

The Veranda Restaurant is open for breakfast, lunch, and Sunday brunch. Casual, yet elegant, this 110-seat restaurant offers dining in a light, airy white wicker setting with poolside views. The menu offers a variety of soups, salads, sandwiches, and lunch plates. The Waterford Marriott Hotel is especially well known for its Sunday Jazz Brunch in The Veranda Restaurant, which features live jazz along with an elaborate brunch buffet.

An evening of fine dining in The Waterford Dining Room offers contemporary cuisine in a 19th-century English country setting. The menu consists of regional and continental gourmet specialties—from a smoked salmon appetizer to a "from the grill" dinner of a T-bone steak with grilled onions.

The Waterford Marriott Hotel put its culinary excellence in the spotlight when inviting the community to the grand reopening of the hotel following extensive renovation. During this event, guests were invited to "taste the difference" with culinary masterpieces from the Chef and colleagues from two sister hotels in Cambridge, Massachusetts, and St. Petersburg, Florida.

Guests were able to get a taste of the hotel's food and beverage selection while also taking in the interior and exterior enhancements to the hotel.

The level of excellence for the Waterford Marriott Hotel is sustained through attention to hotel aesthetics. A four-month, $3.3 million renovation was completed in March of 1996. This phase of the project included a new front desk

and lobby area, redesigned guest rooms, corridors, ballrooms, cafe, meeting rooms, and health spa. The addition of a porte-cochere and an enclosed terrace rounded out the renovation.

Every guest room was stripped to the dry wall, then refurbished with new paint, carpet, wallpaper, bedspreads, and curtains. Although much of the furniture remained the same, many pieces were refinished. The previously dark color schemes were exchanged for a lighter color strategy. Other luxurious room features of marble bathrooms and dressing areas remained intact.

Engineering and structural issues were also addressed. Along with improvements made that are not readily apparent, guests can enjoy a state-of-the-art phone system equipped with guest voice mail and in-room data ports.

Although the hotel has undergone major physical restoration and improvements, the Waterford Marriott Hotel's service and tie to the local community extends beyond the hotel walls. Contributions of time and resources benefit the Greater Oklahoma City Chamber of Commerce, literacy programs, Big Brothers/Big Sisters, plus other charitable and nonprofit organizations.

The hotel brought back the popular Sunset Serenade from the 1980s to raise funds for the Oklahoma City bombing relief fund. The event brought in $8,000 for donation to the fund and was such a success that the Sunset Serenade series has been reinstated Sunset Serenade is now a bi-weekly event from May to July in an after-work happy hour environment that features live jazz music.

The hotel also provided a generous number of hours of service and thousands of dollars of food products to specific events. In addition, the hotel, in one calendar year, donated over $16,000 of individual gift certificates to assist with fund-raising efforts and volunteer motivation programs.

The Waterford Marriott Hotel has been heavily involved in the Redbud Classic, an annual race that features bicycling, running, and walking events with proceeds benefiting various organizations

The Waterford Mariott Hotel features 197 guest rooms, including 32 suites.

across the city. The event begins with thousands of participants lined up on North Pennsylvania, just west of the hotel, ready to begin the race. The

traditional and festive "Pasta on the Pond" is held Saturday evening before the race around the pond on the Waterford Complex grounds.

The Waterford Marriott Hotel is uniquely situated in a multiuse complex.

Along with the hotel, Waterford Properties oversees more than 155,000 square feet of rentable footage in several office buildings that ring the outer north portion of the complex. Other property owners within the complex include Fleming Companies, one of the world's largest food distributors, and a branch office of Boatmen's Bank. Additionally, the Waterford Condominium Complex is managed by the Homeowners' Association.

Common areas for the complex include the boulevard, parking lot, and the pond, which is home to families of ducks and elegant black swans. Sitting at the point of the pond area is a large statue titled *The Three Madonnas* which was dedicated in 1984. The statue's artist is Norma Penchansky Glasser.

A standing plaque along the boulevard pays tribute to the Oklahoma Baptist Orphans Home, which moved to this site in 1907. The home was founded by the Rev. J.A. Scott in March of 1903 "under the auspices of the Oklahoma and Indian Territories Baptist Conventions" and cared for more than 4,000 children in more than 20 buildings throughout the 75 years on these grounds. The history plaque continues the story: "The Baptist General Convention of Oklahoma in 1936 changed the name to Oklahoma Baptist Children's Home. Moved to 16301 South Western in June of 1982."

The Waterford Marriott Hotel is especially well known for its Sunday Jazz Brunch in The Veranda Restaurant.

include the Colonial Inn in La Jolla, California; Hotel Atop the Bellevue in Philadelphia; The Charles Hotel in Cambridge, Massachusetts; Don CeSar Hotel, St. Petersburg, Florida; and the Goodwin Hotel in Hartford, Connecticut. Interstate also manages several prestigious resorts across the country.

The Waterford Marriott Hotel in Oklahoma City has earned the honor of being a part of this distinguished group. The Total Quality Management (TQM) process, being implemented over a five-year time frame, guarantees the hotel's commitment to unparalleled service into the future. The progress of this TQM implementation has been rewarding. Associates are empowered, so decision making is encouraged on the guest contact level. Due to the feedback from the TQM process, a full-sized iron and ironing board are now found in the closet of each guest room.

Since the hotel staff is responsible for the safety, comfort, and security of their guests, all associates work together for total quality. They strive to give visitors the finest of everything.

So whether the grandeur is designated by four stars or four diamonds, there is no doubt that the Waterford Marriott Hotel is making its mark of excellence on the Oklahoma City community. ✪

From the full-time florist on staff to the large swimming pool flanked by four lion fountains, all facets of the Waterford Marriott reflect a unique style and service.

Although the Waterford Complex is tied by name, the entities occupying the complex are independently owned and managed. In 1997, the Waterford Marriott Hotel became an affiliate of Host Marriott, Washington, D.C. This new partnership offers guest conveniences such as Marriott's Rapid Rewards program, the best frequent traveller program in the industry, and an extensive central worldwide reservations system.

Interstate Hotels, an international hotel management firm, was selected to oversee the renovation, plus market and manage the property. For four years in a row, both *Hotel Business* magazine and *Hotel Motel Management* Magazine, have ranked Interstate as the largest management company when it comes to generating revenues for its owners. Interstate was also ranked first in overall number of rooms and number of properties managed.

Interstate operates hotels in virtually every major market in the United States. Sister luxury independent hotels to the Waterford Marriott

Culinary delights served from The Waterford Dining room reflect a tradition of excellence from renowned chefs.

Smicklas Chevrolet

John Smicklas

When traveling along highways in the metropolitan Oklahoma City area, large Smicklas Chevrolet billboards featuring the dealership and its "Friends for Life" animal adoption program reflect a company committed to its business and community. The combination of automotive dealerships, the program to help abandoned animals, and additional equestrian ventures solidify this company's genuine devotion to provide a good product to customers, together with beneficial programs for the community.

MOVING TO OKLAHOMA CITY

Smicklas Chevrolet's diverse and active business and community agenda originated with John and Barbara Smicklas. Since moving to Oklahoma City in 1982, the Smicklases and their company have had a positive impact on Oklahoma City on many levels. Their life in Oklahoma City began with a love for horses and Oklahoma City's predominant position in the horse arena.

The Smicklas family has a deep and abiding affinity for horses. Through participation in the sport of polo, they began a friendship with Bob Moore, an Oklahoma City auto dealer. It was Moore who convinced John and Barbara to leave their thriving Denver, Colorado, dealership to take up new challenges in Oklahoma City. The first of these challenges was to become, perhaps, the most well known—the Smicklas Chevrolet GEO dealership. Today, it is the number one Chevrolet dealer in the state of Oklahoma.

THE SMICKLAS AUTO BUSINESS

The Smicklas company began by buying the Buck Morris Chevrolet dealership. This 10-acre compound, at 50th and north May, is today a very successful complex of new, used and program cars, trucks, and vans as well as a busy service department, and a highly regarded parts department with a more than $3 million inventory of GM parts, Oklahoma's largest.

The Smicklas approach to sales and customer service offers many unique attributes. Among these attributes is the Smicklas "Lemon Protection Plan." This is a written guarantee that one will not buy a "lemon" when buying a used car from Smicklas. Another attribute is the "$500

(l to r) Steve Gates, General Sales Manager, John Smicklas and Brad Smicklas.

We Won't Be Undersold" certificate. With literally hundreds of vehicles available on the lot and an aggressive policy of customer satisfaction as the foundation for every sale, Smicklas has become the number one Chevrolet dealer in several metro and state categories. With annual sales topping the 100 million mark, Smicklas Chevrolet GEO has been a boon to the economy of Oklahoma.

Standing still, even on success, has never been part of the Smicklas philosophy. So expansion is once again under way. The Smicklas Automotive Group has been formed, with Brad and John Smicklas and Steve Gates, Smicklas general sales manager. The initial purpose is to create an innovative dealership named Autoright. Now open in Edmond, Autoright provides a new way

for customers to buy used cars. It is an exciting concept and plans are under way to extend this one location into several in the near future.

HORSE RACING AND POLO

The Smicklas decision to become Oklahomans has also brought an infusion of enthusiasm, as well as excitement, for both thoroughbred racing and polo in Oklahoma.

In the Norman area, alongside the Broad Acres Polo Club, John and Barbara run the Smicklas Horse Farms. While John no longer rides in polo matches, his older son, Dale, continued to ride and was a member of the world champion Michelob team. He was also a respected trainer of polo ponies. Now retired from polo, Dale has joined the Smicklas Automotive team.

Robyn Morgan, Friends for Life coordinator, with two "friends".

The Smicklas colors have also many times graced the winner's circle at Remington Park and other race tracks across America on such champions as "No More Hard Times" and "Belle of Cozzene."

"FRIENDS FOR LIFE" CAMPAIGN

While the Smicklas automotive dealerships and the Smicklas farms have given, and continue to give, employment and economic opportunities to hundreds of Oklahomans, it may well be that the greatest impact the Smicklases are making on the culture of Oklahoma is through an entirely selfless effort called "Friends for Life."

This program was founded to help educate the public about two important animal needs: the need to spay or neuter pets to prevent litters for whom no homes can be found and the need to find as many homes as possible for animals abandoned in shelters.

"Friends for Life" is publicized through newspaper ads, television, and radio commercials, as well as outdoor billboards. The entire campaign is paid for by Smicklas. Why such an extensive campaign? John and Barbara explain it quite simply: they love animals and feel a responsibility for their care and protection.

This responsibility extends from their own home where special kennels house animals waiting for new owners, to their dealership where a special assistant runs an office dedicated to finding shelter for any animals in need.

While successful at business with Oklahoma's #1 Chevrolet dealership, the Smicklas family is known for its outstanding

John Smicklas and the Smicklas Chevrolet office staff.

loyalty and uncommon devotion in the areas of abandoned animals and equine ventures. Oklahoma City gained a strong business with a good product, along with a dedicated family and a commitment to make a positive impact, when the Smicklases came to town. ✺

Remington Park

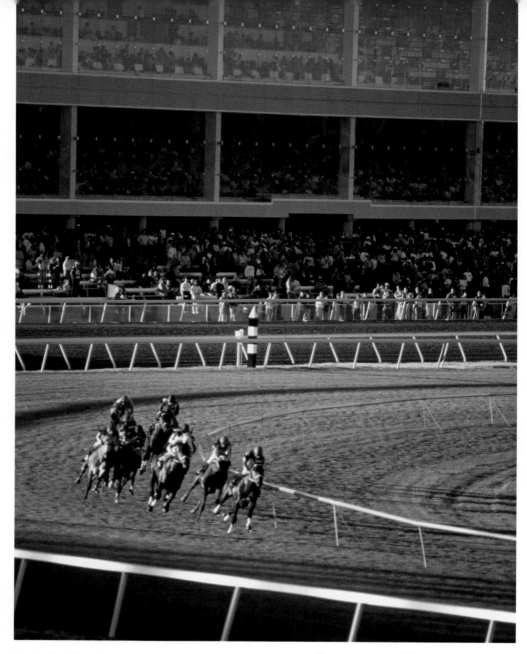

Through the centuries, artists have captured the spirit of the American West through paintings and sculptures that inevitably included horses. World-class Remington Park continues to seize the beauty and brilliance of these powerful creatures throughout the year by featuring Thoroughbred and Quarter Horse racing at its impressive facility in northeast Oklahoma City. As the Horse Capital of the World, it is no wonder Oklahoma City touts a crown jewel horse racing facility like Remington Park.

HORSE POWER

Remington Park is a powerful factor in the entertainment, growth, and prosperity of Oklahoma City. As the starting gate opened for the first time on September 1, 1988, a new era for Oklahoma horse racing began.

* Remington Park is Oklahoma's first major pari-mutuel racing facility. This $97 million, state-of-the art facility encompasses 375 acres.

* Remington Park is Oklahoma's number one tourist attraction.

* From 1988 to 1996, Remington Park hosted 10 million fans who won $1.3 billion dollars.

* Remington Park directly employs as many as 1,000 people with an annual payroll in excess of $1 million.

* Remington Park promotes a world-class image for Oklahoma through national exposure on ABC, NBC, ESPN, CNN, *Sports Illustrated, USA Today, Forbes and The New York Times.*

* Remington Park's impact on Oklahoma City's economy in its first five years was more than $1.5 billion according to the Greater Oklahoma City Chamber of Commerce.

The late Edward J. DeBartalo developed Remington Park through the vibrant Edward J. DeBartolo Corporation. DeBartolo was recognized worldwide for his accomplishments in real estate development and management, and he gained additional acclaim when pioneering this amazing horse racing facility.

MODERN FACILITY

From the stable area to the glass-enclosed, climate-controlled stadium, Remington Park possesses some of racing's most modern and clean facilities. Whether enjoying the all-you-can-eat gourmet buffet at the Eclipse Penthouse Restaurant, or a family outing at the Trackside Cafe, the stadium's eight levels (including three available to the public) are brimming with opportunities.

The Track Level features the track apron, free bench seating, Racing Information Center/Video Library, two gift shops, Trackside Cafe, 11 concession stands, and a video arcade. Privately owned box seating with television sets, Silks Restaurant, reserved seating, Paddock Terrace Restaurant, and five concession stands are a part of the Clubhouse Level.

The elegant Penthouse Level is lined with 20 private suites, the Remington Suite, Eclipse Penthouse Restaurant, Penthouse Lounge, Business as Usual, Jockey Club Lounge, and incredible art exhibits in the Equine Art Gallery.

Patrons do not miss a racing moment at Remington Park. Along with the 8' x 17' full-color viewing screen positioned above the paddock, more than four-hundred 27-inch color monitors located throughout the stadium and infield show races live and on instant replays, odds updates, views of the walking ring, Winner's Circle, and starting gate.

The technology and design of this racetrack are developed to focus its energies on service for both patrons and horsemen.

SHOWCASING EXCELLENCE

Horses and horsemen are the stars of the show at Remington Park. This facility allows guests to have a close-up look at the strategy and race preparation between distinguished owners, trainers, jockeys, and horses who come to Oklahoma City to race at Remington Park.

Horse racing fans can observe the horses from the moment they arrive in the Paddock Gardens to gear up for the race, until they are led through the glass tunnel and out onto the track. As the bugle sounds and the thundering hooves blaze across the 1-mile dirt track or 7/8 mile turf course, horses give it their all to win.

Champions such as Clevor Trevor, Brother Brown, and Silver Goblin established their careers at Remington Park and have since raced into national prominence. Remington Park's first world record was set by Silver Icon, a grey gelding trained by Ron Goodridge, when it equaled the world record for 6 1/2 furlongs.

Patrons are entertained by and reap the rewards of this racing competition. An unidentified woman made racing history in February of 1990, when she collected a track record Pick Six payoff of $1,070,482.40.

three, and off-track wagering has skyrocketed in popularity.

Legislation passed in 1996 now allows Remington Park to offer more product to its fans. By bringing racing from other jurisdictions into Remington Park through simulcasting, patrons can watch and wager on races taking place throughout the United States and across the world via satellite.

Other ambitions for Remington Park include striving to improve the quality of horses competing and working to step up the awareness of the horse as an athlete and horse racing as a professional sporting event.

There is no doubt, world-class Remington Park racetrack and the horses who are the stars of the show will remain the cornerstones of the warm-hearted, ever-proud Horse Capital of the World. ✪

Yet Remington Park's prosperity is recognized by more than those who enter the turnstiles on race days.

LET YOUR HEART RACE

Since it began operation, Remington Park has taken an active role in charitable events and raised more than $1.5 million for local charities. Philanthropic involvement ranges from direct contributions to hosting events such as the Muscular Dystrophy Association Easter Egg Hunt, Juvenile Diabetes Foundation Walk for the Cure, and the Mini Grand Prix for the Arthritis Foundation.

Figures through 1995 reflect Remington Park has paid more than $40 million in direct taxes to the state, county, and city. In fact, its tax dollars have helped rebuild Oklahoma City's Millwood public schools through remodeling, addition of computer systems, and an upgraded library.

Remington Park does back the slogan "Let Your Heart Race."

RUNNING STRONG

Remington Park is constantly working to improve its product. In addition to live Quarter Horse and Thoroughbred racing, Remington Park has added two more dimensions to racing.

To provide the fan base easier access to its product, Remington Park now offers off-track wagering. In the fall of 1995, following the passage of Senate Bill 450 in April, Remington Park transmitted its signal to three satellite wagering sites for the first time. A number of additional sites have been added to the original

Penn Square Mall

An excellent mix of specialty shops, four department stores, a comfortable atmosphere, and the desire to meet the everchanging needs of customers are the reasons Penn Square Mall has continued its trend as Oklahoma City's premier shopping mall.

Approximately 45 percent of the merchants in Penn Square Mall are exclusive to the Oklahoma City market, and some are even exclusive to the state—a true invitation for shoppers. With a convenient central location, the mall welcomes more than 14 million visitors per year.

When Penn Square Mall opened in 1960, it was billed as the largest "mall-type" shopping center in the area between Chicago, Dallas, and Denver. This open-air shopping center boasted 575,000 square feet and 46 retail shops.

A grand reopening of the mall in 1988 transformed the single-level center into a two-story Victorian landmark. Reminiscent of the turn of the century amusement park that once stood on the property, the open and light Victorian park design reflects the community and memories of the past. Square footage of the mall increased to 1.15 million square feet. It featured 130 specialty shops, a festive food court called Picnic Square, Oklahoma City's largest cinema complex, and three major department stores: Dillard's, Foley's, and Montgomery Ward.

Further expansion and renovation was completed in October, 1995. This expansion added JCPenney department store and increased the number of specialty shops to 140.

In addition to providing great shopping opportunities, Penn Square Mall has strived to provide the community with educational, yet entertaining, opportunities. The State Arts Council of Oklahoma awarded Penn Square Mall the Business and the Arts Award in 1994 for supporting local arts organizations through successful concert series scheduled throughout the year. A variety of groups perform during these series, including classical ensembles, jazz ensembles, folk groups, and vocal groups including school choirs. Musicians are chosen for these concerts based on their musical abilities and how they communicate with the audience.

Entertainment which provides an educational element for children is also a priority for the

A grand reopening of the mall in 1988 transformed the single-level center into a two-story Victorian landmark.

mall. Various programs include the "Penn Square Players," which produce and perform original plays for children and "Club Santa," which offer children's entertainment every Monday night in December and features music, theatre, and puppet shows. Other events which bring education and entertainment together include the Dinamation Whales and Dinosaur exhibits. Field trip guides and curriculums are sent to area schools to encourage teachers to visit the exhibit with their students. During the 1996 "Whales" exhibit, students from three to four schools per day, for a month, ventured to see the species from the ocean lining the mall walkways.

Penn Square Mall stays focused on pleasing the customer and meeting public needs. One-of-a-kind specialty shops, holiday valet parking, excellent customer service and its strong community ties continue to maintain Penn Square's position as the premier shopping center in Oklahoma City. ✪

Hudiburg Auto Group

The Hudiburg Auto Group is home grown and proud of the grassroots efforts that have made it one of the largest auto groups in the country. As an extremely people-oriented company, fair treatment in employee and customer relations have solidified the company's place in the state and national automobile marketplace.

Six locations and 14 franchises make up the Hudiburg Auto Group. Hudiburg Chevrolet, Hudiburg Pontiac GMC, and Hudiburg Toyota are in Midwest City. Hudiburg Nissan-Buick is found on I-240 and Shields in Oklahoma City. The Tulsa locations include Riverside Chevrolet and Riverside Nissan. Riverside Autoplex in McAlester includes Dodge, Jeep Eagle, Chrysler Plymouth, Honda, and Mazda.

As a 16-year car industry veteran, David Hudiburg is president of the Hudiburg Auto Group and employs approximately 1,000 people in Oklahoma. He also has partners at each location. "My business philosophy is people. If you hire good people, train them well, pay them well, motivate them, and treat them fairly, then they will do a good job for you," said Hudiburg.

David Hudiburg began his business career in 1982, when he became an active partner in the family-owned Hudiburg Chevrolet. "My dad started with very little and built an extremely large dealership because he was an aggressive trader with a great business mind," said David.

Six locations and 14 franchises make up the Hudiburg Auto Group.

"I am fortunate he gave me the start I had with this business and the opportunity to take the company to the next level—as one of the top 50 dealership groups in the country."

David's father, Paul Hudiburg, launched his business career when he was 22 years old by establishing a Kaiser-Fraser automobile and Case farm equipment agency in Prague. This business later became a Pontiac dealership. Two years later, a brother, K.H., joined Paul in business. The brothers bought another Pontiac dealership in Okemah, which Paul operated for five years. The brothers also bought a Buick dealership in Prague and Okemah.

Dissolving the partnership in 1952, Paul retained the Pontiac and Buick dealerships in Okemah. He bought an additional Buick dealership in Freeport, Texas, kept it for a short time and then sold it. In April, 1957 Paul came to Midwest City, buying the Chevrolet dealership from the Atkinson family, which he directed to unprecedented success. The firm is one of the largest dollar volume dealerships in Oklahoma, a record it has maintained for years.

The company record also shows long-term support for the success of the community and its citizens. Hudiburg Auto Group is a member of several chambers of commerce. It gives to the Mid-Del Public Schools Foundation and the Chairman's Club of the YMCA. David's leadership includes 1997 president of the Midwest City Chamber of Commerce and member of the board of directors for the Oklahoma City Economic Development Foundation. He was instrumental in securing the Metro Auto Dealers' substantial contribution to the committee formed to ensure Tinker Air Force Base continues to thrive amidst the federal government's Base Closure and Realignment Committee (BRAC) studies.

The future of the Hudiburg Auto Group is to remain strong and keep expanding when opportunities arise. David points out that customers are the main key to success: "Treat them fairly. That's what we try to do every day." ✪

The Hudiburg Auto Group is home grown and proud of the grassroots efforts that have made it one of the largest auto groups in the country.

The State Fair of Oklahoma

More than a million people enjoy The State Fair of Oklahoma each year. Photo by Dale Metzler.

For 17 days each September, The State Fair of Oklahoma comes to life! It brings free family entertainment, educational exhibits, exciting competitions, tasty fair food, and the colors of the midway, with more than 100 rides, games and attractions to Oklahoma. Considered "Oklahoma's Largest Attraction," more than a million people enjoy The State Fair of Oklahoma each year. The State Fair of Oklahoma consistently ranks as one of the top five fairs in the country. It is run by a volunteer board, comprised of prominent businesspeople from across the state. To promote clean family fun is one objective of The State Fair of Oklahoma. Some of the many events taking place during the fair include Walt Disney's World On Ice, State Fair Circus and the PRCA Championship Rodeo. The State Fair of Oklahoma is proud to host the largest international trade show of any fair in America.

Oklahomans have been entertained and educated through The State Fair of Oklahoma since 1907. This tradition of family values and service to the public began at Northeast 10th and Martin Luther King Boulevard (Eastern Avenue). The people of Oklahoma inaugurated the new and expanded fairgrounds at Northwest 10th and May Avenue at the 1954 State Fair. This location remains the permanent home for The State Fair of Oklahoma.

State Fair Park is one of the largest, most versatile, fair facilities in the country. It plays host to more than 2,000 event days a year. The year-round schedule includes 12 national-caliber horse shows, craft shows, motor sports events, boat and auto shows, rodeos, concerts, antique shows and two major fairs. Besides The State Fair of Oklahoma, the Spring Fair and Livestock Exposition was launched in 1994. This event indicates the fair's commitment to agriculture, youth and education and is highlighted by the 4-H and FFA Junior Livestock Show.

Spread over 435 acres and boasting 29 permanent buildings, State Fair Park houses 300,000 square feet of air-conditioned space. The livestock and barn complex, State Fair Arena, exhibit halls and State Fair Grandstand are the four main components of the State Fair Park building network.

State Fair Park is one of the largest, most versatile, fair facilities in the country. Photo by Dale Metzler.

Extensive remodeling of the State Fair Arena is part of the Metropolitan Area Projects (MAPS). Construction and remodeling of barn number six and an additional show arena, plus infrastructure work on five barns to fortify the excellence of the equine facilities, will complete the MAPS construction on the State Fair Park grounds.

Outside, more than one million square feet of exhibit space is framed with arches, the Conoco Space Tower, and the AT&T Wireless Services Monorail.

To augment the full range of activities on the State Fair Park, the Food Service Division serves approximately two million customers annually, during the interim (non-fair) event days. OK Tickets is charged with the responsibility of selling tickets for State Fair Park events and a variety of events and activities held throughout the Oklahoma City area.

Each year, the goal is to make The State Fair of Oklahoma bigger and better than the previous year. With an obligation to the people of Oklahoma and a desire to live up to nearly a century of expectations, The State Fair of Oklahoma continues to stride down the colorful midway of success! ✪

Medallion Hotel

From the grand entrance hall's sophisticated flourishes to the five-star Concierge Level Presidential Suites, the Medallion Hotel celebrates a renaissance of elegance in the heart of Oklahoma City's downtown area. Located at One North Broadway, the Medallion's top-to-bottom $9 million transformation is the first major renovation of an entire hotel in recent local history and introduces a new, upscale level of hospitality to the metro area.

"Since we began operating the hotel in 1994, our approach has been to implement the

The two 1,200-square-foot Presidential Suites are calculated to revive spirits through a calming harmony of comfort and refinement. The five-room suites showcase hand-crafted furnishings, sunken oversized bathtubs, marble wet bars, and a view of Bricktown from the hotel's top floor.

A full-service business center located on the Concierge Level offers access to a computer, e-mail, Internet, and facsimile equipment. In-room amenities include a dataport, voice mail accessible off the premises, complimentary fax service upon request, and personal delivery of

coffee with the morning newspaper. Life Cycle exercise equipment is also available for delivery to the Concierge Level suites.

"We are giving our guests maximum return on their dollar in service and style," Browne says.

Throughout the hotel, guests experience a revival of cosmopolitan artistic detail. A floor blend of Italian marble, domestic ash wood, and English-made carpets welcome guests to the lobby, which is accented with full-lead crystal accessories, dramatic pottery, and original artworks. A full-length painted mural displays foliage in motion.

The enclosed board room and nearby intimate wine cellar, with more than 100 selections, are adjacent to the Aria Grill. This restaurant serves gourmet specialties created by a maestro chef with international five-diamond restaurant and resort experience. It also offers a contemporary backdrop for dining on fresh, naturally flavored High Plains cuisine unique to this region.

"With the MAPS program and Oklahoma City's economic development campaign, Medallion recognizes the city's commitment to becoming a leading national business city," says John Steinle, vice president of New York-based Medallion Hotels, Inc. "Our vision is to form a long-term partnership with Oklahoma City's business community that energizes downtown and fosters civic and cultural opportunities." ✪

Medallion guests are welcomed with a classic contemporary style that artistically blends elegance, luxury, and sophistication into a relaxed, comfortable atmosphere. **Travel Weekly** *has recognized the Medallion for devoting care and attention to both its properties and staff.*

Medallion standard. It raises the quality of guest services in the marketplace to a heightened level of hospitality. In each of our six facilities, we incorporate a 'let me show you' attitude of service and guest-friendly property to meet the needs of executives," says Douglas V. Browne, general manager. Other Medallion properties are in Texas, Kentucky, and Ireland.

Travel Weekly magazine noted the Medallion has achieved success by devoting care and attention to both its properties and staff, resulting in high morale and improved customer satisfaction.

Renovation of the 15-floor, 400-guestroom Oklahoma City Medallion features totally refurbished guest rooms and suites at an above-standard cost of $6,500 each. Aria Grill, the Mediterranean-style restaurant with private dining rooms, a distinguished corporate board room, and a spacious 18,000-square-foot conference area, are designed to meet every business and personal need.

Home to Oklahoma City's two largest Presidential suites, the Medallion's five-star Concierge Level anticipates the needs of today's traveling executive—privacy, security, luxury, and efficiency.

18

High Tech & Energy

Smith Cogeneration plant at night. Photo by Jack Hammett.

Devon Energy Corporation

From its modest beginnings in the early 1970s to its present status as one of the nation's largest independent oil and gas producers, Devon Energy Corporation is one of Oklahoma City's outstanding success stories. Devon employs over 300 people, more than 200 of whom reside in Oklahoma City. The company has grown to become a major operator of oil and gas wells in Texas, New Mexico, Oklahoma, Wyoming, and western Canada. Devon's oil and gas reserves now exceed 150 million barrels and annual revenues exceed $100 million. In addition, the aggregate value of Devon's common stock exceeds a billion dollars.

Devon's co-founder and chairman of the board, John W. Nichols, has long been a force in the energy industry. In 1950, he pioneered the first oil and gas drilling program registered with the Securities and Exchange Commission. In the 1960s, prior to the start of Devon, he established several successful oil and gas-related businesses around the world.

One of the companies founded by John Nichols was an oil service company in Libya. When he sold the assets of this business in 1969, he was left with a corporate shell. While formalizing the sale in a barrister's office in London, John was asked for the name he had selected for the corporate shell he was retaining. Not having anticipated the situation, John glanced up at the wall to a map of England. His eyes landed upon the county of Devon and he replied, "Devon—Devon Energy Corporation will be the name." And so it was.

Following the sale of the service business, Devon had but one remaining asset—a $5 million tax loss carry-forward. In 1971, John's son, Larry, joined him and the father-son team set out to put that asset to use. They combined the tax loss with the financing horsepower of major European investors and created Devon International Royalties Limited.

From 1971 to 1977, Devon International Royalties Limited acquired over $70 million in oil and gas properties, gas processing facilities, and transportation systems. The company established the world's first international royalty fund—and a legacy of financial creativity.

From 1978 to 1982, Devon again exhibited this creativity as the company formed a series of business

Flame in the background reveals another successful well in Devon's Northeast Blanco Unit in northwest New Mexico. This is a geologically unique area where natural gas is produced from coal deposits (coal seam gas) located some 3,000 feet beneath the surface of the earth.

Devon Senior Executive Officers: (Seated L-R) H.R. Sanders, J. Larry Nichols and J. Michael Lacey. (Standing L-R) H. Allen Turner, Darryl G. Smette and William T. Vaughn.

ventures that raised an additional $88 million in investment capital. During the mid-1980s, Devon was one of the first oil companies to roll-up its public drilling funds into a master limited partnership that traded on a major stock exchange.

In 1988, Devon International Royalties Limited and the master limited partnership were combined into a new publicly held company. This union, Devon Energy Corporation, was destined for a period of dramatic growth. Since the beginning of 1988, the company has increased its oil and gas

reserves more than twenty-fold. The company's oil and gas production, revenues, and earnings have all grown dramatically. All in all, Devon has grown to become one of the largest and most profitable independent oil and gas producers in the United States.

Key to Devon's success has been the company's narrow focus on building oil and gas production and maximizing the value of that production. Devon increases oil and gas reserves and production through the acquisition and development of

A helicopter lowers equipment for use in a 3-D seismic survey of a Devon property. Devon's geoscientists use this technology to "see" underground structures where oil and gas deposits may lie.

While Devon has come a long way in the 25 years since John Nichols first selected its name, the company has no intention of stopping now. As this publication goes to press, Devon has just closed the largest transaction in the company's history—over $300 million in value. And while the ink has hardly dried on this deal, Devon is debt-free and busy looking for the next opportunity. Devon Energy Corporation and its employees are looking forward to continuing their partnership in growth with Oklahoma City. ✪

producing oil and gas properties as well as through exploring for undiscovered oil and gas.

Since the beginning of 1988, Devon has drilled more than 800 oil and gas wells with a success rate of over 95 percent. Devon is now one of the most active drillers of oil and gas wells in the United States.

Devon began one of its most successful drilling projects in the early 1990s. Along with a few other oil and gas companies, Devon pioneered a method of economically producing natural gas from coal deposits (coal seam gas). Using this technology, Devon developed the company's fabulously successful Northeast Blanco Unit in New Mexico. Arguably the world's best coal seam gas property, the Northeast Blanco Unit has recoverable gas reserves of around one trillion cubic feet. This singular property with just 102 wells produces enough natural gas to satisfy the needs of a city of several hundred thousand people.

Acquisitions have also played an important part in Devon's growth. The company has completed 16 major acquisitions of oil and gas properties or companies in the last nine years. The acquisitions, when coupled with Devon's exploration and development activities, have earned the company a reputation for building per share value—year after year.

As a result of the company's growth, Devon's shareholders have been richly rewarded. But Devon's shareholders are not the only ones who benefit from the company's success. Devon and its employees have been recognized for their contributions to various charities, the arts, civic organizations, and to the community at large.

Roughnecks prepare to make another in a series of drillpipe connections on a well drilled on Devon acreage. Devon is one of the most active drillers of oil and gas wells in the U.S.

Buttram Energies, Inc.

The story of Buttram Energies, Inc. (BEI) begins with a profile of its founder, Frank Buttram, who was voted Oklahoma City's Most Useful Citizen in 1926.

"Frank Buttram was to become one of the most successful oil men in the United States, as well as a rancher, banker, cattle breeder, citrus grove owner, civic leader, philanthropist, and art collector and would be named one of Oklahoma's outstanding businessmen of all time," reads a description in the book, *One Man's Footprints: The Story of Frank Buttram.*

Dorsey Buttram, current president of Buttram Energies, talked about his father: "Whatever he did was always the best."

MASTER GEOLOGIST AND INDUSTRY LEADER

Born in Oklahoma Territory in a log cabin, Frank was a man who knew hard work and the land. A member of the first geology class at the University of Oklahoma, Frank graduated with a master's degree in geology within an accelerated time frame.

Dorsey Buttram and Preston Buttram

Frank then took a job with the Geological Survey as a chemist in 1911. One of Frank's three bulletins published by the Geological Survey within his first year, *The Cushing Oil Field*, directed his career in the oil industry and placed new-found emphasis on geologists in the work force.

"So accurately did Buttram's bulletin chart the Cushing Field and so closely did the actual development bear out his scientific arguments that within a year virtually every oil company added a geologist to its staff," quoted *The Story of Frank Buttram.*

The rest is history. With Frank's help in soliciting investment funds from the East Coast, Fortuna Oil Co. was born. Frank's accurate geol-

ogy took him to general manager of the company. *The Story of Frank Buttram* described that Fortuna Oil Co. "...turned out to be one of the outstanding success stories in the oil business, and he (Frank) became a millionaire before Fortuna was sold in 1918."

In 1920, Buttram Petroleum Corporation was formed operating mainly in Oklahoma and Texas. Frank's geological expertise led to the continued success of this company through discovery of many fields and prolific wells.

Frank was one of the five originators of the Independent Petroleum Association of America (IPAA) and served as chairman of this organization from 1939 to 1943. He was the only independent oil man chosen by President Franklin D. Roosevelt for the Petroleum Industry War Council in 1942.

The leadership Frank exhibited in the oil industry was extended into many other areas.

DEVOTION TO COMMUNITY

Frank Buttram made an indelible mark on the Oklahoma City landscape. Frank served Oklahoma City by leading the committee to establish a city manager form of government, playing an instrumental role in resolution of the near-crisis flood situation and working diligently to remove the railroad tracks that paralyzed downtown traffic.

In 1926, he was appointed chairman of the Oklahoma City branch of the Federal Reserve Bank. He served as chairman of the Oklahoma City Chamber of Commerce from 1936-1939, and was honored as the only native-born chairman.

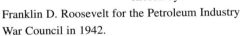

Frank Buttram

Frank was inducted into the Oklahoma Hall of Fame in 1940, and his wife, Merle, was honored with membership as well. They were the first couple to be included in this prestigious Hall of Fame. Frank was also inducted into the National Cowboy Hall of Fame in 1964.

THE BUTTRAM TRADITION CONTINUES

Dorsey Buttram joined Buttram Petroleum Co. in 1945 after earning an engineering degree from Cornell University and taking graduate courses in geology at the University of Oklahoma. Serving as an Air Force pilot and training instructor during World War II, Dorsey's deep interest in the aviation industry nicely complemented his father's geological work.

As a father-son team, they worked well together. Dorsey would take his father skyward for a first-hand look at surface geology.

President of Buttram Oil Co. since 1964, Dorsey has held extended terms on the board and executive committees for both the Oklahoma Independent Petroleum Association and the IPAA.

Dorsey also traveled extensively throughout the United States as a member of the IPAA Speaker's Bureau and Energy Advocators Group

Brangus breed of cattle originated by Frank Buttram.

A BEI subsidiary company was also created to engage solely in real estate projects. This diversification into a new industry is met with optimism that it will be an excellent counterpart for BEI's oil and gas endeavors.

The cattle industry is another area of focus for BEI. The Brangus breed of cattle was originated by Frank Buttram. Dorsey continues his father's work in this area and, as a past president of the International Brangus Breeders Association, remains active in the association. The company maintains a herd of Brangus in Ulman, Missouri.

Through Buttram Energies, Inc. and in honor of family tradition, past commitment, and success, the characterization of Frank Buttram still applies today: "Whatever he did was always the best." ✸

to advise the consumer states and media on energy policy. Along with his U.S. travels, he ventured to England and Europe selling the U.S. and Oklahoma oil.

Dorsey was invited to hold a prestigious membership in the American Association of Petroleum Geologists. He was appointed to the American Heritage Panel and the National Anti-Metrics Committee chaired by Congressman Eldon Rudd. He was also selected by Congressman Edwards to serve on the Air Force Academy Selection Committee and chosen as Honoree of the Year for the Sirloin Club, which promotes 4-H and FFA youth.

Dorsey was and continues to be active locally through many civic and cultural committees.

Dorsey's son, Preston, is vice president in charge of operations for BEI. Educated in Oklahoma, Preston is on the Board of Governors for the National Association of Royalty Owners. He is also a member of the American Association of Professional Landmen, Oklahoma City Association of Petroleum Landmen, National Association of Division Order Analysts, and a charter member of the Los Angeles Association of Lease and Title Analysts.

Preston's leadership responsibilities with BEI include review, evaluation, and recommendation of oil and gas prospects submitted for BEI's consideration, along with management and oversight of BEI's day-to-day operations, staff, budget, and financial matters.

Preston is also founder, president, and owner of Buttram Oil Properties, Inc., a consulting company specializing in revenue and property management of oil and gas properties.

BUTTRAM ENERGIES, INC.

BEI continues to be a 100 percent family-owned oil company committed to the exploration and development of oil and associated energy-related activities. Over the past years, BEI initiated a program for the acquisition of proven reserves that complements its exploration activities.

One of Buttram Energies, Inc.'s early discoveries was the Powell Field near Coriscana, Texas which ultimately produced 65,000 barrels of oil per day.

Innovention Systems Integrators

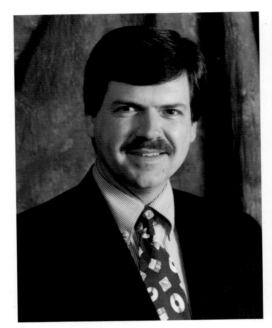

Robert Metivier, President

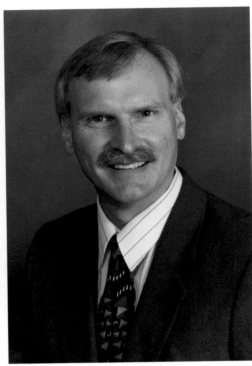

Richard Hesselgren, Vice President

In this increasingly complex world it is difficult for anyone, professionals included, to determine which of today's technologies will become the cornerstones of tomorrow's. Businesses, especially, may not be able to dedicate enough personnel or resources to keep up with all the newest advancements. It falls to the leadership of these businesses to choose a partner that will help them make decisions that ensure their technology dollars are well invested.

In Oklahoma and the Southwestern United States, Innovention Systems Integrators is that technology partner for many companies. They have provided guidance for some of the Southwest's largest companies. Innovention is not simply a computer sales company, but a provider of solutions to enhance a company's productivity and bottom line.

Innovention was born in 1993, founded by people with long experience in the computer and technology arenas. Knowing that Oklahoma needs a true systems integrator, one who can efficiently make different types of computer and communications "talk" with each other, Robert Metivier and Richard Hesselgren came together with a vision to help Oklahoma develop the cutting edge technologies that are the bridge to our future.

But it is no longer just computer systems that are hooked together in a network. Based on observations that many of the communications in use today (telephone, fax, electronic mail, paging, cellular, data, electronic document imaging, accounting, and Internet) are merging and interacting with each other, the company determined a need to help business find the most cost- efficient ways of integrating these state-of-the-art technologies. This integration is nothing less than the dynamic and ever changing future of human communication. Using this "unified computing model" as a framework for solution building, Innovention Systems Integrators has become a technology leader not only in Oklahoma, but is known to technology professionals across the United States.

PARTNERSHIP

Innovention's unique approach to serving its clients' needs rests on a commitment to partnership, not just a vendor-customer relationship. This commitment involves a substantial investment in training for Innovention personnel, an understanding of the client's business and business processes, and recognition of the need to teach the client to manage the technology. Innovention Systems Integrators thus becomes a partner in the truest sense, someone upon whom the client may depend for sound advice, guidance, and performance. As new methods and technologies develop, Innovention is there to show how they can be applied in the customer's workplace. By staying abreast of cutting edge developments, Innovention Systems Integrators is a valuable addition to any company's resource base.

GLOBAL VISION

As the third millennium begins, the global marketplace has become reality. Innovention realizes that individual companies are no longer isolated in their particular geography, and technology has made every company, large or small, part of a system that has grown much larger than any state or nation. Innovention Systems Integrators can prepare every client to become a competent, thriving participant in the world's economy. From a simple presence on the World Wide Web to a sophisticated Electronic Data Interchange system,

Innovention can give virtually any business the technology assistance to compete in their market.

Major technical advances in communication happen more frequently than the average person realizes. The explosive growth of nationwide fiber-optic networks, satellite feeds, and other means of moving digital content has lifted many of the old constraints on the exchange of large amounts of data. In many cases, this means a location near a geographically defined major market is no longer a particularly relevant consideration in a business' plan for growth and operation. Helping businesses design their best "Window on the World" is Innovention Systems Integrators' specialty. Through a thorough business process analysis, Innovention will help a customer determine the network architecture that is right for them. This could be anything from a simple modem and telephone hookup to a far more complex distributed global network.

It is this independence from geography that makes Oklahoma City a prime location for companies seeking to expand or relocate. Ideally situated midway between the East and West Coasts, as well as between our NAFTA trading partners Canada and Mexico, Oklahoma's low cost of living and reputation for good schools and friendly people make it a natural hub for communications systems which will facilitate the development of tomorrow's businesses. In the coming years, the air above Oklahoma will be as valuable as the oil below the ground. Just as Interstate Highways 40 and 35 are the transportation crossroads of America, the wired and wireless networks that will define business in the next century come together over Oklahoma City, and Innovention Systems Integrators will be there assisting companies old and new in their use and development of them.

Innovention Systems Integrators actively supports the efforts of civic and business leaders to attract technology companies to the Oklahoma City area. With the realization that ours is now an information-based economy, Metivier, Hesselgren, and company strongly believe that an aggregate of high-tech firms is good for the community's health and solid foundation for its future.

INTERACTIVE SERVICES AND INTERNET

The explosive growth of the Internet in recent years has meant that the demand for highly qualified networking engineers has skyrocketed, and Innovention's technical staff is unrivaled in network design, installation, and maintenance.

Innovention is also a leader in the development of applications designed to enhance a company's productivity in a wide-area network or Internet environment. While recognizing that technology is a very dynamic field, Innovention's current areas of expertise also include database design and management, document imaging systems, work flow analysis and implementation, and accounting systems. Their client base includes customers from Oklahoma's petroleum industry, as well as medical manufacturing, real estate, biomedical, public utilities, communication companies, and government at all levels.

DYNAMIC SOLUTIONS

Innovention Systems Integrators is one of the Southwest's premier technology companies, and it is proud to share its vision of tomorrow with Oklahoma City. The challenge of determining the direction of tomorrow's technologies and working with today's is what keeps Innovention employees enthusiastic and eager for more business. Innovention wants to help Oklahoma City companies enter the twenty-first century with dynamic, practical solutions for use in the global marketplace. ✦

Kerr-McGee Corporation

Kerr-McGee is a multinational energy and chemical corporation. The strength of this Fortune 500 company's history, corporate beliefs, flexibility, areas of focus, and vision ensure that the fortitude and strategic alliances that built the company will continue to serve the organization well into the next millennium.

Three solid businesses of oil and natural gas exploration and production, inorganic chemicals, and coal represent the core of Kerr-McGee. The company ranks as a leading producer in each of these fields and each area has good potential for growth.

The company's belief statement serves as the corporate theme. Respect for the individual, ethical business dealings, safe work practices, responsible corporate citizenship, responsible care for the environment, and continuous improvement mark the standards of Kerr-McGee.

Intertwined in this philosophy are nearly 4,000 employees, the company's stockholders, customers, and thousands of citizens across the country who benefit from Kerr-McGee's charitable giving.

HISTORY

The company's vision is to be an innovative, respected global energy and chemical company, recognized as outstanding by employees, investors, customers, and the public, and to be the standard by which other businesses are measured. This vision is solidly founded in Kerr-McGee's history.

The story of Kerr-McGee began in Ada, Oklahoma, in 1929, when Robert S. Kerr and his partner organized the Anderson & Kerr Drilling Co. This company's search for oil and gas called for a skilled geologist. Dean A. McGee accepted the challenge. After several years and name changes, the company became Kerr-McGee.

The historical significance of this company includes its founding by two men who became well-known Oklahomans and its pacesetting business standards.

In 1947, Kerr-McGee made history when it launched the offshore oil industry. Its Rig 16 drilled the first commercial well out of sight of land on Ship Shoal Block 32 in the Gulf of Mexico. The company continued to show its progressive colors in 1965 when Transocean No. 1, the first offshore drilling rig designed for North

Kerr-McGee Corp., a Fortune 500 energy and chemical company, conducts its worldwide business activities from the company's headquarters in Oklahoma City.

Kerr-McGee coal from Jacobs Ranch Mine in Wyoming's Powder River Basin supplies energy needs for electric utilities.

Sea operations, began work in the North Sea's German sector.

Kerr-McGee observed its 50th anniversary in 1979, celebrating the steady leadership, innovative nature, and success of the company.

In 1983, Dean A. McGee stepped down as chairman after 46 years of leadership. Frank A. McPherson succeeded McGee as chairman and chief executive officer. McPherson's successor, Luke R. Corbett, is only the fourth chairman and chief executive officer in Kerr-McGee's history.

"Frank McPherson has led Kerr-McGee through some challenging times and positioned us well for future growth," said Luke Corbett, chairman and CEO.

OIL AND NATURAL GAS

A strong exploration and production team conducts Kerr-McGee's domestic and international oil and gas operations. The company's inventory of exploratory and producing properties is balanced between oil and natural gas. Areas of activity are in the Gulf of Mexico, the North Sea, China, Southeast Asia, and the Middle East.

Kerr-McGee announced in 1996 that Kerr-McGee China Petroleum, Ltd., a wholly owned subsidiary, signed a petroleum contract with the China National Offshore Oil Corporation to explore an area of approximately 756,000 acres near the mouth of the Pearl River, South China Sea in China. In March of 1997, Kerr-McGee

Kerr-McGee's Breton Sound Central Facility produces crude oil and natural gas from the Gulf of Mexico.

signed production sharing agreements in the Republic of Yemen to explore approximately 10 million acres in that country.

Careful analysis of environmental factors precedes exploratory drilling in new areas. Through continuous improvement in safety and environmental protection, yesterday's standard of excellence becomes today's expected performance level in Kerr-McGee's exploration and production activities.

CHEMICALS

Major chemical products produced by Kerr-McGee include titanium dioxide pigment, electrolytic, and specialty chemicals and pressure-treated forest products. Kerr-McGee's people and plants are among the safest in the chemical industry.

The company is a global producer and marketer of titanium dioxide pigment, which is used extensively as an opacifier in paint and plastics. Titanium dioxide plants are in the United States, western Australia, and Saudi Arabia.

Electrolytic products are in strong demand. Manganese dioxide is a fine black powder used in alkaline dry-cell batteries. Kerr-McGee is North America's largest producer of manganese dioxide and the third-largest in the world. Another electrolytic product is sodium chlorate. This odorless and colorless chemical resembling salt is used in an environmentally preferred pulp-bleaching process.

COAL

Kerr-McGee ranks seventh among U.S. coal producers in production tonnage. The company produces low-sulfur coal from Jacobs Ranch Mine, a large surface mine in Wyoming's Powder River Basin, and relatively low-sulfur coal from Galatia Mine, a modern underground mine in southern Illinois. Jacobs Ranch is the nation's fourth-largest coal mine.

Safety and environmental stewardship are top priorities for Kerr-McGee's coal operations, which continue to earn awards and praise from regulatory agencies for environmental and land reclamation programs.

COMMUNITY

Kerr-McGee's community spirit and vision are well known. Across from the company's worldwide headquarters, McGee Tower, Kerr Park was dedicated and presented to the people of Oklahoma City in 1974.

At Columbus Elementary School in Oklahoma City, Kerr-McGee employees spend 60 to 100 hours per week tutoring students during the school year. The company provides transportation to and from the school for involved employees. The company also contributed 30 computers to establish a computer lab at Columbus.

In the area of environmental preservation, the donation of 1,300 acres in Florida to The Nature Conservancy preserved 300 scenic acres along the Suwannee River. Sale of the remaining acreage raised $1.5 million to help fund Conservancy projects in Oklahoma, including a significant portion of the Tallgrass Prairie Preserve in Osage County.

This community spirit is felt in many nonprofit arenas and locations where Kerr-McGee has a presence.

After nearly seven decades of successful operations, Kerr-McGee looks ahead to continued growth in its principal businesses and in the company's value for its shareholders and community. Because of the corporate beliefs and the long-standing vision to be the standard by which other businesses are measured, Kerr-McGee will remain a respected multinational energy and chemical company well into the future. ✪

In Hamilton, Miss., Kerr-McGee produces titanium dioxide pigment used in coatings and plastics.

Chesapeake Energy Corporation

Chesapeake Headquarters — Oklahoma City.

The February 27, 1997, edition of *The Wall Street Journal* sums up Chesapeake Energy Corporation's track record of success as an independent energy producer. The headline reads: "America's Best Performing Companies." The following article says: "A combination of high technology and old-fashioned promotion made Chesapeake Energy Corporation the best-performing stock among the 1,000 in Shareholder Scoreboard over the past three years."

This distinction is impressive for any company, especially one started in 1989 with only a $50,000 investment. Aubrey McClendon serves as chairman of the board and chief executive officer, while Tom Ward is president and chief operating officer. The men were informal partners in the oil and gas industry from 1983 to 1989 before incorporating under the name of Chesapeake Energy Corporation.

COMPANY INTRODUCTION

Chesapeake Energy Corporation, an independent energy producer, is headquartered in Oklahoma City. Additional district offices are in Lafayette, Louisiana; College Station, Texas; and Lindsay, Oklahoma. The company employs approximately 350 people, with 250 of those employees located in Oklahoma City.

Since its inception in 1989, Chesapeake has drilled or participated in 700 wells of which approximately 90 percent have been commercially productive. The company focuses on utilizing advanced drilling and completion technologies to develop new oil and natural gas discoveries in major onshore producing areas of the United States.

Chesapeake became a publicly-owned company in 1993 and is acknowledged as one of the most successful and fastest-growing companies in the country. Its common stock is publicly traded on the New York Stock Exchange under the symbol CHK.

Chesapeake roustabouts.

TRACK RECORD OF SUCCESS

This dedicated and aggressive company has proven that its growth through the drillbit business strategy works. Figures from the first quarter of 1997 show Chesapeake has increased annual revenues to $255 million, increased annual earnings to $42 million, and increased annual operating cash flow to $150 million—all from zero in 1989! The company's annual oil and gas production is now more than 12 million barrels of oil equivalent, an increase of 20-fold from its IPO level, while its stock price has only increased 12-fold from the IPO.

Chesapeake ranked in the top one percent of all public companies in return to shareholders from 1994 to 1996. The stock had an average annual increase of approximately 274 percent during this three-year period.

As the fourth most active land-based operator in the country, Chesapeake is currently using 30 drilling rigs in its U.S. exploration program. In addition, Chesapeake is drilling the deepest and most technically challenging wells in the United States, with an average depth of more than 16,500 feet. More than 50 percent of these active wells utilized horizontal drilling technology, exhibiting Chesapeake's leadership in horizontal drilling.

"Chesapeake has drilled nearly 80 horizontal wells deeper than 13,000 feet, more than twice the number of the next most active horizontal driller, confirming Chesapeake's status as the recognized worldwide leader in deep horizontal drilling expertise," McClendon said on April 2, 1997, announcing the success of its Brown #1-H well.

Drilled in the Washington County, Texas, portion of the Deep Giddings Field, this world-class natural gas well may be one of the most prolific wells drilled in the past decade. "Based on initial flow rates and according to petroleum information data, the Brown is both the most productive

Chesapeake rig — Oklahoma.

Giddings is one of the most active fields in the United States and generated approximately five percent of all onshore drilling activity in 1995 and 1996. Natural gas equivalent reserves of more than one trillion cubic feet (TCFE) have been developed in this area, and drilling continues at an aggressive pace.

COMMUNITY INVOLVEMENT

Chesapeake Energy Corporation contributes more than simply developing oil and gas reserves for the company and its shareholders. The company's community development efforts support a variety of worthwhile causes. Its strongest and most focused giving is in education, as Chesapeake believes in financially supporting both the public and private educational community in Oklahoma City and the state.

"We believe that good things are more likely to happen in today's economy whenever smart people congregate. We have attempted to build an intellectually challenging environment at Chesapeake and hope to have a meaningful impact on our community through continuing to build our company," stated McClendon in a recent speech in Oklahoma City. "We remain committed to our goal of building one of the premier North American exploration companies. We are now well-positioned to continue executing our business strategy to achieve this goal." ✪

horizontal well ever drilled and the most productive well of any type drilled onshore in the United States during the past 10 years," he added.

Chesapeake has generated its impressive track record by developing a clear and focused strategy to create competitive advantages over its peers and enhance shareholder value. These five competitive advantages include:
* Identifying high-potential, geologically complex oil and natural gas reservoirs.
* Developing creative engineering solutions to profitably extract hydrocarbons from these reservoirs.
* Establishing dominant leasehold positions to create effective barriers to entry.
* Generating high returns through an accelerated production profile and aggressive cash flow reinvestment.
* Owning a 40 percent equity position, thereby aligning management interests with those of the company's other shareholders.

"We demonstrate our confidence in our strategy by having one of the highest management equity ownership interests of any public oil and gas company," McClendon said.

PRIMARY OPERATING AREAS

Chesapeake's three major operating areas are the Louisiana Austin Chalk Trend, the Giddings Field of south Texas, and the Knox, Golden Trend, and Sholem Alechem Fields in southern Oklahoma. The company is also active in New Mexico, Montana, North Dakota, and Canada.

For the next several years, Chesapeake will focus its advanced-drilling expertise in the expanding Louisiana Austin Chalk Trend, a 5,000-square-mile area that has potential to be

the largest onshore oil and gas exploration play in the past 25 years. The promise of significant oil and gas recoveries in this project area has attracted Chesapeake, Union Pacific, and Sonat, 3 of the 10 most active onshore operators.

Chesapeake is now the largest leasehold owner in the Louisiana Trend with approximately 1,300,000 undeveloped acres. This project inventory of up to 500 prospective drillsites has the potential to more than double the proven reserves of Chesapeake over the next several years.

Chesapeake — Glenmora, Louisiana.

Ted Davis Enterprises

Ted Davis has come a long way since he started a small business in his garage. Four companies later and the sale of two companies to the Siemens AG giant of Germany have made Ted Davis Enterprises a storybook success. Yet great achievements have not swayed the values or priorities of this native Oklahoman, who still uses ma'am and sir when greeting a stranger.

Oklahoma has been good to Ted Davis, and Davis has not forgotten. "My goal as an Oklahoman is certainly to tell all the advantages of Oklahoma City and Oklahoma. That is my main theme in life," says Ted Davis, president and CEO, Ted Davis Enterprises.

"The entrepreneurial spirit is alive and well in Oklahoma. If a person is determined to succeed, and has a vision to do something, the state allows him to do that," says Davis. "I grew up in a very poor farm family, but the reason I have been effective is because I am willing to spend time to learn and work to become a successful businessperson. Oklahoma's spirit has allowed me to do this."

TED DAVIS ENTERPRISES

The history of Ted Davis Enterprises (TDE) is easily traced through four companies: Ted Davis Manufacturing (TDM), Visionary International Products, Surface Mount Depot, and Progressive Stamping, Inc. (PSI).

TDM was Davis' first company started from the garage of his home in 1983. Its product is electronic assemblies for the disc drive industry. Visionary International Products was established in 1994 as the sales and marketing arm of TDE. In 1996, these two companies were sold to VAC-UUMSCHMELZE GmbH, a division of Siemens AG, headquartered in Germany. Davis stayed on as president and CEO.

Surface Mount Depot, a company specializing in electronic assembly and modification, was formed in 1989 by Davis. To continue the family entrepreneurial tradition, Davis' daughter, Stacey, became president and majority owner in 1992. Progressive Stamping, Inc. was purchased in 1993 and is run as a division of Ted Davis Enterprises.

Davis has assembled a group of companies that provides full service to all of their customers worldwide. Each company is successful in its own right.

VACUUMSCHMELZE CORPORATION

With the June, 1996, announcement of the acquisition of Ted Davis Manufacturing by VACUUMSCHMELZE GmbH, a subsidiary of Siemens AG, Germany, came accolades from Oklahoma Governor Frank Keating, Oklahoma City Mayor Ron Norick and chairman of the Greater Oklahoma City Chamber of Commerce.

This acquisition included the products, facilities, equipment, and research development activities of Ted Davis Manufacturing, which now operates as VACUUMSCHMELZE Corporation (VAC). It is the center of operations for Siemens' permanent magnets business and classified as a Siemens Corporation operating company.

"The merging of two world-class companies, Ted Davis Manufacturing in magnet assemblies and VACUUMSCHMELZE, GmbH in high-tech rare earth magnets, will allow for increased production capacity for magnet assemblies, increase our sales and marketing staff, and expand the research and development capability of the company," says Davis. "The new company will continue to reduce time to market and provide more design expertise for magnetic assemblies in a continuing effort to serve our customers."

At the time of acquisition, TDM employed approximately 200 people in Oklahoma City as well as 75 people in Asia, including Malaysia and Singapore. This company has grown to 1997 figures of $50 million in sales and number of employees close to 500 people worldwide.

BENEFITS OF OKLAHOMA CITY

Davis indicated there are two areas of concentration that continue to make the companies great: customer service and process control. All companies are internationally certified with an ISO 9002 designation. "We are always willing to go the extra mile and add the personal touch," says Davis. "We are all very much involved—ready to roll up our sleeves and do the job." Davis outlines the specific reasons he believes starting and growing his companies in Oklahoma City is also a major benefit.

"People are loyal and productive in Oklahoma. They will give it 200 percent when they know everyone is equal and all are important to the future of the company," says Davis. "In regard to employees, it makes no difference what the difference is in title. Everyone is key to the whole team accomplishing goals." Along with the work ethic, Davis suggests the cost of doing business, supply of water, real estate values, and low-cost

SURFACE MOUNT DEPOT

Surface Mount Depot is a woman-owned, minority-owned small business specializing in electronic assembly and modification. The firm's 38,000-square-foot facility in Oklahoma provides competitively priced, high-quality, surface-mount, thru-hole, or mixed technology circuit board assemblies and subassemblies representing 5,000 different products.

Nearly 200 employees work to make these products, which are used in everything from televisions, washing machines, computer systems, and medical applications to down-hole oil field monitoring equipment and garage door openers. Oil and gas applications are currently the largest portion of the business, although industry customers range from communications and electronics to medical and military.

"The relationship we have with our customers begins at development with prototype assemblies and doesn't end until we provide them with a completed assembly which is functionally tested when it arrives at their plant. This has helped us to become a full service supplier and offer more value added to the customer. The commitment we have to our customers starts at the top of our organization starts with Ted Davis and is something each employee takes ownership in," says Ted's daughter, Stacey Davis. "He is a man of great vision that we believe in, because he not only believes in his vision, but his employees as well."

PROGRESSIVE STAMPING, INC.

Some of the most quality-conscious customers

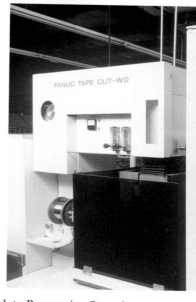

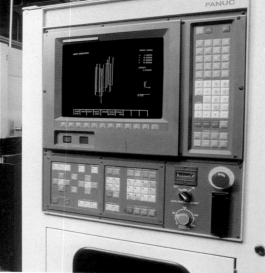

in the world look to Progressive Stamping, Inc. (PSI) for their stamping, tooling, and plating needs. The company's ability to serve both domestic and international customers clearly indicates its success in continually improving quality and productivity. PSI offers medium to high-volume production capabilities and takes great pride in meeting close tolerances and exact specifications. The 50 employees of PSI work precisely to stamp metal into specific shapes, sizes, and specifications. These stampings can be found in such products as seat belts, computer disc drive components, road-building equipment, compressors, electronic components, and Computer Numerical Control (CNC) machining.

utilities are assets of the city. "Another big key to our success is the fact we have the best vo-tech system in the United States," Davis adds. "They have provided us all the training for our employees—many times free of charge."

One Oklahoman and his daughter, coupled with a superb work force and all the benefits of doing business in Oklahoma City, are making their mark locally and in the global marketplace with a unique mix of companies. It is a storybook success, and Ted Davis believes Oklahoma gave him this chance to succeed. ✺

Louis Dreyfus Natural Gas

"LD" on the New York Stock Exchange stands for Louis Dreyfus Natural Gas, an independent energy company headquartered in Oklahoma City. The company name reflects a recent French influence, which is coupled with a strong Oklahoma City heritage. This combination produces an energetic business finding the leading edge in technology, risk management, and environmental quality.

CONSERVATIVE GROWTH

In 1990, an Oklahoma City-based oil and gas company was purchased by the French company Societe Anonyme Louis Dreyfus & Cie. Within a six-year time frame, Louis Dreyfus Natural Gas has multiplied the reserve base 16 fold, spent more than $670 million acquiring oil and gas properties and drilling for oil and gas. On a natural gas equivalency, the company reserves are nearly a trillion cubic feet.

The long-term marketing of these reserves has set Louis Dreyfus apart. To provide steady and predictable cash flow in the volatile natural gas industry, the company began supplying gas under long-term, fixed-price contracts. This removal of price risk variables resulted in increased profits for the company and funding for future development activities and acquisitions.

This conservative approach has now brought the company into a 50 percent hedged and 50 percent spot market sales strategy.

FOCUS AND TECHNOLOGY

The company focuses its operations in four core regions where it possesses considerable expertise. The geographically concentrated operations include the on-shore Gulf Coast region in south Texas, Permian region of west Texas and southeast New Mexico, the Mid-Continent district, and the Sonora area of west Texas.

Natural gas represents 85 percent of the reserves and production, with the remaining percentages attributable to oil.

Louis Dreyfus continues gas and oil exploration and drilling with the leadership of the company's skilled team of geologists and engineers. The company owns and uses technology-driven tools relevant to the industry such as three-dimensional seismic and three dimensional moduling for completion and reservoir characterization in all phases of operation.

Health and safety issues out in the field are also paramount for Louis Dreyfus. Photo by Jeff Corwin.

The company focuses its operations in four core regions where it posesses considerable expertise. Photo by Todd Flashner.

ENVIRONMENTAL HEALTH AND SAFETY

Ongoing company efforts exemplify the Louis Dreyfus philosophy as it relates to the environment and health of employees. With a strong commitment to these issues, two full- time employees plus support staff are dedicated entirely to implementing leading edge programs to not only comply with environmental and safety laws, but to exceed requirements.

The company takes pride in its high standards. Louis Dreyfus has received two awards from the Environmental Protection Agency.

Health and safety issues out in the field are also paramount for Louis Dreyfus. The company's record for employee safety is among the best in the industry. This record is attributed to the emphasis on health and safety highlighted during monthly safety meeting, focus groups, quarterly district meetings, plus an annual banquet that provides awards, cash prizes, and recognition for safety.

From safety to technology, the character exhibited by Louis Dreyfus Natural Gas is a compelling demonstration of the company's commitment to its shareholders, employees, and corporate citizenship. This style concisely reflects the conservative, leading-edge, and energetic business known as "LD." ✪

Organon Teknika

Organon Teknika's Oklahoma City facility was established in 1977 as CCI Life Systems to manufacture and distribute a sorbent dialysis system, using technology that was developed during the Apollo manned space flights. In 1978, CCI was acquired by Akzo N.V., Arnhem, the Netherlands, and the name changed to Organon Teknika.

Organon Teknika Corporation (OTC), a division of Akzo Nobel N.V., develops, manufactures, and markets a wide spectrum of diagnostic health care systems used in today's most advanced medical environments worldwide. Akzo Nobel is composed of a number of companies, some of which date back to 1792 and were founded by the distinguished scientist Alfred Nobel.

Organon Teknika is located on the south side of Oklahoma City (airport property) in a 80,000-square-foot building and employs 175 people. The company has a strong reputation for building customer satisfaction through world-class quality systems, delivering products and services across the globe. OTC's operation is ISO 9001 certified. The latest methods for inventory control, team involvement, cost containment, and quality systems are used. Materials used to manufacture the instruments are procured from local suppliers whenever possible.

"Organon Teknika is committed to Oklahoma. We attribute our success to dedicated people with the strong Oklahoma work ethic, strong working relationships, and our partnerships with local suppliers," says Joe Shimkonis, director of operations for the Oklahoma facility.

The plant manufactures several automated and semiautomated diagnostic systems used in medical laboratories worldwide. A premier system is the BacT/Alert®, an automated instrument for the rapid detection of bacteria in blood and other body fluids. Using patented technology to detect bacterial growth, this system enables timely initiation of specific patient therapy.

Organon Teknika also manufactures the MDA™ 180, a fully automated coagulation analyzer providing a wide range of diagnostic assays. The MDA™ 180 requires minimum operator intervention and provides maximum flexibility in the selection of assays to be performed. Additionally, three other coagulation instruments are manufactured to cover the broad range of throughput requirements, from large hospitals and laboratories to lower volume doctor's offices.

The next generation of products is based upon the patented NASBA technology and these

MB/BacT™, a non-invasive, walk-away system.

products are known as NucliSens™. Development of this technology has resulted in a system for the selective amplification of ribonucleic acid (RNA) and the automated detection of the amplified material. Nucleic acid amplification is one of the most sensitive diagnostic tools in the clinical laboratory today. Prime examples of its utility are for the detection and monitoring of infectious agents such as hepatitis C and human immunodeficiency viruses.

COMMUNITY INVOLVEMENT

Organon Teknika's contribution to the Oklahoma City community is a way to say thank you for the company's success. The company had

100 percent employee participation in the United Way of Oklahoma City resulting in the highest per capita contribution rate for the Oklahoma City area for the second year in a row. At least twice a year, Organon Teknika employees actively contribute to the Oklahoma Blood Institute blood drive. Support is also provided for Big Brothers and Big Sisters through bowling competitions and other fund-raisers.

"We understand the need for each person to be involved and contribute to the success of their community by whatever means is suitable for them. A number of our employees contribute their time and knowledge as coaches for kids' sports, scout leaders, science fair judges, and as leaders in their community," says Shimkonis. "Organon Teknika is happy to be in Oklahoma City." ✪

Organon Teknika's diagnostic health care instrument manufacturing facility, located at 5300 S. Portland Avenue.

BacT/Alert®, the totally automated blood culture system.

Seagate Technology, Inc.

Seagate's Oklahoma employees have greatly contributed to the company's outstanding success in the high-end disc drive marketplace. One of their contributions has been the Barracuda™ product family, providing users with "predatory performance" and unmatched reliability.

The computer is the most revolutionary communications tool since the telephone. Not only is it becoming as commonplace as the TV set, it's beginning to resemble one, too. And with the emergence of the Internet, people all over the world are beginning to rely on computer technology to work, play, and interact with one another, regardless of location or time of day.

Throughout the Information Age, Seagate Technology and its Oklahoma City Operations have had an extraordinary impact on the computer's transformation. As the world's largest supplier of computer disc drives and related components, the company provides the industry with some of the most advanced data storage technology available, giving computer users just the information they need, when they need it, and how they want it.

The Oklahoma City Operations (OCO) is the city's third-largest manufacturer, employing more than 2,000 local residents and purchasing materials from more than 500 Oklahoma suppliers. OCO designs and develops high performance, high capacity hard disc drives for mid-range to high-end computer applications. Considered the "heart" of the computer system, the disc drive is a highly complex peripheral storage device used to store data received from the computer's main memory and retrieve it later at the request of the system's user. From Hollywood film and TV production to NASA space shuttle testing, OCO's Hawk™ and Barracuda™ product line families can be found throughout the world's most data-intensive computing environments.

"Throughout its more than 30-year history, the Oklahoma operation has been an achievement-driven organization, producing a number of industry 'firsts.' These accomplishments are attributable to a highly talented team of individuals," said Don Colton, senior vice president, Worldwide Product Line Management. "Our work force—with an average length of service approaching 25 years—takes pride in Seagate's continued success. This is indicative of Oklahoma's strong work ethic and our employees' commitment to doing the best job possible."

Seagate and its Oklahoma employees have a long, proud history, not only in their contributions

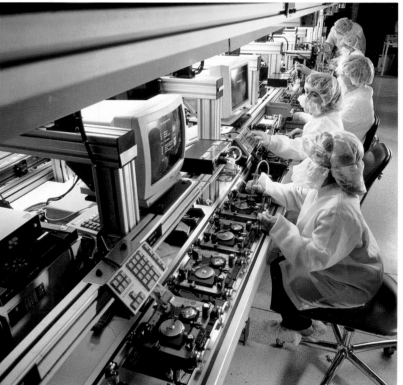

The assembly of Seagate's hard drives begins inside Class 100 "clean rooms" in which the air is continuously monitored and controlled by sophisticated computers to ensure proper cleanliness levels. This air filtration, which prevents damage to the internal components of a disc drive, results in an environment that is about 100 times cleaner than a hospital surgery room and 3,000 times cleaner than a typical household.

to the computer industry, but also in their contributions to and involvement in the local community. Areas such as education, economic development, and human and health services receive significant company support through financial, product, and surplus equipment donations. Employees volunteer thousands of hours each year to participate in numerous community projects sponsored or hosted by Seagate. And through their annual Federated Fund Drive, initiated in 1964, OCO employees have contributed an approximate grand total of $3.5 million in earnings to nearly 150 statewide, non-profit organizations.

Today, Seagate is building the highest quality precision electronic device in the world, outside of the military. Its new generation of disc drives achieve the engineering equivalent of an F16 fighter jet flying at MACH 813 just 1/62nd of an inch above the ground . . . and landing on a *blade* of grass. And through the evolution of data storage technology, there's a little bit of Oklahoma inside every one of these remarkable products. ❂

America Online Oklahoma City Member Services Call Center

America Online, Inc. (AOL) is the leading Internet online service in the exciting new world of cyberspace with more than eight million members around the globe—and growing. AOL Networks oversees the America Online consumer online network which offers its subscribers such services as electronic mail, conferencing, software, computing support, interactive magazines and newspapers and online classes, as well as easy access to the Internet. Founded in 1985, AOL has a global workforce of more than 7,000 people.

One of the five AOL Member Services call centers across the country, the Oklahoma City call center handles an average of 40,000 telephone calls and 20,000 electronic mail messages daily. It is the primary support center for AOL's Canadian members. Work shifts run 21 hours each day, 365 days a year.

The first call came into the Oklahoma City call center on May 5, 1996. Nearly 800 AOL employees now work at the center. By the end of 1997, the Oklahoma City operation is expected to employ up to 1,400. The center's permanent training department provides employees with 120 hours of training along with continuous performance improvement and update training.

"Our role is to provide our members with the technical and other member services support which they need to maximize their interactive experience," said Michael J. Ritonia, general manager of the Oklahoma City call center. "Our educated, friendly and articulate employees here in Oklahoma City is critical to AOL's mission."

The member Services call center communicates with AOL members via telephone, mail and online to provide support in four main areas:
1. Technical Support: Furnishing technical support to members in accessing and using the service;
2. Acquisitions and Payments: Answering general set-up questions, ordering AOL software shipments to prospective members, and providing billing information;
3. Member-Save Program: Reaching out to members calling to cancel with an educational assistance program;
4. Community Action Team: Handling complaints from members about other members who have violated the terms of their membership.

The Oklahoma City call center devotes itself to a variety of community projects, including free inter-activity classes for teachers. Our College Associate Program offers a reimbursement plan for the college tuition of employees attending college full-time.

"Our call center must always be flexible and reliable in handling its changing and challenging assignments," added Ritonia. "There is no telling how our industry will evolve in the future, but we will be prepared to provide all the support necessary for our members to enjoy a productive, convenient and entertaining time on AOL." ✪

AOL Networks oversees the company's consumer online network which offers subscribers a wide variety of services and entertainment including: Electronic mail, Instant Messages, and Buddy Lists. Photos by Jack Hammett.

Bibliography

Born Grown, by Roy P. Stewart. Fidelity Bank, Metro Press, Inc., Oklahoma City, 1974, 352 p.

Heart of the Promised Land: Oklahoma County: An Illustrated History, by Bob L. Blackburn. Windsor Publications, Woodland Hills, California, 1982, 264 p.

Oklahoma City: A Centennial Portrait, by Odie B. Faulk, Laura A. Faulk, and Bob L. Blackburn. Windsor Publications, Northridge, California, 1988, 253 p.

Stalwart Sooners, by Anson Burton Cambell, Paul and Paul Publishers, 1949.

Tinker Air Force Base: A Pictorial History. Tinker Air Force Base, Oklahoma City Air Logistics Center, Office of History, 1983, 163 p.

Enterprises Index

Index